ENERGY

Sources/Applications/Alternatives

by

Howard Bud Smith
Executive Editor—Technology
Goodheart-Willcox Co., Inc.

South Holland, Illinois
THE GOODHEART-WILLCOX COMPANY, INC.
Publishers

Library of Congress Catalog Card Number 84-21105
International Standard Book Number 0-87006-956-X

1 2 3 4 5 6 7 8 9 10   93   97 96 95 94 93

Library of Congress Cataloging in Publication Data

Smith, Howard S. (Howard Sylvester)
  Energy.

  Includes index.
  1. Power resources.   2. Energy conservation.
  I. Title.
TJ163.2.S63   1985   621.042   84-21105
ISBN 0-87006-956-X

INTRODUCTION

Whatever moves or changes in our world is made to do so by the use of a force we call energy. Even the human body must depend upon energy for the ability to talk, walk, and work. Indeed, without it, our marvelous technological society would not exist.

Even so, until the energy shortages of 1973, little thought was given to the importance of a ready supply of energy. Nor was much thought given to the finite nature of our fossil fuels—that one day they would be gone, never to be replaced.

Energy: Sources, Applications, Alternatives deals with many aspects of our energy sources. It explains how we extract, process, convert, and use them to power our technology. It also deals heavily with the alternative sources—where they come from and how we can use them to conserve and eventually replace dwindling fossil fuels.

Further, this text examines how and where we use our energy, where demand is the heaviest and what the outlook is into the year 2000 and beyond. It explains the various "use sectors" and what demands each makes on available energy supplies.

From a base of knowledge which includes an understanding of the various alternate energy sources, this book then deals with the need and the tactics of conservation from one sector of society to another. Final chapters present a view of the impact the changing energy picture will have on society, what the future holds as energy sources evolve, and what the career opportunities might be.

Each chapter includes a list of words to know, test questions, and suggested outside activities. A workbook/laboratory manual includes drawings and directions for various activities and experiments in alternate energy and energy conservation as well as activities to gauge the knowledge you have drawn from the text.

Howard Bud Smith

CONTENTS

1 CHAPTER

ENERGY AND ENERGY SYSTEMS

The information given in this chapter will enable you to:
- *Define energy.*
- *Explain how energy is part of a system.*
- *Tell why a study of energy is important.*
- *List and describe the six forms of energy.*
- *Describe how light and heat travel.*
- *Define the term inertia.*
- *List two forms of energy conversion that occur in nature.*
- *Explain, in simple terms, the two laws of thermodynamics.*
- *Explain and give an example of entropy.*
- *List the sources of energy available for use today.*

Energy is the ability to do work. We are all familiar with the idea of work. We could say that anything that causes movement has energy. That would be true, but there is much more that is caused by energy than just movement. We really need to define work more exactly.

Work is that which causes change. This would account for objects being moved, for cold objects becoming warm, or for darkness to be lighted. However, it would not account for the energy used, for example, to prevent objects from falling or scattering. Thus, we can say that work involves change, movement, or support.

Looking around us we can see a great deal of energy being used and changes occurring. To give only a few examples:
- The sun lights and heats the earth, Fig. 1-1.
- Winds cause trees to bend, flags to flutter, and wind gauges to spin. See Fig. 1-2.
- Water, Fig. 1-3, moves from higher ground to lower ground. It can move so slowly that it hardly dis-

Fig. 1-1. The sun is the greatest source of energy we know. Every day, winter or summer. Top. During the day, it lights up the earth. Bottom. At night, with no sunshine, the earth is dark and streets must be lit by other means.

turbs a floating leaf. It may move rapidly so that waves can change shorelines by moving sand about, as in Fig. 1-4.

HUMAN ENERGY

People exert energy, too. Leaping, running, and pulling all require use of energy within us, Fig. 1-5. Our actions create an energy that all moving bodies have.

Fig. 1-2. Wind is the movement of air. We see its effect on flags, weather vanes, and the anemometer above a coastguard station.

Fig. 1-4. Waves like these are energized by the wind. They have enough energy to move sand and boulders, changing the shorelines of lakes and oceans.

Fig. 1-3. Water has energy as it moves from high ground to lower ground as in rivers. Even slow-moving water with little energy will carry a leaf, a stick, or a boat downstream.

Fig. 1-5. People too have energy and can expend it by child's play or sports activities.

Power is sometimes confused with energy. It is related to energy but is not the same. A later chapter will define this term.

ENERGY AS A SYSTEM

A system is an organization of parts, or sometimes of individuals, that operates in concert (together) to achieve a certain purpose. Each component complements (supplies something lacking in) the other components in the system.

Our entire world is made up of systems. The human body is a system made up of a brain, blood, tissue, organs, and many subsystems. Its purpose is to be a thinking, acting, person.

The furnace that heats your home is a system. A burner turns fuel into heat. A plenum collects heated air. The hot air is distributed through ducts by a fan. The ductwork carries the moving air to different parts of your home. A thermostat turns on the furnace when it senses that the home is too cool. When the home is warm enough, it signals the furnace to stop producing heat.

We also refer to the sun, planets, and stars as our solar system and we speak of a system of government with many departments. Truly, our lives are intimately tied to systems of many different types. Some are natural. Others are fashioned by humans.

All systems have the same components, Fig. 1-6. There are inputs, process, output, and feedback. An **input** is also called a resource. The inputs for all technological systems are energy, people knowledge, materials, capital (tools, machines, and buildings), finance, and time.

The **process** of a system is the change produced or the work required to give value to what is produced. In transportation the process is moving people or cargo. The **output** is the new location of people and cargo. **Feedback** is any signal or information that might adjust the process or initiate it.

One way of thinking of energy is to consider it as a system. The input of an energy system is matter. The process is the change that takes place in the matter. The output is a force that can produce a change which we know as work.

WHY STUDY ENERGY?

It is important to understand energy because it drives our world and all that is in it. To use energy without understanding it is to run the risk of wasting it. Whatever we do requires energy. The type of energy you choose will have an effect on the environment.

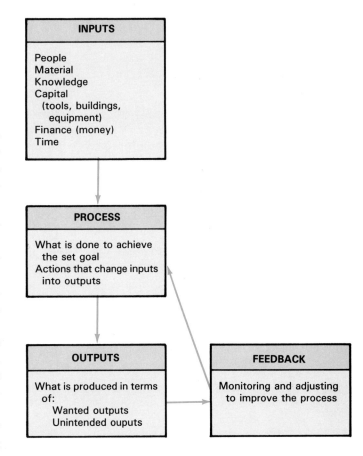

Fig. 1-6. Energy is part of a system with all the parts shown in this diagram.

When we study technology and how it benefits humankind, we must understand that energy is the driving force behind technology. All other inputs contribute to the outputs, but without energy, processes cannot go into action.

If you were to study the technologies of Manufacturing, Construction, Communication, and Transportation, you would discover that energy is a primary input in all of them. For example, a manufacturing process may require heat and electric power. Both of these are outputs of energy.

Another example: this book was written on a computer that used electricity. The source of the energy might have been the burning of a fossil fuel such as coal. The paper it is printed on required energy to harvest the wood and turn pulp into paper. It was printed on huge presses that were run by electric motors. Vehicles powered by fuel-burning engines delivered the books to your school.

PROPERTIES OF ENERGY

Energy is contained in all matter (molecular energy). It can be readily changed from one form to another.

Energy in matter is continually changing. It exists in substances as heat energy and chemical energy.

EFFECT OF ENERGY

You can easily see the effects of energy when it causes motion such as a windmill turning or an automobile moving down the street. It is even quite easy to realize that energy causes light and heat. You can see the light and you can feel the heat. It is not as easy to realize that energy is causing plants and trees to grow. See Fig. 1-7.

While you may know it is happening, you cannot see the sun lifting water into the air as vapor. You may see the effects of it in clouds or rain, Fig. 1-8.

Nature sometimes releases great forces. Sometimes the forces cause great damage. A high wind will destroy buildings and trees, Fig. 1-9. Fires that are not controlled can destroy forests, Fig. 1-10.

As you can see, energy is available in different ways. It produces many different effects. Since this book will be concerned with the single subject of energy it is important that you become aware of the types and forms of energy.

TYPES OF ENERGY

Science classifies energy in a physical sense as:
• Potential energy.
• Kinetic energy.

Fig. 1-8. Clouds are made up of moisture vaporized and drawn into the atmosphere by the work of the sun.

Fig. 1-9. Sometimes the natural energy in a storm destroys buildings and trees. A tornado moving across midwestern prairies did this damage.

Fig. 1-7. Energy from the sun is stored in all growing things. When you eat fruit or vegetables consider how you change its chemical energy into motion.

Fig. 1-10. Lightning is a spectacular example of the electrical energy present in nature. Occasionally, it causes fires that destroy property and burn forestlands.

find it easier to consider radiation as particles of energy.

These particles or **photons**, are bundles of energy that travel through space and give up their energy upon striking a surface through which they cannot pass. It is believed now that the wave and quantum theories complement (help) each other.

Heat/light differences

Though both light and heat are the result of radiant energy, heat energy acts differently from light energy. Light exists outside of matter and does not enter into matter that is not transparent or translucent. (**Transparent** means you can see through it; **translucent** means it transmits light but you cannot see through it.) Heat exists outside of matter, too, but it will enter matter and cause its temperature to rise. The atoms of matter that heat has entered begin to move about more rapidly. The higher the temperature of the material, the more rapid the motion of the atom.

Also, heated atoms must have more room to move about. This accounts for the well-known fact that heated liquids or gases expand. For example, if water or air is heated in a closed container, expansion may cause the container to bulge or burst, Fig. 1-16.

HOW HEAT TRAVELS

While light travels by way of radiation only, heat can move from place to place by three methods:
• Radiation which was just discussed under "electromagnetic radiation."
• Convection.
• Conduction.

Convection

Convection is movement within liquids or gases. The heat travels by circulation of a heated liquid or air. Circulation will continue until all of the material is the same temperature.

Water illustrates this action. Because of expansion, hot water is lighter than cold water. Therefore, cold water sinks to take up space vacated by warmer water which tends to rise, Fig. 1-17.

Likewise, when air is heated, it becomes lighter and rises. Cooler air sinks and takes its place. See Fig. 1-18. A room may be heated on this principle. Warm air from the furnace rises as it enters the room. Cold air sinks to the floor and is drawn back to the furnace through the cold air return and reheated. See Fig. 1-19.

Conduction

Conduction is heat travel produced by molecular action. Heated molecules of a substance move about so rapidly that they collide with one another. Warmer

Fig. 1-17. Convection taking place in a liquid. Warm water is less dense since its molecules are father apart. Being lighter, heated liquids rise while cooler liquids sink to the bottom of the container.

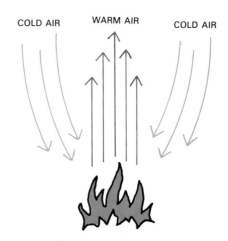

Fig. 1-16. Fluids in closed containers will expand as they are heated and may even burst their containers.

Fig. 1-18. Convection also occurs in the air above a fire. The warmed air rises while cooler air descends to take its place.

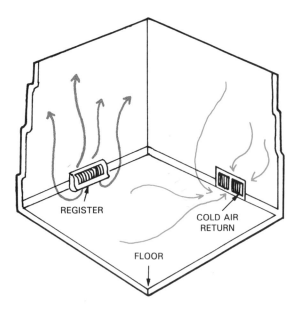

Fig. 1-19. Home heating uses the principles governing convection. Warm air rises from the floor-level register, heating the room. Cool air enters an opeing called a cold air return that takes cooler air back to the furnace to be heated.

molecules bump into cooler ones and give some of their heat to the slower moving cool molecules. This action is repeated until all molecules in the substance are the same temperature. See Fig. 1-20. This principle is applied to cool an automobile engine and to cook food.

MECHANICAL ENERGY

Of all the forms of energy, **mechanical energy** is the one we usually think of as "doing work." Mechanical energy is the same as kinetic energy which you learned about earlier. However, the term "mechanical" is

Fig. 1-20. Heat will travel along a solid material, such as an iron bar, as heated molecules bump into cooler slower moving molecules. Through this contact, heat is passed along until the entire bar is hot.

generally used wherever the energy is harnessed to do work.

Though its name suggests machines, mechanical energy is not limited to the work output of machines. All moving bodies have it. Bodies in motion do work because they resist forces exerted to stop them. This tendency to stay in motion is called **inertia**. The heavier an object is the more inertia it has.

CHEMICAL ENERGY

Chemical energy is locked away in the molecules of many kinds of substances. Growing plants collect radiant energy from the sun, carbon dioxide from the atmosphere, and minerals and water from the soil. They combine all these chemically to form new molecules that make up the pulp and fiber of the plant. The stored energy can be released by being eaten or burned.

For example, fire creates a chemical change in wood through the process of burning. Burning can also be called "rapid oxidation." The output of this process is a change in the wood from fibers to heat and ashes. The heat energy may power a steam engine, heat a building, or cook a meal. See Fig. 1-21. The burning rearranges the molecules, releasing heat and light in the form of radiant energy. The ashes left behind are

Fig. 1-21. Wood or coal, when burned, change chemically and give off heat which can be used to power a steam engine.

an entirely different form of matter because of the chemical change.

Gasoline, oil, and natural gas are other examples of chemically stored energy. A battery also stores energy as a chemical and releases it as electrical energy, Fig. 1-22.

ELECTRICAL ENERGY

Electrical energy results when negatively charged particles are attracted to positively charged particles. The negatively charged particles are called **electrons**. The positively charged particles are **protons**. Attraction between them causes a flow of electrons. This electron flow is known as electrical current. Special conducting metals, like copper and silver, allow electrons to flow easily from one molecule to another. A path that allows current through copper wires to produce energy is called a circuit. Electrons can be made to produce light and heat when they flow through the filament (fine wires) in a light bulb. They will also create a magnetic field and cause an electric motor to rotate and provide power for turning machines. Electric current can also produce heat when traveling through certain types of wires, Fig. 1-23.

Another quality which makes electrical energy so popular is the ease with which it moves from one place to another. It can be produced at one location and moved long distances before it is used.

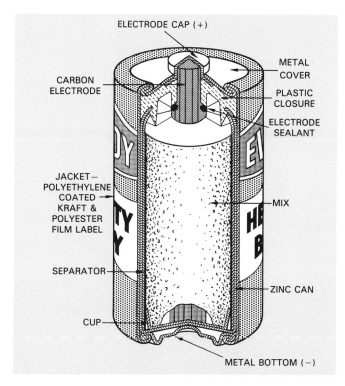

Fig. 1-22. Cutaway of a small cell. It stores electrical energy in a chemical paste which is called a "mix." (Union Carbide Corp.)

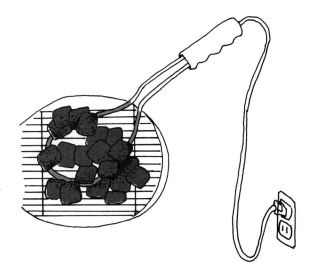

Fig. 1-23. An electrical charcoal lighter heats up and starts charcoal afire. Wires in the lighter heat up as heating coils resist the electric current passing through them.

NUCLEAR ENERGY

Atomic or **nuclear energy** is released when certain kinds of atoms are split. The release of the energy can be swift as in the explosion of an atomic or hydrogen bomb. It can also be controlled or slowed down to produce huge quantities of heat with a very small quantity of material. The heat of fission is used to heat fluids. The fluids boil to create steam that drives a turbine. Connected to a generator or a propeller the mechanical energy in the turbine produces electricity or propels a ship.

As with electricity, nuclear energy depends on the action of atoms. The center of an atom is called a **nucleus.** All nuclei (plural for nucleus) are made up of positive particles called protons and neutral particles called **neutrons.** Nuclear energy is produced when the nucleus of an atom of uranium is changed. (Uranium is used because its atoms have many protons and neutrons in their nuclei). The process which changes the nuclei of uranium is called **fission.**

Fission

One way of changing an atom is to bombard it with a particle such as a neutron. This causes the nucleus to split into two parts each with the same number of protons and neutrons. This changes the mass (weight) of the nucleus and causes it to give up tremendous amounts of heat energy. See Fig. 1-24.

If other nuclei are struck by neutrons flying out from the first fission, the splitting or fission will continue. This is called a **chain reaction.** When controlled inside a reactor, the release of heat takes place at a slower rate than the exploding of an atomic bomb. It is,

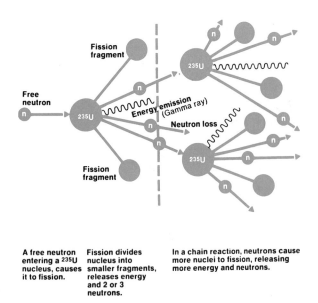

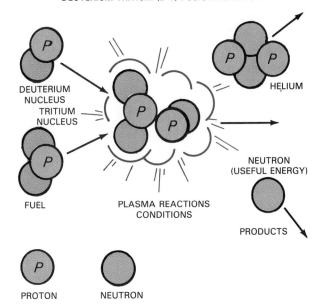
DEUTERIUM-TRITIUM (D-T) FUSION REACTION

Fig. 1-24. Nuclear energy is released as molecules of a certain kind of uranium are split. It creates other kinds of matter and releases large quantities of heat. (Atomic Industrial Forum, Inc.)

Fig. 1-25. Fusion is another form of nuclear energy. Two atoms combine as energy is released in the form of heat.

therefore, possible to put the heat energy to use. Usually it heats water for steam energy.

Fusion

Another way of tapping the energy of the atom is to combine two light atoms into one heavy atom. This process is called **fusion**. It propels two nuclei together. This is difficult because the nuclei are both positively charged and tend to repel one another. This tendency is overcome by heating the fusion fuel to a high temperature. The nuclei bounce around rapidly. Then they slam into each other with such great force that they overcome their natural repulsion and remain together. Fig. 1-25 is a simple drawing of how fusion takes place. As with fission, fusion releases great quantities of heat which can be used to create other useful forms of energy.

ENERGY CONVERSION

One characteristic of energy is that it is easily changed from one form to another. It happens so often in nature that we hardly ever stop to think about the conversion.

Some of the common conversions occurring in nature are caused by solar energy. For example, solar radiation is responsible for several conversions which occur quietly all around us.

• The sun, like a giant pump, lifts water from the earth's surface. Some is collected from lakes, oceans,

and rivers. Part of it comes from the evaporation of moisture from other surfaces. Every year, 100,000 cubic miles of moisture are drawn into the atmosphere by the sun. It is later released as rain, sleet, dew, or snow to replenish the flow of rivers. This flow is mechanical energy which can be captured and put to use producing other forms of energy.

• Plants collect solar radiation, along with nutrients from the soil and store it in matter which makes up leaves, stems, and stalks. This process is called **photosynthesis**. The energy will remain locked in the plant as chemical energy until some outside agent transforms it into still another form of energy. Let us consider some of these transformations:

• Plants eaten by animals or humans supply the energy needed to maintain body temperature and nourish muscles which control the movements of the body.

• Burning the plant changes its chemical energy into heat energy.

• Plant life buried and subjected to heavy pressure, forms **fossil fuels,** Fig. 1-26. These stored forms of energy are rich in carbon and are useful for providing heat when burned.

LAWS OF ENERGY CONSERVATION

Energy follows certain basic behavior characteristics known as the **Laws of Thermodynamics**. The first of these laws is: Energy cannot be destroyed. It can be changed from one form to another, but it never ceases to exist.

Fig. 1-26. This chunk of bituminous coal came from a coal mine in Kentucky.

Fig. 1-27. Top. Child on a swing uses energy to pump the swing. Bottom. Eventually the swing will stop when no more energy is put into making it continue to swing. Friction and wind resistance bring it to a stop.

At first glance, this seems like a wonderful arrangement. If it cannot be destroyed, can we continue to use it over and over again?

The answer is "no" because of the Second Law of Thermodynamics. It states that is it is impossible to make heat flow from a cold body to a warmer one without using another source of energy. The effect of the law is that, once used, energy loses its ability to do work even though it is not destroyed.

A child on a swing, Fig. 1-27, shows how this principle works. Mechanical energy in the child's body was used to set the swing in motion. When no more energy is put into pushing the swing, it will gradually stop. It would seem that all the energy is gone. However, if you had a sensitive thermometer, you could measure that the mechanical energy had changed to a low heat. It would show an increase in the temperature:
• Of the swing parts and the swing supports.
• Of the air around the swing.

The heat is not of a high enough temperature to be useful. It is passed off to cooler surroundings. In such cases the energy is said to be **degraded** (weakened) to a point where we are not able to use it.

ENTROPY

The child on the swing is an example of what **entropy** means. The swing slowed down and came to a stop. The energy was not all gone, but it was so scattered that it could do no more work. The principle of entropy is this: eventually all energy, as it is used, becomes so random (scattered) that it loses the ability to do work.

Let us consider another example. Suppose that you were to heat up water to operate a small model of a steam engine. When the water becomes steam, it will run the engine as long as you continue to heat the water. However, when you turn off the burner supplying the heat, the steam will cool. The engine will stop running. The water is still hot but it no longer has the energy to run the steam engine.

SOURCES OF ENERGY

What does it take for a form of energy to be useful as a source? Several things are required.
• It must be plentiful and easy to acquire.
• It must be easily converted into a form which our machines can use.
• It must be easy to transport.
• It must be easy to store.

Up to now, only a few energy sources have been developed that fit all these requirements:
• Hydrocarbons or fossil fuels. They are: natural gas, coal, and petroleum.

- Nuclear energy.
- Hydroelectric power.

Other sources are available which may not now meet all the requirements as well as those just listed. However, they could be developed to the point of meeting the requirements better. This group includes:

- Solar energy. The radiation of the sun, discussed earlier, can be used directly to produce power and heat buildings.
- Forms of kinetic energy, including tidal, wave, and wind energy.
- Thermal energy. This includes geothermal (heat from the center of the earth) and the heat stored in ocean waters.
- **Biofuels.** These include a group of gases and liquids which can be extracted from organic wastes, seaweed, grains, and a variety of plant life.

RENEWABLE, NONRENEWABLE, AND INEXHAUSTIBLE

Energy resources also fall into one of three groups that will greatly affect their future use. Renewable resources are those that can be grown and collected or harvested time after time. The supply cannot be exhausted. Nonrenewable energy sources have a limited supply. When this supply is gone, it cannot be replaced. Inexhaustible energy resources exist in such large quantity that they will never run out.

Fossil fuels, since the process which formed them took millions of years, are considered nonrenewable. When the last oil and gaswells runs dry and the last coal mine is stripped, the fossil fuels are gone forever. The matter from which they were formed was limited. There is neither time nor material to produce more.

Renewable resources are those like plant matter which can be grown and used any number of times. The energy of the sun, wind, and water are considered inexhaustible. As long as the sun shines they can be made to do work. All of these energy sources will be discussed fully in later chapters.

SUMMARY

Energy is the ability to do work. While we cannot always see energy doing work, we see and feel its effect on our world. There are several ways to classify energy. One is by whether the energy is at rest or doing work. Energy waiting to do work is called potential energy. Energy at work is known as kinetic energy.

Another way to classify energy is by its forms.
- Light energy or radiant energy. The sun is the primary source of radiant energy.
- Heat energy. Closely related to light energy it also comes from radiant energy.
- Mechanical energy. This is the energy that is found in any object in motion. This form allows objects in motion to continue in motion and to actually create movement in other objects.
- Chemical energy. This is energy stored in molecules. Energy is released by change taking place in atoms.
- Electrical energy. It is the result of the movement of electrons from one atom to another.
- Nuclear energy. Another form of molecular energy, it occurs as atoms are split with a resulting release of huge quantities of heat.

A useful characteristic of energy is that it can quite easily be changed from one form to another. This is called energy conversion.

Physics teaches us that while energy can be changed and used it cannot be destroyed. Another property of energy is that once used it becomes so random that it can no longer do work. These two physical principles are know as the Laws of Thermodynamics.

To be useful as a way of performing work, energy must be plentiful, easily converted, easy to transport, and easy to store. Today, our main sources are fossil fuels, also known as hydrocarbons; Nuclear (atomic) energy; and hydroelectric (water power to electricity).

DO YOU KNOW THESE TERMS?

biofuels
chain reaction
chemical energy
conduction
convection
degraded
electrical energy
electromagnetic radiation
electron
energy
entropy
feedback
fission
fossil fuels
fusion
gravity
input
intertia
kinetic energy
Laws of Thermodynamics
mechanical energy
nuclear energy
nucleus
neutron
orbit

output
photon
photosynthesis
potential energy
process
proton
quantum
radiant energy
translucent
transparent
wave theory
work

SUGGESTED ACTIVITIES

1. Secure books from your library and research the quantum theory of energy. Prepare a written report for your class.
2. Build or bring in a toy or device which demonstrates that heated fluid, such as air or water, rises.
3. Construct and/or demonstrate a toy or device which illustrates a principle learned in this chapter. Ask your instructor for suggestions and help.
4. Place an open-topped glass container of water in direct sunlight. Cover the top with a piece of clear plastic or glass. Observe at intervals. What happens to the underside of the cover? Report your observations to the instructor or class. Discuss what happens and give reasons why it happens.

TEST YOUR KNOWLEDGE

1. What is energy?
2. Energy and energy sources are able to produce (check correct answers):
   a. Heat.
   b. Light.
   c. Motion.
   d. None of the above.
   e. All of the above.
3. _____ energy is the type of energy that is waiting to do work.
4. Define kinetic energy and given an example.
5. List the six forms of energy and briefly describe each one.
6. Heat moves from one place to another in three ways: _____, _____, and _____.
7. Chemical energy is locked away in the _____ of many kind of substances.
8. How is nuclear energy produced?
9. Give two examples of energy conversion which take place in nature.
10. The first Law of Thermodynamics states that energy (can/cannot) be _____.
11. List three fossil fuels.
12. Fuels made from plant life or organic wastes are said to come from a (renewable, nonrenewable) source.
13. Entropy means that eventually all energy, as it is used, becomes so scattered that it (complete the sentence) _____.

What type of energy is represented by a small boy going down a slide?

2
CHAPTER

HARNESSING ENERGY
THEN AND NOW

The information given in this chapter will enable you to:

- Trace development of energy conversion from primitive to modern times.
- List and describe a number of energy converters including windmills, water wheels, turbines, internal and external combustion engines, electric motors and generators, and fuel cells.
- Discuss modern systems of energy conversion and usage.
- Assess the importance of energy systems to the economy.
- Discuss the impact of energy on the environment.
- Discuss the role played by the design process in developing early energy-using machines.
- Name and discuss modern conversion systems.
- Assess the importance of energy systems.

Fig. 2-1. Primitive humans used fire to provide warmth and to cook food. Wood was probably the first energy material (fuel).

Our ancestors did not know much about using energy. At first they depended upon the natural warmth of caves to shelter them from the cold and to keep them safe. Then they learned to use fire. They found that it helped to keep them warm, Fig. 2-1. During thousands of years of simple existence, early humans had to depend on their own muscle power to travel, find food, and house and protect themselves.

PRIMITIVE ENERGY USAGE

In that time they discovered that muscle power could hurtle a rock at an animal, bringing it down to provide a bigger supply of food. Then they learned to

shape the rocks into better weapons.

The heat energy from fire was found to be useful in making spears, knives, and arrows. Cold water, dropped carefully on the edges of heated stones, would break away small chips until good cutting edges were formed. See Fig. 2-2. Thus, our ancestors could make their muscle (mechanical) power work better. When they used these improved tools, they could enlarge their supply of food and ensure better protection against animals and enemies.

The first communities were developed as families and tribes began to take up more or less permanent possession of lands. These people gathered around the valleys of the Nile, the Tigris, and the Euphrates rivers in Egypt and what today is Iraq.

There is also evidence that humans living in southern California 70,000 to 100,000 years ago were fashioning crude tools. Their newly acquired tool-making skills allowed them to grow food from the soil and kill more

Fig. 2-3. Animals were trained to pull farm implements.

Fig. 2-4. Sails were found to be better than muscle power to drive boats.

Fig. 2-2. Later, fire was a source of heat for toolmaking. Top. A special kind of rock was heated. When cold water was dripped onto the heated stone, part of the stone flaked off to produce a cutting edge. Bottom. The shaped stones were used as arrow or spear heads as well as for crude knives.

wild game. Animals, which they had tamed, provided both a source of food and mechanical energy to help them power crude implements and work the soil, Fig. 2-3.

As civilizations developed the need for travel, the water power of flowing rivers was used to carry rafts downstream. Later still, fixed sails were added, Fig. 2-4, to capture the kinetic (mechanical) energy of the wind.

DESIGN: THEN AND NOW

How did labor-saving, energy-using tool and machine designs come about? Did early designers follow the same steps in design as modern designers? Since paper and writing/drawing tools did not exist in prehistoric times we can only guess. No doubt, the pro-

cess, then as now, started with recognition of a need. Observation told them that fire and wind had energy that they could use. All they had to do was find a way to use it. From this point on, trial and error probably developed improved ways of using heat and different designs to capture the energy of wind. Through their experiments, early inventors built up a knowledge base consisting of designs that worked best. It is also possible that they traced out their designs in the dirt. Possibly a crude system of measurement was worked out.

The most inventive minds said: "What if" and thought of different ways to solve the problem and fill the need. By experimenting early inventors tried and discarded designs until they found one that filled the need. An early hunter discovered that a rock, thrown hard enough could kill small game. Later the hunter saw a need for a way to throw the weapon harder. From this need the bow and arrow developed. So it is with development of modern energy-using machines. Now, however, the design process is developed to a high degree. It follows several steps:

1. Statement of the design problem. This is called the design brief. It tells what the product must do but not how to do it.
2. The designer quickly sketches out all the ideas that occur to him or her.
3. The most promising sketches are refined.
4. The designer presents the ideas to the client who selects one for further development.
5. A working model (prototype) is built and tested.
6. Using the test results, the designer either improves the design or sends it on for manufacture.

DESIGNING LABOR-SAVING MACHINES

The first attempts at applying engineering principles to labor-saving machines probably came around 300 B.C. Greek scientists developed a number of these machines but regarded them as little more than toys and curiosities for the amusement of the rich.

However, Archimedes, who lived from 287 to 212 B.C., is said to have put one of the inventions to use. To defend a Greek city from a Roman invasion, he used large curved mirrors to reflect concentrated sunlight that burned up the sails of the Roman ships.

The inventors of the Middle Ages were much better at making machines that were more efficient as well as labor-saving. Both animal and human muscle were better used. See Fig. 2-5 and Fig. 2-6. For the first time, human muscle and even horses were replaced with another source of mechanical energy. The first water wheels captured the energy of falling or flowing water.

Fig. 2-6. Animals were later trained to power machines.

EARLY WINDMILLS

The first windmills were thought to be a Persian invention. They were popular in the desert where there are heavy winds blowing most of the time. Fig. 2-7 shows an early design. The wooden vanes for catching the wind were mounted on a vertical shaft. The rotor looked somewhat like a modern revolving door.

The shaft was connected to a wooden disc which had pegged teeth fastened along its outer edge. These meshed with a slotted drum. The drum operated a long, endless rope ladder which had dippers attached to it at regular intervals. The ladder, called an elevator, was

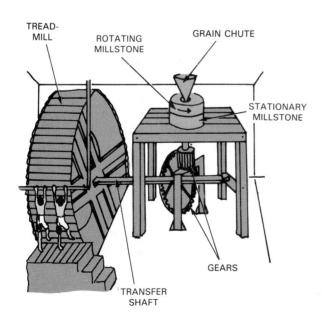

Fig. 2-5. Human effort was often used to provide power for running grain mills.

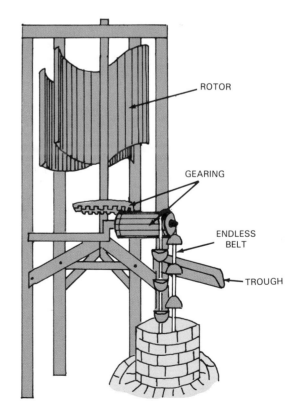

Fig. 2-7. Sketch of what an early Persian windmill might have looked like. It could catch the wind from any direction.

simply an endless rope. It extended downward into the well. As the drum turned, the dippers picked up water which was brought to the surface and dumped into a trough as the rope traveled over the drum, Fig. 2-8. The top of the tower was open so that the vanes might catch the winds from every direction.

EUROPEAN WINDMILLS

The **windmill** first began to appear in Europe in the 12th century. The first ones were called post mills and looked like the sketches in Fig. 2-9. The vanes looked like four huge paddles. They were wooden frames with

Fig. 2-8. The Persian mill was hooked up to gears and a lift so that buckets of water could be drawn from the well.

Fig. 2-9. Later, European mill designs improved upon the Persian mill. Left. A post windmill. Right. Smock mill.

canvas stretched over them. The vanes were supported by a post and an open structure like a treehouse. This allowed the mill to turn into the wind. Sometimes the open structure was closed in for storage.

The smock mill, which came later, was an improvement over the post mill. It had a second rotor (wheel) which kept the large rotor turned into the wind. **Smock mills** were huge structures, often standing 120 ft. tall. Their rotors (often called "sails") could be as much as 80 ft. long. A series of wooden gears and shafts connected the wheel with grinding stones or devices to pump water.

WINDMILLS IN AMERICA

Wind power became popular on American farms in the early 19th century as a means of providing water for farm families and their livestock. Stretching the railroad to the West Coast was made possible, in part, by the windmill. It pumped and stored water needed for the steam-powered locomotives. Earlier designs had wooden towers and large wooden wheels, Fig. 2-10. A vane (fan-like piece), extending behind the wheel, kept the wheel turned directly into the wind. Later, towers and wheels of galvanized steel replaced the wooden mills. Though often abandoned due to the coming of cheap electric power to the farms, many of the towers still dot the midwestern and western farmlands, Fig. 2-11. The wheel rotated an offset shaft or crank which lifted a long shaft in a pumping action. The shaft was connected to the piston of a lift pump. As the wheel revolved, the piston of the pump was made to move up and down to draw water from the well.

Fig. 2-10. Early mills in America, both wheel and tower, were made of wood. Without such mills settlement of the West would have been extremely difficult.

Fig. 2-11. Later windmills were made from metal. Many of them still stand. Few are still used for pumping water, having been replaced by other forms of energy.

RISE AND DECLINE OF WIND POWER

Though relatively cheap, wind power was never highly dependable, nor was the windmill as powerful as the water wheel. It was more valuable, of course, in lowlands which did not have water power. By 1810 it had become an efficient source of power and was used for irrigation, drainage, and, to some extent, for grinding corn and small grain. In 1890, Denmark built the first wind generator. Eventually other European countries adapted them for generating electric power. The excess electricity was stored in batteries for use during calm days.

HARNESSING WATER POWER

Primitive travelers might have been content with letting the power of moving water carry their boats downstream. Later generations found another way to use water energy. The waterwheel, which was invented about 200 B.C., captured the energy of flowing or falling water and converted it to power for grinding grain. This was a big improvement over old methods of grinding corn with a stone bowl and a smooth rock.

The first waterwheels were designed to let the water flow push against paddles from below. This is known

as an **undershot waterwheel**, Fig. 2-12. During the Middle Ages, the undershot waterwheel had been perfected to the point where a single wheel could develop 40 to 60 horsepower.

In 1750, John Smeaton, a British civil engineer, designed the **overshot waterwheel**. This was a big improvement over the undershot wheel. The water flowed over the top of the wheel on its downward side. Thus, the weight of the water, added to its motion, made the wheel much more powerful, Fig. 2-13. It could work harder and power heavier loads.

TURBINES

Waterwheels did not get enough power out of the mechanical energy of the water. This led to development of the more efficient water turbine. One of the early successes was designed by Benoit Fourneyron, Fig. 2-14. It had curved vanes and was suspended on a vertical shaft. The shaft was hollow and the water entered the turbine through the shaft. As it entered the turbine it was forced to move outward, pushing against the blades before it ran off the edge of the turbine.

The modern water turbine is mainly used to generate electricity. Power is supplied by dammed up water. The kinetic energy is supplied by the water's long drop from the top of the dam to the turbine below. This vertical distance is called "head." The water is channeled through a sluice or penstock (a special waterway or tunnel formed in the dam). It strikes the turbine and causes

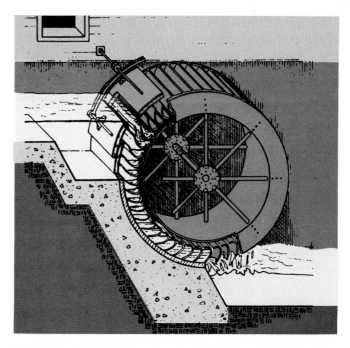

Fig. 2-12. Undershot waterwheel. In this early design, the water flowed under the wheel. The water pushed against the paddles attached to the wheel.

Fig. 2-13. A model of an overshot waterwheel. Because the weight of the water was combined with its kinetic energy, this type of waterwheel was more efficient.

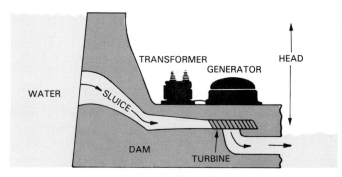

Fig. 2-15. Modern water turbine. Kinetic energy is supplied by the "head." This is the distance the water falls before it hits the turbine.

USING HEAT TO PRODUCE MOTION

Cave dwellers built fires to create more comfortable surroundings. Early metalworkers burned wood and coal to melt their materials for casting and forging. Heat, however, was not used for creating motion until someone invented a machine which could contain and convert the heat energy. This invention was the heat engine.

HEAT ENGINES

Devices which can contain the energy of heat and make it do work are called **heat engines**. The first working model of one was developed about 130 B.C. The inventor was Hero, a scientist, writer, and mathematician of Alexandria — part of present-day Greece. His machine, Fig. 2-16, hardly resembles a steam engine

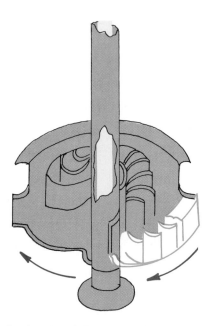

Fig. 2-14. A sketch of Fourneyron's water turbine. The cutaway shows stationary vanes in black and movable vanes in color. Water coming down the hollow shaft flowed out through the stationary (not moving) vanes and struck the movable vanes on the outer rim of the wheel. Its downward force was converted into rotational motion.

Fig. 2-16. The first model of a heat engine was a Greek toy. Steam escaped from the kettle-like container below. It traveled up the pipes on either side to the hollow sphere above. As pressure built up, steam escaped through the spigots, causing the sphere to spin.

it to spin rapidly as it rushes through the fins or blades. Being connected by a shaft to a generator, the turbine supplies the power to generate electricity. Fig. 2-15 is a cross section of a modern hydroelectric power plant.

or steam turbine as we know them today. It looked like a kettle with a large ball mounted on top of it. The lower section of the "kettle" held water. It had a tight-fitting cover to contain the steam. Two pipes extended upward from the flat top and were connected to a hollow ball. The ball, which could rotate on the pipes, had two nozzles. These were angled to face in opposite directions. When water was boiled in the kettle, the steam escaped through the pipes into the hollow sphere. As pressure built up in the sphere, the steam escaped through the nozzles. This caused the ball to rotate.

You can easily see the force of steam created by boiling water if you observe a covered pot in which food is being prepared. The pressure will actually lift the lid until some steam can escape and relieve the pressure.

What do you suppose would happen if you were to attach a whistle so that the steam would have to travel through it to escape? This principle is used on the whistling teakettle, Fig. 2-17. Is the steam doing useful work by blowing the whistle?

Hero had no idea how important his invention was. If further attempts were made to try to harness steam, they were not recorded until Leonardo da Vinci invented a gun which used steam to fire bullets. That was in 1495.

STEAM POWERED WATER PUMPS

Little came of da Vinci's invention and little more was done with steam as a power source until English miners needed a better way to pump water out of the peat mines. In 1698 Thomas Savery designed and built a steam-operated pump that worked on pressures above 150 psi (pounds per square inch or 1034 kPa). It marked the beginning of the modern era of power development which saw the employment of fossil fuels (coal, petroleum, or natural gas) as chief source of energy.

Savery's engine was further improved when he went into partnership with Thomas Newcomen. The improved engine shot steam into a hollow cylinder to move a piston. The piston was connected to one end of a rocking beam. The other end was connected to a pump rod. As the steam worked the piston, the beam rocked up and down like a seesaw. See Fig. 2-18.

HARNESSING THE FOSSIL FUELS

The earliest practical steam engines were designed to burn fossil fuels. A Scottish instrument maker, James Watt, developed the first successful reciprocating steam engine, Fig. 2-19. Other designs based on the same principle would drive boats and locomotives. See Fig. 2-20.

HOW THE STEAM ENGINE OPERATES

The steam engine depends on the pressure of super hot expanding steam. The steam is let into a large cylindrical chamber that has been fitted with a movable cap called a piston. The pressure of the expanding steam pushes against the piston causing it to move along the

Fig. 2-17. A "singing" teakettle demonstrates the power of expanding gases. Fuel from the stove top provides the energy to create steam. The force of the steam will blow the whistle. (DACOR)

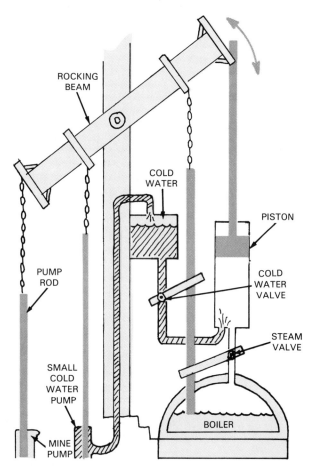

Fig. 2-18. Newcomen's engine was used to pump water from coal mines in England. The steam piston rocked the beam up and down to move the pump piston in a pumping motion.

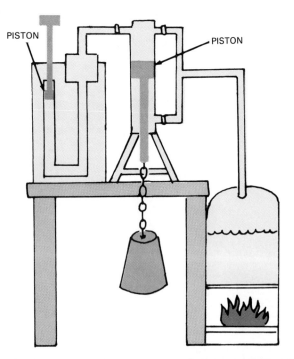

Fig. 2-19. Watt's steam engine invention is much simplified here. It did not work too well and used huge quantities of fuel. The problem with his design was having to cool and reheat the walls of the cylinder after each cycle.

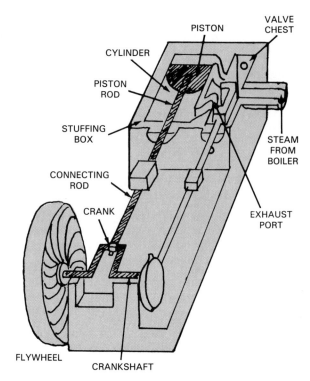

Fig. 2-20. A later version of Watt's steam engine. He learned to pack insulation around the outside of the cylinder walls. This is the area known as the "stuffing box." The engine was eventually adapted to boats, tractors, and locomotives.

cylinder. The piston is attached to a long rod which turns a crank. The crank will turn whatever is connected to it.

STEAM AND GAS TURBINES

The **turbine** is a more efficient mechanism for converting heat energy and kinetic energy of steam or gas to mechanical energy. It is able to use the pressurized steam and convert it directly into rotary motion. The steam engine converts linear (straight-line) motion to rotary motion. In the process, some mechanical energy is wasted because of friction, inertia, and nonproductive motion.

Another reason for the efficiency of the turbine is that energy can be applied to it continuously. While the piston of a reciprocating steam engine returns to the top of the cylinder it cannot use the energy of the steam. That is, its conversion of energy is only intermittent (off and on).

TYPES OF STEAM TURBINES

In the **impulse turbine,** high pressure steam is piped to the turbine blades. It escapes through nozzles and is moving at very high speed as it strikes the blades. See Fig. 2-21. In modern turbines, the speed may be faster than 4000 ft. (1219 m) per second.

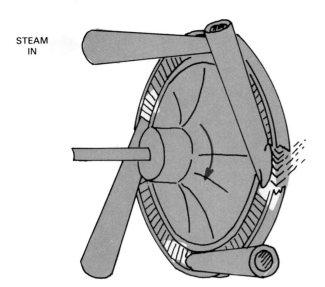

STEAM IN

Fig. 2-21. The de Laval steam turbine is an example of the impulse type turbine. High-pressure steam is piped to the turbine blades.

REACTION TURBINE

In a **reaction** type **turbine,** Fig. 2-22, the steam passes through several sets of blades. The result is a stronger "push." All of the blades are on the same shaft. As the steam leaves one set of blades it is redirected so it strikes the next set at an angle. This angle causes great pressure even though the pressure of the steam has been reduced.

COMPOUND TURBINE

A third type is known as the **compound turbine.** Actually, it is a series of turbines. The steam passes through one after the other. Each is designed to make the best use of the steam as its pressure and temperature

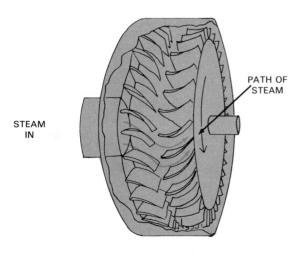

Fig. 2-22. Single element of a Parsons reaction steam turbine. Steam had to pass through several sets of blades.

drop, Fig. 2-23. This design is used on very large turbines where the large blades of the reaction type would be too large.

INTERNAL COMBUSTION (IC) ENGINES

Combustion engines harness energy by capturing the force of rapidly burning fuels. Steam engines are called external combustion engines because the fuel is burned outside the engine. See Fig. 2-24. **Internal combustion engines** burn the fuel inside the engine, Fig. 2-25. The two popular types are the gasoline engine and the diesel engine.

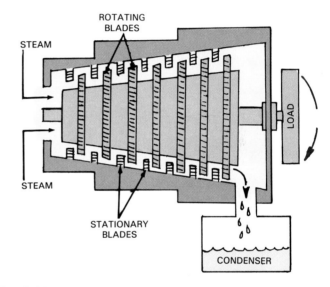

Fig. 2-23. A compound turbine is actually a series of turbines. This type of turbine makes the best use of steam as temperature and pressure of the steam drop.

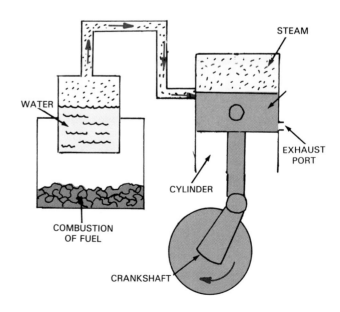

Fig. 2-24. Steam engines are external combustion engines because the fuel supplying the energy is burned outside the engine.

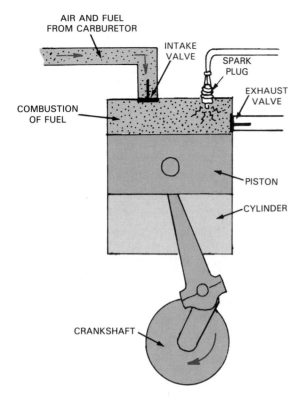

Fig. 2-25. The internal combustion engine is basically a free-moving piston moving up and down in a cylinder. A crankshaft converts up and down motion into circular motion. One port allows fuel to enter; a second one allows spent fuel to escape.

HOW THE INTERNAL COMBUSTION ENGINE DEVELOPED

Experimental models of four-stroke internal combustion engines were developed in both England and Europe around 1800. The first fuel was a mixture of coal, gas, and air. About 60 years later, the French engineer, Etienne Lenoir, built a two-stroke engine that ran smoothly but did not develop very much power. Even so, several hundred were sold over the next few years.

Alphonse Beau de Rochas, another French inventor, developed the engineering principles that led to the development of the first successful four-stroke engines. He listed the sequence of operations:
1. "Suction during the first outstroke to bring in the fuel-air mixture."
2. "Compression of the mixture during the following instroke."
3. "Ignition of the compressed charge . . . by means . . . of a spark from an induction coil, with the expansion of the combustion products and the performance of useful work during the next outstroke."
4. "Expulsion of the burned gases during the next instroke."

Fourteen years later, Nikolaus A. Otto, a German engineer, applied these principles in the first four-stroke engine. Even today, most internal combustion engines operate on these basic principles. It is often called the "Otto Cycle."

Improvements and modifications continued. These included Charles Curtis' velocity-compound turbine introduced in New York in 1897 and Zoelly's multistage impulse turbine. The latter was introduced in 1900. It had fewer stages than the Curtis turbine and was very simple.

Otto's engine was manufactured in Germany and in the United States. Nearly 50,000 of the engines were sold between 1877 and 1894.

Improvements in the internal combustion continued. In 1855, Gottlieb Daimler, the German inventor and engineer, designed and built an engine that burned gasoline. Another German engineer, Rudolph Diesel, built the first **diesel engine.** It was designed to operate on low-grade fuels. See Fig. 2-26.

HARNESSING IS CONVERSION

You have just studied the history of human effort to shift the burden of work from themselves to machines. First they were able to train animals to do the work. Then they designed and adapted machines for this purpose.

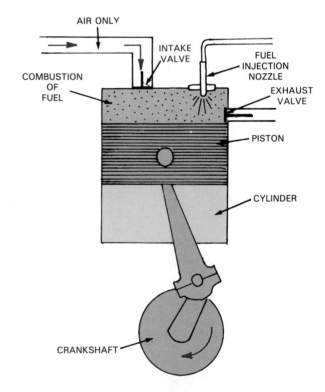

Fig. 2-26. Diesel engines burn low-grade fuel inside the cylinder depending upon heat of compression to ignite the fuel-air mixture.

One common element in these efforts was the use of some agent that could convert energy. One form of energy was usually converted to a different form so it could do what needed doing. Thus, the devices or beasts that were harnessed to do the work came to be called **energy converters** and the process, **energy conversion**.

MODERN CONVERSION SYSTEMS

As you learned in Chapter 1, energy can be readily converted from one form to another. It is common to convert to a new form to make the energy usable. Consider the following:

- Mechanical to electrical. Examples are generators and alternators. They may be coupled to an engine, a water or steam turbine, or a windmill.
- Electrical to mechanical. Examples are an electric motor used to turn a fan or a solenoid that controls a switch.
- Chemical to electrical. A typical example is an electric cell that provides power for a radio or a toy.
- Electrical to chemical. An alternator produces electricity and stores it in a cell or a battery.
- Light (or radiant) to electrical. A solar cell collects sunlight and changes it into electricity.
- Molecular to heat. Increased movement of molecules in a substance produces heat that increases the temperature of water so it can be used to do work.

It is not unusual for energy to be converted several times before it is used to do work. It is desirable, however, to reduce the number of conversions since some energy is lost in each conversion.

Many of the ancient conversion devices are still with us in improved designs. The gasoline and diesel engines, together with the steam turbine, supply most of the mechanical energy used in the world for transportation and electric power. Hydroelectric turbines provide electric power.

Generators convert mechanical energy to electrical energy. The electric motor converts the electrical energy back to mechanical energy.

OTHER ENERGY CONVERTERS

Use of fossil fuels to provide heat for cooking and comfort, began, as we saw, with campfires to heat caves and cook meat that was turned on a spit. The open fire soon was replaced with crude fireplaces and then stoves, Fig. 2-27. Modern homes are heated with furnaces or electrical heating devices.

Likewise, the use of open fires for fashioning tools was replaced by blast furnaces and process heat. (Pro-

Fig. 2-27. An old wood burning stove. It was once popular as it was a more efficient heating system than an open fire or a fireplace.

cess heat is heat energy used to manufacture products of metal, plastic, and other materials.) Factories today use electric power to run machines.

In the 20th century, jet engines and rocket engines joined the growing list of propulsion systems. These powerful engines are designed for the aircraft and aerospace industry.

LOOKING AHEAD

At some time in the future we can expect that more efficient energy converters will be perfected. More than likely, many of these will be devices that can convert energy directly from one form to another with few or no moving parts. Some of these inventions are already in the development stages. Because power systems and energy converters must always work together so that energy may be put to work, understanding them is important for an understanding of energy. For this reason, Chapter 3 will be devoted to power systems and Chapter 5 to modern energy conversion.

SUMMARY

Energy usage in the early cultures was greatly limited. Caves provided shelter. Skins of animals were crudely shaped as clothing. Muscle power provided a mode of getting from one place to another.

What about food? If you could call it technology, it consisted of flinging crude weapons to bring down wild game.

Eventually, animals were tamed to provide transportation. Crude sails captured wind power for moving rafts or boats.

It would seem that after muscle power, thermal energy and kinetic energy were among the first forms of energy to be harnessed. Windmills were probably developed in Persia and used to draw water from wells. Europeans refined the Persian's invention.

Waterwheels captured the energy of falling or flowing water. A later refinement of the waterwheel was the turbine. It was more efficient than the waterwheel. Today, the turbine, albeit many times refined, is a major device for the generation of electric power. It utilizes both water and steam as a primary energy source.

The idea of using heat to produce motion did not take hold much before the first century. An inventor, scientist, and mathematician living in what is present-day Greece, developed the forerunner of the modern heat engine. Even to him it was no more than a toy for the amusement of his friends.

While the first heat engines ran on steam produced outside the engines, modern heat engines nearly always burn fuel inside the engine. Engines power all types of land vehicles, water vessels, aircraft, and spacecraft.

Machines that convert some form of energy to a form that produces power are called "converters." Energy may be converted many different times before it is actually used to perform work. Even the human body is a converter. Think about it!

DO YOU KNOW THESE TERMS?

compound turbine
diesel engine
generator
heat engine
impulse turbine
internal combustion engine
overshot waterwheel
post mill
reaction turbine
smock mill
turbine
undershot waterwheel
windmill

SUGGESTED ACTIVITIES

1. Prepare a report on an invention of the last 100 years which harnessed energy or improved upon the usage of energy and made life more pleasant for people. Collect your information from your school's resource center and by interviewing an older person who has experienced change because of an invention.
2. Construct a working model of a waterwheel and demonstrate it to your class. Explain how and why it works.
3. Build a model of an early windmill, working with metal or wood. Use descriptions and illustrations in the text to assist in the design.
4. Sketch a device that can convert steam into motion. Suggest or describe how you think it can be built.

TEST YOUR KNOWLEDGE

1. Name two early forms of mechanical energy.
2. Give the steps involved in designing.
3. The _____ was invented to make use of the energy in falling or moving water.
4. The _____ was the first invention that captured wind for pumping water and grinding grain.
5. The invention of the machine in No. 3 was credited to:
   a. The Dutch.
   b. Settlers in America.
   c. The English.
   d. The Persians.
6. Describe the importance of the windmill to American farms.
7. The waterwheel gave way to the _____ because the latter made better use of the energy of the falling water.
8. What is the meaning of the word "head" in relation to the energy of falling water?
9. Devices which can contain the energy of heat and make it do work are called _____.
10. Make a sketch of Newcomen's steam engine.
11. The steam _____ uses the energy of steam and changes it directly to rotary motion.
12. Describe the difference between an external combustion engine and an internal combustion engine. Give examples of each.
13. When a device or machine is designed to change one form of energy to another which can be used to do work, it is called:
    a. A machine.
    b. An energy converter.

3
POWER SYSTEMS

The information given in this chapter will enable you to:
- Define a power system.
- List four basic power systems.
- List and describe the role of each essential element of a power system.
- Define the energy-related terms: work, power, torque, velocity, head, pressure, Btu, efficiency, and quad.

In Chapters 1 and 2 you learned about the nature and sources of energy and how they can be harnessed to do work. In this chapter you will study power systems and their relationships. You will also learn how energy is measured.

POWER SYSTEMS

A power system is important to the harnessing of energy so that the energy can work. The system converts, transmits, and controls energy in such a way that the energy performs useful work. A farm tractor, Fig. 3-1, is a good example of a power system. The tractor is very powerful and can do the work of many horses. There are other power systems, such as a lawn mower that are much smaller and have fewer parts. On the other hand, some power systems are very large. They take up a great deal of space and have many parts. An

Fig. 3-1. A commonly seen agricultural power system. Tractors convert fossil fuel to do farm work.

electric power plant would be a good example of a large power system. See Fig. 3-2.

BASIC POWER SYSTEMS

There are four basic power systems. These four systems provide power for all of the technology systems anywhere in the world. They are:
- Electrical systems.
- Mechanical systems.
- Fluid systems.
- Heat systems.

No matter how small or big they may be, all power systems have certain basic elements that have the same function. The elements are:
- A source of energy. Often, this is a fuel of some type, but it could be another form of energy, such as waterpower or wind.
- A conversion method. There must be a way of changing the energy to a more usable form so it can do work.
- A method of transmission. This is a way of moving the energy to the load where it can do the work for which it is designed.
- A control system. Like a leash on a dog, a control mechanism keeps the energy in check so that it does useful work and no more.
- Measuring devices. These may be gauges, meters, or indicators. Their purpose is to give the user information about the power system that is needed to keep it working properly.

- An output or load. This is what the power acts upon. On a tractor, the load is the weight of the tractor in addition to the implement (tool) it is pulling.
- Losses. Not all the energy released is used. Some of it is passed on to the atmosphere as heat. Friction of moving parts rubbing against one another rob some of the energy.

If you compare a tractor or a lawn mower with a power plant you will see that all three fit the definition of a power system. See Fig. 3-3.

ENERGY SOURCE

From previous chapters you already know that an energy source is a supply of energy which can be tapped when there is work to do. Some energy sources are said to be **constant.** They are always available when needed. Others are called **changeable** sources because availability is not certain.

Constant Energy Source

A **constant energy source** is one that is always on hand day or night. Nothing needs to be done to it. The energy is ready for use when it is needed. Nearly all constant energy sources are the result of storage. The storage may have been accidental or purposeful. See Figs. 3-4 and 3-5. Fuels such as coal, crude oil, and natural gas resulted from natural or accidental storage. Natural processes caused them to be trapped in beds and pools. This accidental storage makes them a constant source.

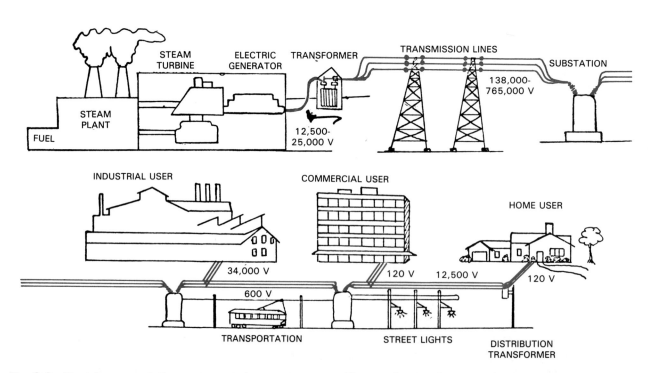

Fig. 3-2. Electric power stations are a complex power system. They are large, take up much space, and have many parts.

ELEMENT	LAWN MOWER	POWER PLANT	EXPLANATION
Energy Source	Fuel and tank to contain it.	Piles of coal, natural gas pipeline, nuclear rods, force of falling water.	Gasoline is fed continuously to a running engine; power plants are supplied fuel on continuous basis—it is available night and day.
Conversion Method	Engine burns fuel and uses expanding gases created by combustion.	Fuels and fission create steam to rotate turbine. Falling water also spins turbine which operates electrical generator.	Gasoline engine converts chemical energy into heat energy so engine can create spinning motion (kinetic energy). Power station uses chemical, nuclear, or kinetic energy to create electrical energy.
Transmission Path	Piston and crankshaft	Electric power lines	Kinetic or mechanical energy travels from expanding gases to piston, to crankshaft to cutting blade. Electrical energy travels through wires to run electric motors and lights.
Control System	Throttle lever handle for steering	Switches Solenoids	Mower's speed and direction are controlled. Switches and solenoids turn electric power on and off.
Measuring Devices	Oil pressure gauge Gasoline tank gauge	Electric meters	In both systems, gauges tell operator operating condition of power system. Fuel, current, and voltage can be measured.
Load	Cutting blade	Electric motors Electric lights Electric heating Electronic equipment	Load is the work done by systems; mower cuts grass; electric power runs machine, provides light.

Fig. 3-3. Comparison of large and small power systems. Both have all the requirements of a power system.

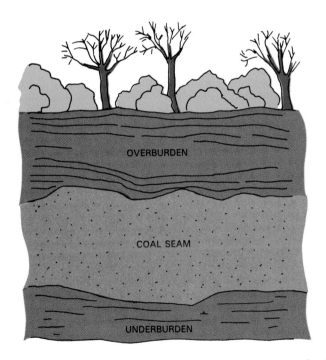

Fig. 3-4. Accidental (or natural) storage of fuel source. Coal beds formed by nature lie stored until the coal is mined.

Fig. 3-5. Automotive fuel is readily found at filling stations and can be stored in a vehicle's fuel tank until it is needed. This vehicle is being fueled with compressed natural gas (CNG) which reduces emissions of nitrogen oxides and hydrocarbons. (The Coastal Corp.)

Water stored behind a dam to provide a continuous source of mechanical energy is purposeful storage. Electricity, however, is one constant energy source which is hard to store. Certainly, small quantities can be captured in batteries but, usually, electricity must be used as it is generated.

Most power systems are designed to use a constant source of energy. Systems built around the use of electric power, natural gas, and petroleum are the most common. Electric power operates lights, electric motors, and appliances, Fig. 3-6. Sometimes it is used for heating and cooling.

Natural gas provides a fuel source for cooking, heating homes, office buildings, and factories. Petroleum provides gasoline to fuel internal combustion engines.

Fig. 3-6. Electricity is a constant source of power. It can be used to produce light, heat, or mechanical energy. This modern cook stove uses electricity for heat. Electricity operates the combination convection and microwave oven through a programmable electronic controller. (Creda Inc.)

CHANGEABLE SOURCES

Some energy sources are available only some of the time. These types are called changeable sources. The wind is a good example of such a source. As you learned in an earlier chapter, it can provide power to run many different kinds of machines cheaply. However, it is not available day in and day out. Sometimes there is no wind or it is not strong enough to do work.

Solar energy is also a changeable source. The sun delivers large quantities of energy to the earth. It can be made to do a great deal of work. In fact, it helps to produce many other sources of energy. Still, solar radiation cannot be depended upon entirely. On cloudy days there is little sunlight and at night none at all. At best, it only shines about 12 hours out of the day in most parts of the world, Fig. 3-7.

Tides and ocean waves are another form of changeable energy. Like the wind and sun, they are not always available to provide dependable energy.

As conventional constant sources are used up, changeable sources will become more and more important. Technology is, therefore, being employed to find new ways for converting and storing the energy from these sources. Then the stored energy will work while the source is not available.

CONVERSION OF ENERGY

As you know from Chapter 2, energy may need to be converted from one form to another before it can do work. Some power systems may make more than one conversion. The engines in lawn mowers and automobiles, for example, cannot use fuel as a direct power source. The fuel must first be converted to hot gases. Fig. 3-8 shows transfer of energy from expanding hot gases through a piston to a crank.

A coal or oil-fired power generating station makes several energy conversions. First it converts stored

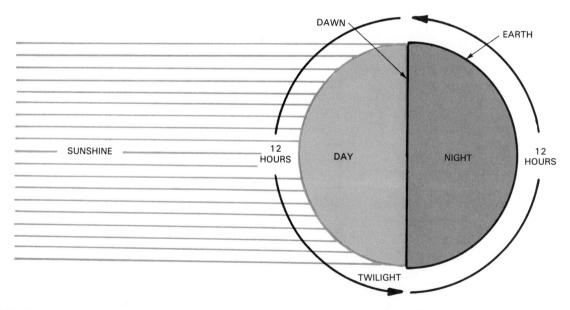

Fig. 3-7. Because of the earth's rotation, sunlight is available only about half of every 24 hour period. The drawing shows why this is so.

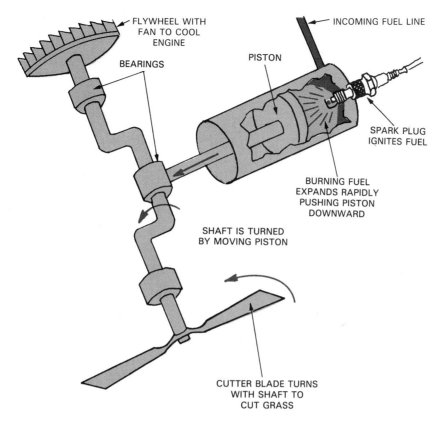

Fig. 3-8. Lawn mower engine converts fuel (chemical energy) to hot gases which create mechanical force to turn the mower blade so it will cut grass.

chemical energy into heat energy. Then the heat energy is converted to mechanical energy. Finally, mechanical energy is changed into electrical energy. See Fig. 3-9.

TRANSMISSION OF ENERGY

The third element of a power system, transmission, is the act of moving power from the source to the load.

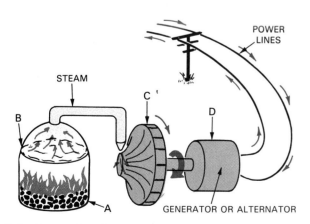

Fig. 3-9. Converting the energy of coal to electricity takes several steps or conversions. A—Coal is burned to produce heat energy. B—Heat raises water temperature until it turns to steam. C—Heat energy in steam is converted to mechanical energy as it spins the turbine. D—Energy of whirling turbine drives generator. The generator converts mechanical energy to electrical energy.

Three basic methods are employed for this purpose:

- Fluid linkage. Fluids or gases are forced through pipes or tubes to the load. The fluids are driven by the force of the converted energy. The brake system on an automobile operates this way. Pressure from the driver's foot or a pump driven by the engine forces brake fluid through tubes to push the brake shoe against the brake drum. See Fig. 3-10. Hoists are another example of fluids transmitting power from one point to another, Fig. 3-11.

  Industry often uses compressed air to move switches and operate controls on assembly lines. The air is under pressure and moves through tubes and channels in metal parts called gates.

- Electrical linkage. Moving through wires, electric current carries the energy of the rotating turbine to homes and factories. There it can be used for lighting and to operate appliances. The wires or conductors, known as transmission lines, Fig. 3-12, allow the generating station to be placed nearer its source of energy.

- Direct mechanical linkage. A shaft, a belt or some other linkage moves power from one spot to another. The driveshaft of a truck or automobile takes the rotational motion of the engine and carries it to the wheels, Fig. 3-13. Belts, chains, ropes, or cables also can be used to transmit mechanical energy, Fig. 3-14.

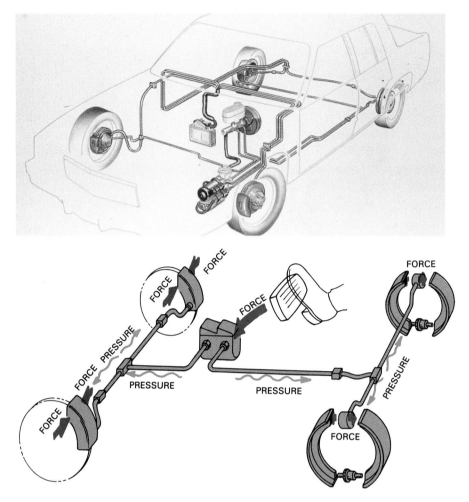

Fig. 3-10. Brakes on an automobile use fluid to transmit pressure. The foot on the brake pedal applies the pressure. The force is moved from the pedal to all four wheels by fluid which is noncompressible.   (Cadillac Motor Car Div.)

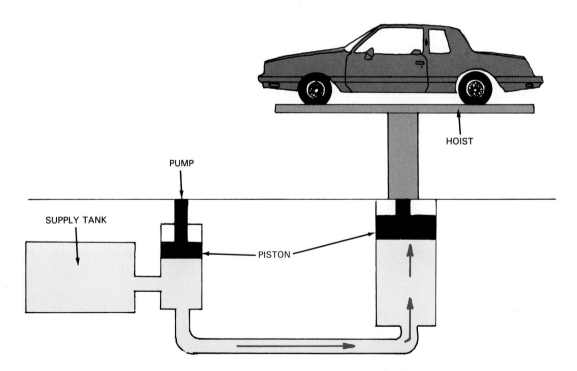

Fig. 3-11. Power to lift an automobile on a hoist is transmitted to the hoist by the mechanical action of a pump. Fluids are forced through pipes and cylinders. The fluid, in turn, forces a piston to move up a cylinder, lifting the automobile.

A

B

C

Fig. 3-12. Electric power can be transmitted long distances and voltages varied for different uses. A—Electric power, generated by a power station leaves the station over high-voltage lines. B—Substations transform the electricity to different voltages and send it out to users. C—Lines shown here carry the electricity to homes.

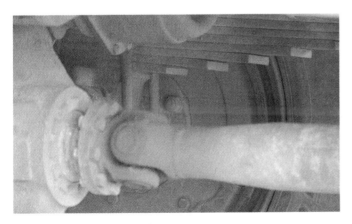

Fig. 3-13. Transmission systems. Chains and driveshafts are two mechanisms used to transfer power from a power source to drive wheels.

CONTROL OF ENERGY

Controls in power systems do three things:
- Turn power on and off.
- Regulate speed, stop movement, or cause some other change in the way power is applied.

Fig. 3-14. Belts and chains as power transmitters. A chain and two sprockets change reciprocating motion to circular and then to linear. Belts on an old thrashing machine transmit rotary motion or torque from one mechanism to another.

• Control direction.

Switches control the electric lights and motors in your home and school. See Fig. 3-15. The automobile has a switch for turning on electric current for starting and operating the engine. Other switches may be used to control lights, windows, locks, and windshield wipers.

Electric power plants have switches and transformers which control the electric current going out to customers, Fig. 3-16.

TYPES OF CONTROLS

Controls can be divided into two basic types:
• Total control. An electrical switch is of this type. When it operates, power is either on or off. The switch on an automobile turns current from the battery on or off.
• Partial control. Controls of this type regulate direction and speed.

As important as "off and on" control is, can you imagine driving an automobile using only the ignition switch? You actually need several partial controls to drive an automobile. The shift lever controls transfer of power from the engine to the drive wheels. The steering wheel keeps the vehicle on the road. The accelerator produces changes in speed. A brake pedal slows or stops the vehicle.

CONTROL OF UNWANTED OUTPUT

Some of the output of power systems cannot be used. Sometimes it must be regulated to make the system safe or to protect it from damage. Outputs which must be controlled include:

Fig. 3-16. Power plants require many kinds of controls. Top. Main control panel in power plants turns power on and off. Bottom. Power plant substation with huge circuit breakers and transformers.

• Friction. This is the resistance between two bodies rubbing against each other, Fig. 3-17.
• Waste heat. Electric power plants and factories cannot use all the heat they generate. Part of it must be carried away and eventually released to the air. See Fig. 3-18.

Fig. 3-15. Switches control electric lights.

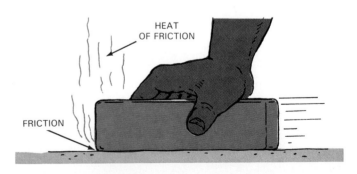

Fig. 3-17. Friction is the resistance of two bodies or surfaces rubbing against each other.

Fig. 3-18. Cooling towers are used by power plants to carry away waste heat.

Fig. 3-19. Instruments are necessary for monitoring the operation of a power system. Instrument panel of automobile tells the operator the operating condition of the charging, cooling, lubrication, fuel supply, and sometimes, the braking system.

MEASUREMENT

Measurement is important in a power system for more than one reason:
- It determines the amount of energy in the system.
- It may determine how much work the system is doing.
- It may also determine how fast work is being done.

METERS AND GAUGES

It is necessary to know about the operating condition of parts of a power system. Indicators such as gauges give this information. The indicators for some systems are very simple. Others are complicated.

A simple gauge to show how much fuel remains in the fuel tank may be all that is needed on a lawn mower. An automobile will have measuring instruments such as:
- A fuel gauge.
- An **ammeter** for measuring current output of the alternator.
- Heat indicator for the engine coolant temperature.
- A speed indicator to gauge how fast the automobile is traveling.
- An oil pressure gauge to indicate whether the engine is getting proper lubrication.

Fig. 3-19 shows the instrument panel of an automobile. The operator needs to "read" the gauges to be sure the vehicle is operating properly.

Power stations have many gauges to show how much electrical pressure (voltage) and how much electricity (current) is moving through power lines. Special meters measure how much electricity is being used by customers, Fig. 3-20.

LOAD

The load of a power system is the useful output. A bicycle is a good example of a simple system. The energy source is the mechanical force exerted on the pedals by the rider. The crank changes the pumping motion to circular or rotating motion. The chain transmits the motion to the rear wheel. The weight of the person and the weight of the bicycle are the load.

LOSSES

Losses are a result of the conversion of energy from one form to another. Every power system has these losses. As explained earlier, it is an unwanted element

Fig. 3-20. Every user of electricity has a watt/hour meter to measure use of electricity for billing purposes.

of energy usage. You will learn about the efficiency of power systems later when you study conservation methods. Power losses are one of the reasons why efficiency is low in some conversions of energy and high in others.

Two important elements mentioned earlier contribute to loss of energy. They are friction and waste heat.

Extra energy must be used to overcome the tendency of parts moving past each other to stick. Even with the use of lubricants such as oil and wax, friction cannot be avoided.

Waste heat is simply energy that has lost its ability to do any more work. It must be passed off to the air, water, or other matter around it. In the automobile, for example, the heat is passed from the engine to the water in the cooling system. The radiator and the heater pass the heat off to the atmosphere.

MEASUREMENT SYSTEMS

In North America, two measurement systems are used to express inputs of work, power, horsepower, torque, velocity, acceleration, and other related values. One is the U.S. conventional and the other is the SI metric measurement system which is used in Canada and almost every nation in the world. The United States uses both systems but has never wholly converted to metric.

POWER/ENERGY TERMS AND MEASUREMENT

As you study energy and power you will begin to use many new terms. These terms are used whenever people discuss energy and its relationship to power, work, and energy sources.

Several terms describe loads. **Work** is the basic unit for output or system load.

Work is force times the distance through which the force acts. A force is anything that causes an object to move.

It is easy to see some types of work being done. For example, when the rider pumps the pedals, a bicycle moves forward, Fig. 3-21. Molecular (movement of molecules or atoms) activity is another matter. We can see the effects of electricity doing work. We cannot, however, observe the progress of electrons in a conductor. Current travels unseen through wiring. While we cannot see it move, we can see the light it creates and feel the heat it causes. We see it in the lighted electric bulb and the rotating of an electric motor. These effects tell us that the electricity is at work.

Fig. 3-21. It is easy to see the application of energy to produce motion in the bicycles of these two young riders.

MEASURING WORK

Work is measured in foot-pounds for conventional U.S. measure. A **foot-pound** is the amount of force necessary to move a 1 lb. load a distance of one foot.

The mathematical formula is:

$$\text{WORK} = \text{DISTANCE} \times \text{FORCE or}$$
$$\text{W} = \text{D} \times \text{F}$$

Problem: To see how this formula works, suppose that the weight of a cyclist is 150 lb. How much work is done if the bicycle is pedaled 500 ft.?

$$\text{W} = \text{D} \times \text{F}$$
$$\text{W} = 500 \times 150$$
$$\text{W} = 75,000 \text{ ft.-lb.}$$

In SI metric, which is being adopted by most countries, work is measured in joules. The unit is named after James Prescott Joule, an English scientist. The **joule** (J) is the work done when a force of 1 newton moves an object 1 meter. The formula is:

$$\text{Work} = \text{Distance (in meters)}$$
$$\times \text{Force (in newtons)}$$

One newton is the force used to accelerate 1 kilogram of mass to 1 meter per second squared.

Problem: How many joules of work are done when a force of 300 newtons (N) is applied for a distance of 150 meters?

Using the formula:

$$W = 150 \times 300 = 45,000 \text{ joules}$$

POWER

Work and power are often confused. Power is used when work is meant and vice versa. It is easier to separate them if you simply remember that work does not concern itself with how long it takes. Only distance and force matter.

Power is the amount of work done in a given amount of time.

To measure power we use the following formula:

$$\text{Power} = \frac{\text{work}}{\text{time}} \text{ or } P = \frac{w}{t}$$

It can also be written:

$$P = \frac{D \times F}{t}$$

Problem: If a rider and bicycle weighing 150 lb. travel 500 ft. in 20 seconds, how much power is developed?

$$P = \frac{D \times F}{t}$$
$$P = \frac{500 \times 150}{20} = \frac{75,000}{20}$$
$$= 3750 \text{ lb.ft. per sec.}$$

HORSEPOWER

In U.S. conventional measure, the standard unit of power is the **horsepower**, Fig. 3-22.

One horsepower (hp) is the energy needed to lift 33,000 lb. 1 foot in 1 minute.

Problem: If 200 lb. are lifted 165 ft. in 1 minute, how much horsepower is developed?

To find the answer:
1. Find out how much work was done.
   Work = D × F
   Work = 165 ft. × 200 lb. = 33,000 ft.-lb.
2. Now, use the formula to find horsepower.

$$Hp = \frac{\text{work}}{\text{time in minutes} \times 33,000}$$
$$= \frac{33,000}{1 \times 33,000} = 1$$

3. The answer is 1 Hp.

In SI metric the unit of power is the **watt** (W). One watt is the power needed to produce work at the rate of 1 Joule (J) per second. A joule is the force needed

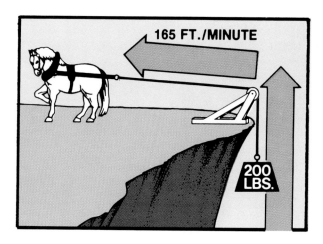

Fig. 3-22. Horsepower is the rate of doing work.

to move 1 kg of mass 1 meter. Since the watt is a small unit, power is usdually measured in kilowatts which is equal to 1000 watts. A kilowatt is about 1.34 hp. The formula for determining power is:

$$\text{Watts} = \text{work (in joules or kilojoules)/time}$$

$$\text{Power (in watts)} = \frac{\text{distance (in meters)} \times \text{force (in newtons)}}{\text{time}}$$

or:

$$\text{Power} = \frac{\text{work (in joules)}}{\text{time}}$$

Problem: If a bicycle and rider with a mass of 70 kg move 300 meters in 20 seconds, how much power is developed?

$$\text{Power} = \frac{\text{distance} \times \text{force}}{\text{time}} = \frac{300 \times 70}{20}$$
$$= 1050 \text{ W}$$

TORQUE

Power has just been described in terms of a force being exerted in one direction. However, not all motion is linear (in a straight line). An engine, for example, or a crank both produce a **turning** or **rotational force**, Fig. 3-23. This twisting force is called **torque**.

Torque is measured the same way as work. Multiply force (in pounds) by distance (in feet). There is one slight difference, however. Work is measured by distance times force. *Torque is always measured using force times distance.* The answer is always given in pound-foot (lb.-ft.) units.

Problem: If a force of 50 lb. is applied to move a load 10 ft., how much torque is developed?

in feet per minute, miles per hour, meters per minute, and kilometers per hour. The formula for determining acceleration is:

$$\text{Acceleration} = \frac{\text{change in velocity}}{\text{time}}$$

An important example of acceleration has to do with gravity. A body falling toward the earth will always be accelerated at 32.8 ft. (10 m) per second. That is, its velocity will increase 32.8 ft. per second *every second*.

CALCULATING POTENTIAL ENERGY

An important aspect of acceleration for engineers is calculating the unreleased energy which exists in an object at rest (but which might fall or could be made to fall). The energy resulting from falling matter is related to acceleration. The formula is:

p = mgh where:
p = potential energy.
m = mass or weight.
g = rate of free fall acceleration (32.8 ft. per sec.).
h = the distance the object or body falls.

To see how it works, try this problem:

A 10 lb. rock sits at the edge of a cliff. It is 100 feet to the bottom of the cliff. What is the gravitational potential energy of the rock?

Use the formula p = mgh

p = 10 × 32.8 × 100 = 32,800 foot-pounds

HEAD

Head is related to potential energy of gravity. It is the difference between the level of a fluid, such as water, and its outlet. A dam holding back a reservoir of water is a good example for illustrating this definition. See Fig. 3-24.

EFFICIENCY

Efficiency is a ratio (percentage or comparison) between the amount of energy put into a power system and the amount of work the system does. The efficiency of an engine can be computed using the formula:

$$\text{Efficiency} = \frac{\text{energy out}}{\text{energy in}} \times 100 \text{ Percent}$$

PRESSURE

Pressure is the force exerted by gas or liquid that is held in a closed container. Its formula is:

$$\text{Pressure} = \frac{\text{force}}{\text{area}}$$

Torque = Force times Distance (F × D)
T = F × D
T = 50 lb. × 10 ft.
T = 500 lb.-ft.

In SI metric, torque is called **moments of force**. The unit of measure is the newton meter (N·m). The formula is the same one used for U.S. conventional measure:

Moments of Force = force × distance

Problem: An engine uses a force of 10 newtons to move a load 2 meters. Compute the Moments of Force.

Moments of Force = 10 × 2
Moments of Force = 20 N·m

VELOCITY

Velocity is the distance traveled divided by the time of travel in one continuous direction. It is used to measure speeds of objects moved by some form of energy.

For example, gasoline is used to propel an automobile at velocities that are measured in miles or kilometers per hour. Measurement can also be in meters per minute, feet per second, or feet per minute.

The formula for velocity measurement is:

$$\text{Velocity} = \frac{\text{distance in the same direction}}{\text{time}}$$

ACCELERATION

Acceleration is related to velocity. It is a change in velocity over a period of time. Acceleration is expressed

Fig. 3-24. Head is the difference in height between the inlet and outlet of water stored behind a dam or in a tower. The greater the head, the greater the force at the outlet.

Pressure is an important force in energy systems. Its applications include:
- Systems which depend upon air for operation: Compressed air may be used to operate some types of brakes; to inflate tires, Fig. 3-25, and to control machines in manufacturing operations.
- Systems which operate with fluids under pressure:
  a. Steam for driving turbines for generation of electricity.
  b. Liquids under pressure in hydraulic systems where they lift heavy loads and exert extremely high forces. An example of such a system is the hydraulic brake system on an automobile. Refer again to Fig. 3-10.

The U.S. conventional unit of measure for pressure is pounds per square inch. It is abbreviated "psi." The metric unit of pressure is the pascal (Pa). A **pascal** is a force of 1 newton per square meter (N/m^2).

BRITISH THERMAL UNIT (Btu)

When heat energy moves from one body to another you can measure it. The British thermal unit (**Btu**) is the U.S. conventional unit of heat measurement. It is the amount of thermal (heat) energy required to raise

Fig. 3-25. Air pressure is an important force in our energy-conscious world.

the temperature of 1 lb. (about a pint) of water 1 degree Fahrenheit. The Btu is used in:
- Calculating the amount of heat in fuels.
- Determining the heat output of mechanisms such as furnaces and boilers.
- Calculating the heat requirements for spaces to be heated.
- Measuring the cooling capacity of an air conditioning unit.
- Measuring heat loss in a home. A chart for converting various quantities to Btus is shown in Fig. 3-26.
- Measuring solar radiation. This is expressed as Btu per sq. ft. per hour.

BARREL

The **barrel** is a unit of measure used in the petroleum industry. A barrel holds 42 gallons. The barrel term is left over from the early days of the oil industry when crude oil was shipped from the oil fields to refineries

ENERGY CONVERSION TABLE

To convert from	to	Multiply by
kilowatt-hour	Btu	3412.8
1 ton bituminous coal	Btu	26,200,000
1 bbl. crude oil	Btu	5,600,000
1 bbl. residual oil	Btu	6,290,000
1 gal. gasoline	Btu	125,000
1 gal. #2 fuel oil	Btu	138,800
1 cu. ft. natural gas	Btu	1031
1 mcf natural gas	Btu	1,031,000
1 therm natural gas	Btu	100,000
1 Btu	kWh	0.000293

Fig. 3-26. Factors for converting various energy units to British thermal units.

in large wooden barrels, Fig. 3-27. Since then, railroad tank cars, huge seagoing tankers, oil barges, and pipelines have replaced the barrel. The term, however, is still used throughout the industry particularly in discussing resources, supply, and demand.

QUAD

A **quad** is a measurement of energy consumption equal to 10^{15} Btu. This amounts to 1 quadrillion Btu, about the same amount of energy released by the burning of a half million barrels of oil daily for a year.

SUMMARY

A power system converts, transmits, and controls energy so that it can perform useful work. A basic power system consists of these elements: an energy source, a conversion method, a transmission path, a control system, measuring system, and a load. An unwanted element of every system is energy losses.

The most valuable energy sources are those that are constant. Among the constant sources are the fossil fuels. Another one is hydropower which comes from water energy usually stored behind a dam.

Other sources of energy are known as changeable, so named because they are not always available. Solar and wind energy are in this classification as are tidal energy and wave energy. As constant sources become scarce, these source will become more important.

Fig. 3-27. Wooden barrels, once used to ship oil, are long gone, but oil is still measured by the barrel.

Before they can be used, most energy sources must be changed to a different form. The most common change is to burn a fossil fuel to produce heat. The heat can be changed to a number of work-producing energies. This change is called "conversion."

It is important to be able to read meters and gauges and to work out measurement problems related to energy gathering and energy usages. These terms and measurements include: load, work, power, horsepower, torque, velocity, acceleration, head, efficiency, pressure, British thermal unit (Btu), barrel, and quad.

There are a number of important measurements related to power systems. It is important to know the formulas for terms such as: work, power, horsepower, torque, velocity, acceleration, head, efficiency, pressure, British thermal unit.

DO YOU KNOW THESE TERMS?

acceleration
ammeter
barrel
Btu
changeable energy source
constant energy source
efficiency
energy source
foot-pound
head
horsepower
joule
moments of force
pascal
power
pressure
quad
torque
velocity
watt
work

SUGGESTED ACTIVITIES

1. Visit a power station where electricity is generated. Ask about job opportunities and the kinds of work done there. Take part in a class discussion on what you heard and observed.
2. Prepare a poster showing the different parts of an electric power system. Label each as to whether it is: energy source, conversion, transmission, control, measurement, or load.
3. Secure a photograph of a small, complete power system from a magazine or old automotive shop

manual. Label different visible parts of the system according to the part's function in the power system.

4. Draw up a list of all the different kinds of power systems in the community where you live. List the source of energy for each.

5. Construct a power system which is capable of moving about with an energy source carried on or within the system.

6. Use a torque wrench and a spring scale to demonstrate torque to the class.

TEST YOUR KNOWLEDGE

1. Define the word "power system."
2. Name the six basic elements or parts of a power system.
3. A(n) _____ energy source is one that is always at hand when needed.
4. A changeable energy source is one which:

a. Can readily be changed from one form of energy to another.
b. Is available some of the time but not all of the time.
c. Neither of the above.

5. List the conversions made by a power station in producing energy.
6. What three things do controls do in power systems?
7. Friction is the _____ between two bodies rubbing against each other.
8. A gas gauge is part of _____ in a power system.
9. Work is measured in _____ in the U.S. conventional measuring system.
10. The _____ is the metric term for measurement of work.
11. What is torque? What term is used in its measurement?
12. Velocity is the distance traveled divided by the _____ of travel.

Top Producing Countries, 1989

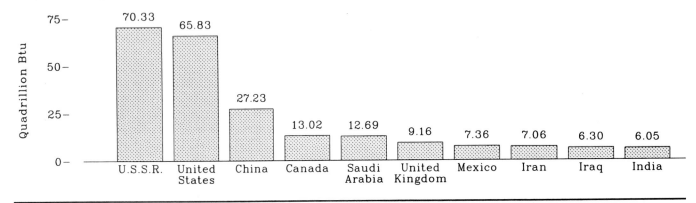

Note: Because vertical scales differ, graphs should not be compared.
Source: See Table 111.

Fig. 4-1. Ten countries supply more than 225 quads of energy every year.   (Annual Energy Review, 1990)

World Total and Leading Producers, 1979-1989

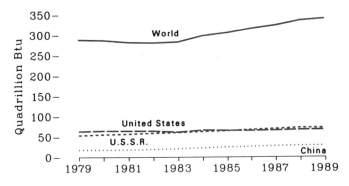

Fig. 4-2. World totals of energy production are graphed for a 10 year period.

World Areas, 1979-1989

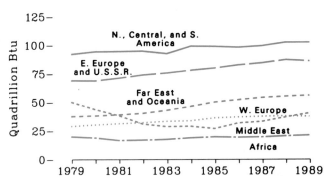

Fig. 4-3. The American continents are the largest energy producers in the world.   (U.S. Dept. of Energy)

4

ENERGY PRODUCTION AND CONSUMPTION

The information given in this chapter will enable you to:
- *Present facts concerning the level of energy production for the world.*
- *Name the major energy-producing regions and countries.*
- *Discuss the usage of energy in the United States and Canada in terms of barrels per day of oil.*
- *Identify the four major energy-using sectors in our society.*
- *Discuss the extent of fossil fuel energy supplies still remaining in the U.S.*
- *Define the terms, proven reserves, and future reserves.*

Because energy drives our daily activities, our economy and our technologies, production of constant energy supplies is a vast worldwide industry. World production of energy amounted to 338.34 quadrillion Btu in 1989, the most current figures available.

While all nations and regions of the world produce a certain amount of energy, 10 countries account for most of the world's energy. See Figs. 4-1 through 4-4 for details.

Energy quantities are figured in two ways:
- British thermal units (Btus). A Btu, as you learned in Chapter 3, is the amount of heat energy needed to raise the temperature of 1 lb. of water one degree Fahrenheit. When discussing energy production and usage, the term, "quad," is used because the amounts of energy discussed are so huge. As you also learned in Chapter 3, a quad stands for one quadrillion. One quadrillion would be written out as 1,000,000,000,000,000 or 1×10^{15} (one times ten to the fifteenth power).

World Primary Energy Production by Area and Country, 1989
(Quadrillion Btu)

Area and Country	1989 [a]
North, Central, and South America	**101.54**
Canada	13.02
Mexico	7.36
United States	65.83
Venezuela	5.58
Other	9.75
Western Europe	**36.66**
France	4.02
Netherlands	2.62
Norway	5.77
United Kingdom	9.16
West Germany	5.87
Other	9.22
Eastern Europe and U.S.S.R.	**85.24**
East Germany	2.88
Poland	5.52
Romania	2.73
U.S.S.R.	70.33
Other	3.78
Middle East	**39.55**
Iran	7.06
Iraq	6.30
Kuwait	4.31
Saudi Arabia	12.69
United Arab Emirates	5.06
Other	4.13
Africa	**20.24**
Algeria	4.21
Libya	2.67
Nigeria	3.80
South Africa	4.27
Other	5.29
Far East and Oceania	**55.12**
Australia	5.78
China	27.23
India	6.05
Indonesia	4.66
Japan	3.07
Other	8.33
World Total	**338.34**

Fig. 4-4. Primary energy production is given by area and country.

- Barrels of oil. A barrel, as you know, is equal to 42 gallons. It is also equal to 5.8 million Btu. When the term is used it often means that total energy from all sources is equal to the energy found in that many barrels of oil.

CONSUMPTION OF ENERGY

Energy consumption in the United States for the year 1990 was 81.44 quads. This means that every day of that year, every person living in the U.S. used 897,000 Btu of energy.

In the year 1989, the entire worldwide consumption of oil was estimated at about 66 million barrels daily. This is nearly 383 quads. In the same year, the U.S. consumed a total of 81.35 quads or 37.9 million barrels a day. Canada consumed 1.76 million barrels a day or 10.2 quads.

CONSUMPTION IN METRIC

The metric unit for energy measurement is the joule (J). A Btu equals 1055.6 J or 1.055 kJ. A barrel of oil is equal to 5,829,000 kJ. A kilojoule is 1000 joules.

ENERGY'S IMPORTANCE TO US

A Btu is hard to envision, but it is roughly the amount of heat given off by one kitchen match. Can you imagine almost 94 quadrillion matches? Fig. 4-5

Energy Unit	Equivalent[1]	
1 Btu of Energy	1	match tip
	250	calories (International Steam Table)
	0.25	kilocalories (food calories)
1,000 Btu of Energy	2	5-ounce glasses of table wine
	250	kilocalories (food calories)
	0.80	peanut butter and jelly sandwiches
1 Million Btu of Energy	90	pounds of coal
	120	pounds of oven-dried hardwood
	8	gallons of motor gasoline—enough to move the average U.S. passenger car about 164 miles (1989)
	10	therms of dry natural gas
	11	gallons of propane
	1.1	days of U.S. energy consumption per capita
	2	months of the dietary intake of a laborer
1 Quadrillion[2] Btu of Energy	45	million short tons of coal
	60	million short tons of oven-dried hardwood
	1	trillion cubic feet of dry natural gas
	170	million barrels of crude oil
	470	thousand barrels of crude oil per day for 1 year
	28	days of U.S. petroleum imports
	26	days of U.S. motor gasoline use
	26	hours of world energy use (1989)
1 Barrel of Crude Oil	15	days of U.S. petroleum consumption per capita
	5.6	thousand cubic feet of dry natural gas
	0.26	short tons (520 pounds of coal)
	1,700	kilowatt-hours of electricity
1 Short Ton of Coal	102	days of U.S. coal consumption per capita
	3.8	barrels of crude oil
	21	thousand cubic feet of dry natural gas
	6,500	kilowatt-hours of electricity
1,000 Cubic Feet of Natural Gas	4.8	days of natural gas use per capita
	0.18	barrels (7.4 gallons) of crude oil
	0.047	shot tons (93 pounds) of coal
	300	kilowatt-hours of electricity
1,000 Kilowatt-hours (kWh) of Electricity	34	days of U.S. electricity use per capita
	0.59	barrels of crude oil[3]
	0.15	short tons (310 pounds) of coal[3]
	3,300	cubic feet of dry natural gas[3]

Fig. 4-5. Energy measurements are more meaningful when given in common, familiar, everyday terms.
(Annual Energy Review, 1990)

shows the relationship of quantities of energy with a number of more familiar units.

Energy, whether in matches or in barrels of oil, is important to our well-being. Without it our industries would not function; our homes and cities would be dark and cold; our planes, automobiles, and trains would be useless. Our means of earning a livelihood would be severely limited. We would be reduced to a level of existence where most work would be done by hand or by using animals.

ENERGY-USING SECTORS

Who uses this energy? Every one of us consumes some of it directly or indirectly every day. Perhaps an electric alarm clock wakes us up each morning. That requires your home to have electric power. The clock itself was produced in a factory. This also consumed some energy. Buses and automobiles take people to work and to school. Fuel, again, was consumed. An automobile or truck manufacturing facility also consumed energy to build these vehicles. Homes, factory activities, and education require buildings. Energy was needed to construct the buildings. Energy is used to heat them in winter and cool them in the warm season. Living, learning, and working consumes energy.

For convenience, energy users are separated into groups or "sectors." There are four major sectors:

- The **nonenergy sector.** This group, Fig. 4-6, includes all those who use fossil fuels as a raw material to produce other products or goods. The energy material—gas, oil, or coal—is used for the manufacture of products such as chemicals, medicines, asphalt, grease and oil, plastics, synthetic fabrics, and steel. In 1990, this sector, by estimate, used about 6.5 quads of energy.
- **Industrial sector.** In this category, Fig. 4-7, are all

Fig. 4-6 Some enterprises, like this asphalt plant, are classified as part of the "nonenergy" sector since they use energy mainly as a raw material for manufacture of other products.

Fig. 4-7. Energy consumers from the industrial sector. Top. Farming today is done with large, fossil fueled tractors. Middle. Heavy machines are used in quarrying operations. Bottom. Oil refinery. This manufacturing operation provides low-sulfur petroleum products.   (The Coastal Corp.)

types of mining, manufacturing, and farming businesses. They use energy in all its forms: gas, oil, coal, and electricity. In the year 1990, this sector consumed 30.17 quads of energy.

- **Transportation sector.** All methods of moving people and materials are part of this group. Automobiles, trucks, trains, ships, planes, conveyors, and pipelines are included. See Fig. 4-8. Since they can use no other energy sources, transport vehicles account for half of the petroleum and a portion of the natural gas used by all sectors. Their share of the energy used in 1990 was 22.06 quads, up about 1 quad from 1980.
- **Residential/Commercial sector.** See Fig. 4-9. Included in this group are all private and public buildings such as stores, office buildings, hospitals, and schools. They consumed 29.22 quads of energy in 1990. This is an increase of about 1.7 quad from the 1980 level.

Fig. 4-8. The transportation sector involves industry that moves people and materials. Transportation involves trucks and cars, trains, airplanes, and pipelines. (The Coastal Corp.)

Fig. 4-9. Residential/commercial sector. Public and private buildings (and homes) use oil, gas, electricity, and sometimes coal or wood to provide heat and light and to operate appliances.

ENERGY SUPPLIES

Meeting the needs of all the energy-using sectors just discussed takes a constant, reliable source of supply. For as long as most of us can remember there have been plentiful supplies of petroleum, natural gas, and coal. Where it came from or how long it would last was of little immediate concern to the average person. It was thought that it would last forever. Then came the Arab oil embargo of 1973 and the organization of OPEC (oil-producing countries surrounding the Persian Gulf). These events served to call national attention to the threat of energy shortages and to the reality that energy would become increasingly expensive.

At the time of the 1973 embargo, the United States was already importing large amounts of oil from foreign countries. Imports included 40 percent of our crude oil, 40 percent of our fuel oil, and smaller percentages of gasoline and jet fuel (kerosene).

The immediate result of the embargo was that imports of fuel fell 3 1/2 million barrels per day below the projected (expected future) demand. The United States had no oil in storage and could not increase production from its own wells. Available supplies had to be spread among all energy-using sectors. In 1973, 46 percent of all energy consumed in the U.S. was oil. To keep industry going, the government allowed it almost all the energy it needed. The burden of cutting back was left mainly to the residential/commercial sector.

Gas station fuel allotments were cut back. Long lines formed at the gas pumps. Commuters began forming car pools to get to their jobs. States imposed 55 mph (89 k/ph) speed limits on highways to conserve gasoline and diesel fuel.

Along with the scarcity of fuel came huge increases in the price of crude oil. In 1973, just before the crisis, the price of a barrel of crude ranged from $3.89 for domestic to $4.08 for imported oil. By 1979, although the embargo was gone and all but forgotten, the price of crude oil had moved up to an average of $21 a barrel. This represented an increase of almost 700 percent in six years. Prices at the gas station pump had moved up to more than $1.00 per gallon. See Fig. 4-10.

The years since 1973 have seen oil prices rise and fall several times but never returning to pre-1973 prices. In 1986, overproduction by OPEC nations caused crude oil prices to decline sharply but briefly to a low of $1.26 per million Btu. In 1989 prices began to slowly rise once more, moving up to $1.32/million Btu that year. In 1989 it rose to $1.40/million. When Iraq invaded Kuwait, uncertainty over oil supplies pushed the price up to $2.63/million Btu. That represented a 21 percent increase from 1989 levels. Such volatile prices have caused international concern for energy costs and supply. Our oil-dependent society continues to be vulnerable to political unrest and the limits of fossil fuel supplies.

LOOKING AHEAD

The crisis of 1973 and the uncertainty since has done two things. It makes everyone realize the importance

Fig. 4-10. Higher prices for crude oil are reflected in gasoline prices which advanced to well over $1 per gallon in the 1970s. Since then prices have fluctuated. In 1979, gasoline sold for $1.35 a gallon at the pump. After the Gulf War, prices hovered around $1.14 to $1.26.

of energy to all sectors of our society. It also draws attention to the disturbing fact that the fuels we depend so heavily upon will one day be depleted, some of them in our own lifetimes. At the same time, the more scarce these energy sources become, the more expensive they will become.

A third effect of the vulnerability to fossil fuel shortages is a growing conviction that nations should be conserving these supplies and supplementing them with alternate sources of energy.

HOW MUCH IS LEFT?

We can only speculate about how long our supplies of petroleum, coal, and natural gas will last. We are able to estimate with some degree of certainty the amount of our reserves. We can also project with somewhat less certainty our annual consumption for some years ahead.

RESERVES

Proven reserves are those fossil fuel supplies which have been located and estimated with reasonable certainty and which can be recovered with available technology. These proven reserves for the entire world are shown in Fig. 4-11. Reserves for the United States and Canada are given in Fig. 4-12.

There is another classification of fuel supplies known as resources or **future reserves**. These are stocks that have been estimated to exist with a lesser degree of certainty. There is a possibility that they can be recovered sometime in the future with technology not now available. See Fig. 4-13.

Figs. 4-12 and 4-13 show that our most plentiful fuel resource is coal. It accounts for 90 percent of all our reserves. At current levels of consumption, supplies could last for about 300 years.

Oil reserves, estimated at 34 billion barrels, may not last out the turn of the century according to some oil industry officials and government experts. Others believe that oil reserves will last to about the year 2020.

WORLD RECOVERABLE RESERVES OF FOSSIL FUELS

SOURCE	AMOUNTS AVAILABLE	MEASUREMENT UNIT
Coal	1,347,142[1]	Million Short Tons
Petroleum	1,002.2[2]	Billion Barrels
Natural Gas	3,989.4[1]	Trillion Cubic Feet

1. World Energy Council
2. Oil and Gas Journal

Fig. 4-11. In 1990 these were the recoverable reserves of fossil fuels. (U.S. Dept. of Energy)

U.S. AND CANADIAN RESERVES OF FOSSIL FUELS

	SOURCE	AMOUNTS AVAILABLE	MEASUREMENT UNIT
U.S.	Coal	232,395[1]	Million Short Tons
	Petroleum	26.5[2]	Million Barrels
	Natural Gas	167.1[2]	Trillion Cubic Feet
CANADA	Coal	5,473[1]	Million Short Tons
	Petroleum	6.1[1]	Million Barrels
	Natural Gas	167.1[2]	Trillion Cubic Feet

1. World Energy Council

2. Oil and Gas Journal

Fig. 4-12. U.S. and Canadian reserves of fossil fuels in 1990. (Annual Energy Review, 1990)

FUTURE RESERVES

ENERGY SOURCE	AMOUNTS AVAILABLE	BTU EQUIVALENT (IN QUADS)
Coal	1.8 trillion tons	37,863
Oil	139 billion barrels	806.2
Natural gas liquids	28 billion barrels	.115
Natural gas	895 trillion cu. ft.	917.375
Total		39,586.49

Fig. 4-13. Future reserves or resources are stocks of fossil energy of which geologists and energy companies are less certain. They may be recoverable in the future. To the fossil fuel future reserves we can add 3.8 billion short tons of uranium supplies.

Oil production in the lower 48 states as was predicted by these same experts, began to decline during the 1980s and 1990s.

Natural gas has been in short supply in some areas for some time. Demand has been outstripping production since 1973. Production from existing reserves will continue to decline into the 21st century.

NUCLEAR ALTERNATIVE

Once heralded as a long-term source of energy that would reduce dependency on fossil fuels, nuclear power has not developed as was once predicted. Concerns for its effect on the environment, the threat of nuclear reactor accidents, and questions of how to safely store spent radioactive wastes have brought development to a standstill. While there were 111 nuclear power stations operating in 1990, no construction permits were pending. There were no units on order.

However, nuclear power generation has increased almost every year since the mid 1960s. A total of 577 billion kilowatt-hours of nuclear-generated electricity were consumed in 1990.

RENEWABLE ENERGY ALTERNATIVES

The history of energy shows that through the centuries, societies have shifted their dependence from one source of energy to another several times. Ancient Rome experienced a shortage of firewood and had begun to turn to other sources. Ruins of ancient buildings also show evidence of passive solar construction. Our own country once depended upon wood for a large part of its energy. Later, the shift was to coal. Still later, the emphasis shifted to petroleum which was more desirable because its by-products were less polluting.

Now the need is urgent to change our energy appetites once again. On a short-term basis, conservation of energy will help stretch out or save the fossil fuel resources remaining. A long-term approach will include the development of renewable energy such as solar, kinetic, and certain thermal resources. In 1990, about 3.2 quads of renewable energy was produced and consumed in the U. S. Most of it was from hydroelectric generating stations. Another estimated 3.6 quads consumed consisted mostly of non-electric utility use of biofuels such as wood, waste, and alcohol fuels. Total consumption of renewable sources has been estimated at 8 percent of total energy consumption.

Geothermal sources contribute to the total electrical energy resources. This source peaked in 1987 at 10.8 kilowatt-hours (kWh). In 1990 production had fallen to 8.6 kWh.

Each of the inexhaustible and renewable resources will be discussed in greater detail in later chapters. Conservation of energy will be the subject of still another chapter.

SUMMARY

Production of energy is a vast, world-wide industry that produced over 338 quadrillion Btus of energy in 1989. Ten nations accounted for most of these supplies.

While one of the largest producers of energy, the United States is also the largest energy user in the world. In 1990 the U.S. consumed 81 of the 338-plus quads produced. By contrast, Canada used around 10 quads in 1990.

The four main categories that consume energy are:
1. Industries that use fossil fuels as a feed stock (raw material) to produce products such as chemicals, medicines, synthetic fabrics, and steel.
2. Manufacturing, construction, mining, quarrying, and farming industries.
3. Transportation which includes privately owned vehicles, trucks, trains, air travel and transport,

ships and other water based systems, and pipelines.

4. Residential and commercial. This is a group that uses energy to provide heat and power used in buildings such as homes, businesses, and stores.

Since 1973, the world and its people have been concerned about the dwindling supplies of fossil fuels. Parallel concerns are the effects of energy consumption on the environment and the danger of over-dependence on foreign countries for fossil fuel supplies.

There are alternatives to be considered as well as conservation measures that will ease the impending shortages. These are taken up in later chapters.

DO YOU KNOW THESE TERMS?

future reserves
industrial sector
nonenergy sector
proven reserves
residential/commercial sector
transportation sector

SUGGESTED ACTIVITIES

1. Prepare a report on how the method of heating homes in the central and northern regions of North America differs from practices in other cold climates. Ask the school librarian to help you find reference materials.

2. Interview homeowners in your neighborhood on the effect of rising energy costs on home heating bills. Try to estimate average rise (by percentage) from 1973 to the present. Report the results of these interviews to your class.

3. Secure an EPA (Environmental Protection Agency) report on estimated fuel consumption for various models of automobiles manufactured. Prepare a graph showing difference in fuel costs between a 4-cylinder auto and an 8-cylinder full-size family auto based on 12,000 miles of driving and an average cost of $1.28 for unleaded gasoline.

4. Draw up a list of the products and services you personally require and indicate which energy-using sector contributes the service or product.

TEST YOUR KNOWLEDGE

1. In 1989 how much energy was produced in the entire world?

2. Name the energy-consuming sectors in your society and define each one.

3. Name the two units used for reporting energy production and usage.

4. There are _____ Btus of energy in a barrel of crude oil.

5. What is meant by the term, "proven fuel reserves"?

6. What is our most plentiful fossil fuel reserve and how long is it estimated to last?

Year	Motor Vehicle Registrations (millions)					Motor Fuel Consumption [1] (thousand barrels per day)		
	Passenger Cars	Motorcycles	Buses	Trucks	Total	Gasoline [2]	Other Fuels [3]	Total [4]
1960	61.7	0.6	0.3	11.9	74.4	3,953	159	4,112
1961	63.4	0.6	0.3	12.3	76.6	4,034	176	4,210
1962	66.1	0.7	0.3	12.8	79.8	4,120	192	4,312
1963	69.0	0.8	0.3	13.4	83.5	4,274	211	4,485
1964	72.0	1.0	0.3	14.0	87.3	4,454	236	4,690
1965	75.3	1.4	0.3	14.8	91.7	4,644	269	4,913
1966	78.1	1.8	0.3	15.5	95.7	4,846	306	5,152
1967	80.4	2.0	0.3	16.2	98.9	5,014	329	5,343
1968	83.6	2.1	0.4	16.9	103.0	5,300	370	5,670
1969	86.9	2.3	0.4	17.9	107.4	5,604	413	6,017
1970	89.2	2.8	0.4	18.8	111.2	5,845	439	6,284
1971	92.7	3.3	0.4	19.9	116.3	6,125	494	6,619
1972	97.1	3.8	0.4	21.3	122.6	6,529	554	7,083
1973	102.0	4.4	0.4	23.2	130.0	6,819	642	7,460
1974	104.9	5.0	0.4	24.6	134.9	6,531	639	7,170
1975	106.7	5.0	0.5	25.8	137.9	6,719	628	7,347
1976	110.4	5.0	0.5	27.7	143.5	7,075	697	7,772
1977	113.7	5.0	0.5	29.6	148.8	7,287	760	8,046
1978	116.6	5.1	0.5	31.7	153.9	7,555	837	8,392
1979	120.2	5.5	0.5	33.3	159.6	7,291	913	8,204
1980	121.7	5.7	0.5	33.6	161.6	6,820	896	7,716
1981	123.5	5.8	0.5	34.5	164.3	6,726	969	7,695
1982	123.7	5.7	0.6	35.3	165.3	6,679	972	7,651
1983	126.7	5.6	0.6	36.5	169.4	6,731	1,043	7,774
1984	127.9	5.5	0.6	38.0	172.0	6,850	1,127	7,977
1985	132.1	5.4	(5)	39.6	177.1	7,020	1,158	8,178
1986	135.4	5.3	(5)	40.8	181.5	7,229	1,202	8,431
1987	137.3	4.9	(5)	41.7	183.9	7,359	1,242	8,601
1988	141.3	4.6	(5)	43.1	189.0	7,405	1,306	8,711
1989	143.1	4.4	(5)	44.2	191.7	7,437	1,385	8,822
1990[6]	145.0	4.3	(5)	45.2	194.5	7,474	1,408	8,882

[1] Includes only motor fuel taxed at the prevailing tax rates in each State. Excludes motor fuel exempt from tax payment, subject to tax refund, or taxed at rates other than the prevailing tax rate. Experience has shown that the total motor fuel consumption quantity cited here equals more than 99.0 percent of gross reported motor fuel consumption.

[2] Includes motor gasoline, aviation gasoline, and gasohol.

[3] Includes distillate fuel oil (diesel oil), liquefied gases, and kerosene when they are used to operate vehicles on highways. Excludes jet fuel beginning in 1962.

[4] Excludes losses allowed for evaporation, handling, etc.

[5] Included in trucks.

[6] Previous-year data have been revised. Current-year data are estimated and will be revised in future publications.

Note: Sum of components may not equal total due to independent rounding.

Sources: •1960 through 1975—Federal Highway Administration, *Highway Statistics Summary to 1975*, Tables MV-201 and MF-221. •1976 through 1986—Federal Highway Administration, *Highway Statistics Annual*, Tables MV-1, MF-21, and MF-25. •1987 and forward—Federal Highway Administration, *Selected Highway Statistics and Charts 1989*.

Fig. 5-I. This table shows the number of automobiles, motorcycles, trucks, and buses in the United States and their consumption of gasoline and other fuels year by year. (Annual Energy Review, 1990, Dept. of Energy)

5 CHAPTER
TRANSPORTATION SYSTEMS

The information given in this chapter will enable you to:
- *Define a system and explain transportation as a system.*
- *Name the transportation environments and discuss the transportation system used in each environment.*
- *List the propulsion systems that have been adapted to different environments.*

Transportation is an important sector of the U.S. economy. As such, it has great impact on the life of every citizen. Without it we would find our lives drastically changed. An enterprise of such importance and magnitude as transportation also has a huge impact on energy usage. According to statistics released by the U.S. Dept. of Energy, transportation consumes 63 percent of all the oil used in the United States. In 1990, transportation's share of oil consumption was nearly 11 million barrels a day. See Fig. 5-1.

TRANSPORTATION AS A SYSTEM

We refer to all of transportation, and to most, if not all, of its components, as systems. Let us refer to Chapter 1 for a definition of a system. If you recall, a system is an organization of components (parts) to perform a specific function.

The vehicle that carries you to school is a system within the larger system called "land transportation." A school bus consists of a structure or body and chassis to carry all parts of the bus. An engine provides power for motion. Wheels support the vehicle and allow motion. Brakes stop the vehicle or reduce its speed. A steering mechanism controls direction of motion.

An oiling subsystem lubricates the moving parts of the engine. A cooling subsystem carries away unwanted heat that could destroy the engine. Another subsystem provides electricity to ignite fuel and provide power and light.

Transportation systems are designed to move people and cargo from one place to another. Modern technology is employed to make this movement as efficient as possible given the time, cost, and care needed by the process. Each year we spend 20 percent of the gross national product moving people and products. So efficient have modern modes of transportation become that transportation today is global in dimension. People have always been curious about new places and wish to explore the nations of the world, Fig. 5-2. At the same time, fast modes of transport bring us products from all over the world, Fig. 5-3. Our own products find markets in many foreign countries.

The term, system, is applied to transportation in different ways. We may use it to mean all of the transportation companies along with their vehicles, tracks or roadways, buildings and the people who keep the companies operating. We may think of transportation as a managed grouping of vehicles to provide efficient movement of people and cargo in a particular environment. We may also use the term to mean the different parts making up one vehicle.

TRANSPORTATION ENVIRONMENTS

Transportation environments are the different mediums in which transportation vehicles travel. Since the mediums are vastly different, the vehicles must be so designed that the environment supports the actions of the transportation vehicle. The term, **modes,** also is used to mean environments. It is also applied to the systems or vehicles designed to suit each environment.

A   B

Fig. 5-2. Foreign travel is a common experience not only for Americans or Canadians but for people of other nations. A—Air travel is common in some countries. B—These people are traveling in Japan. The train shown travels by magnetic levitation (HSST)

Fig. 5-3. It is common to find goods at your local shopping mall that came from a foreign country.

Modern transportation is provided in several environments or modes. They are:
1. Land.
2. Water.
3. Air.
4. Space.

The nature of each of these environments determines the design of the vehicles that can be used in each of them. There are great differences in these vehicles because of the differences in the environment's ability to support the weight of the vehicle and its passengers or cargo.

LAND TRANSPORTATION

Land transportation is any means used on or under the ground to move people or goods from one place to another. There are different modes that can be grouped in different ways. One method is to consider them by what they are chiefly designed to move and the distance of the move.

PEOPLE MOVERS

The first major grouping is known as **people movers**, Fig. 5-4. In this group are bicycles, cable cars, elevators, escalators, scooters, and carts. People movers are generally employed in moving people short distances at relatively slow speeds. Many of them are considered **on-site transportation** because they do not leave the building or area where they are located. For example, elevators and escalators move people between floors of buildings while cable cars move people up steep hills or across gorges. Conveyors move people down walkways or halls in places such as airports. A rail system with small transporters may move airline passengers from terminal buildings to parking lots.

However, usage of any land transport vehicle depends upon the needs of people at any one time. A

Fig. 5-4. Top. People movers in O'Hare Airport, Chicago. The "moving sidewalks" speed passengers walking to and from planes and terminals. (Murphy/Jahn Architects) Bottom. A bicycle is another kind of people mover with people providing the energy.

Fig. 5-5. These conveyances are all examples of people transport.

conveyor, for example, may also be used to move cargo such as sand, gravel, and coal in mining and quarrying operations. Can you think of other places that a conveyor is used to move cargo? Consider shipping rooms and warehousing operations.

PASSENGER TRANSPORTERS

Passenger transporters are also designed generally for the movement of travelers. However, unlike people movers, this class of transportation is intended to move people over longer distances at considerably higher speeds than people movers. In this class are automobiles, commuter trains, limousines, buses, and motorcycles. See Fig. 5-5.

CARGO TRANSPORTERS

Cargo transporters move material and products over long distances. In this class are pipelines, railroads, and trucks. In general, these vehicles and structures are sturdily built to carry heavy loads. See Fig. 5-6.

Pipelines are stationary (do not move) structures. However, various types of material move through them. They are designed to carry gaseous and liquid gravel, other minerals, cement, and coal. Solids are mixed with water before they are moved. Experiments are being conducted to transport other solids, among them sulfur, steel, grain, and, possibly, bulk mail.

An important advantage of pipeline systems is the rapid movement of goods. Once in motion, goods moved unhampered by stoplights, traffic snarls, grade crossings, or bad weather.

Pipelines have been around for more than 165 years. One of the first was constructed at Fredonia, New York, from hollowed out logs butted end to end. The pipeline was built to transport natural gas.

One of the best-known pipelines is the Trans-Alaska which carries oil 762 miles (1227 km) from Alaska's Prudhoe Bay oil fields to the Valdez terminal on the Gulf of Alaska. See Fig. 5-7.

ON-SITE CARGO MOVERS

A fourth class of land transportation is designed to move cargo and workpieces in mines, quarries, warehouses and factories. In this class are off-site trucks, motorized forklift trucks, automated guided vehicles, and conveyors.

A

B

C

Fig. 5-6. Long distance cargo transporters. A—Wagons were once the chief mode of moving goods to the American frontier. B—Railroad cars are an important method of moving goods over long distances. These cars are stored in a rail yard until needed. C—These trucks are hauling logs from a Washington forest to a sawmill.

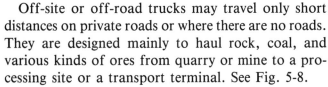

Fig. 5-7. This Alaskan pipeline carries crude oil from the Prudhoe Bay oil fields to tanker terminals in Valdez.

Off-site or off-road trucks may travel only short distances on private roads or where there are no roads. They are designed mainly to haul rock, coal, and various kinds of ores from quarry or mine to a processing site or a transport terminal. See Fig. 5-8.

Forklifts are designed to move and store raw materials and manufactured products. Automated guided vehicles, also known as AGVs, are designed to deliver stock to automated factory machines and carry away machined parts or products to another machine or to storage. See Fig. 5-9.

The pneumatic dispatch tube of the 19th century, found today at drive-in banking facilities, is another example of on-site movement of products. A modern adaptation of pneumatic transport is a pneumatic pipeline at Hamilton, Ontario, Canada. It is designed to move gravel, ore, and similar materials. At intervals along the line, fans create either air pressure or a vacuum. The transporters are gondolas with gasketed ends to seal off airflow. The gondolas are moved along at about 27 mph (about 43.5 km).

Fig. 5-8. Top. Typical off-road truck is powerful and rugged to haul heavy loads on rough terrain. Bottom. These trucks are used to transport ore from a copper mine in Mexico.
(WABCO Creative Services)

FIXED OR RANDOM PATHS

Land transporters can be further classified by the "paths" or routes that they follow. These are either

Fig. 5-9. This forklift and an automated guided vehicle are used to handle stock in a warehouse. The forklift pulls stock off storage shelves and loads it onto the AGV. The AGV is designed to transport the stock to a loading dock or to a new storage area.
(Mannesman Demag Corp. Material Handling Systems)

Fig. 5-10. A dugout could carry more weight than a log of the same size.

fixed path or **random path**. A fixed-path vehicle can only move along one path which has been constructed for its use. A train, an elevator, escalator, or a pipeline follow fixed paths. The vehicle cannot leave the path. Random-path vechicles can be directed onto various routes. A bicycle's path or that of a truck or automobile is random.

WATER TRANSPORTATION

Water-based vehicles are also known as vessels. Since they are intended to travel in water, they are designed differently than land vehicles. Because water is less dense than land, the vehicles must be **buoyant** to stay afloat. (Bouyant means capable of floating in water.) The density of water is 1.00 gram per cubic centimeter at 39°F (4°C). The density of earth is, on average, 5.5 grams per cubic centimeter.

Being less dense than water, wood will float. Ancient humans used logs to carry them and their cargo across or along bodies of water. Later they hollowed out the logs (dugouts) so they could carry more weight, Fig. 5-10. Later, when boats and ships began to be made of sawn timbers and boards, the builders made them hollow so that the vessels, themselves, could carry more people and goods and still stay afloat.

TYPES OF VESSELS

Water transporters are known as **vessels**, boats, and water craft. There are many other classifications and subclassifications. Water transporters, like land transporters, are made to move cargo and people. Another class is designed for recreational purposes. Size also varies as does the loads they are designed to carry.

Cargo transporters

Water cargo transporters include bulk freighters, general cargo freighters, tankers, container ships, LASH (Lighter Aboard Ship) and Seabee (Sea Barge) vessels.

Bulk freighters like the one shown in Fig. 5-11 transport dry and liquid bulk (loose) material. Their cargo may be oil, coal, grain, gravel, gypsum, or iron ore. This cargo is contained in holds located below a single deck.

Fig. 5-11. This bulk freighter operates on the St. Lawrence Seaway. It is taking on a load of coal at a Great Lakes port.

General cargo freighters and containerships are large vessels designed to operate on the open seas. These ships are fitted with derricks and booms to facilitate loading and unloading. General cargo might include cars, heavy equipment, palletized products, lumber, fruit, vegetables, and manufactured goods. A few years ago ships were also fitted with refrigerated compartments to accommodate frozen foods and perishables. Containerships are designed to handle products preloaded into large containers. The containers are loaded into cells especially designed for them. The cells are fitted with guides to hold the van-type containers. Decks are also fitted with hold-downs for storage of containers on deck. Some containerships are capable of handling more than 1000 containers.

A special kind of containership are the RO/RO (roll on/roll off) vessels. They are built to carry large cargoes of land vehicles that can be driven onto and off the vessels. Other cargo too large for containers but without wheels is placed on a special chassis and driven on and off board.

Tankers carry only liquids. Tank-shaped holds are designed to contain petroleum products and other liquid fuels, wine and molasses, and various kinds of liquid chemicals.

Towboats and tugboats, Fig. 5-12, are the workhorses of water transportation. Tugboats are small vessels with powerful engines. They are not intended to carry cargo but to tow or push larger vessels around in harbors or in close quarters. Similarly, towboats are not suited to carrying cargo. They are, however, employed to push or pull barges loaded with cargo. Towboats are found mainly in inland waters. Examples are the towboats that push barges up and down rivers and lakes. See Fig. 5-13.

People transporters

Passenger liners, cruise ships, and ferries transport people traveling by water. Ferries, however, also carry

Fig. 5-13. Towboat on an Illinois river pushes a string of barges.

a limited number of vehicles such as automobiles, trucks, and even trains. See Fig. 5-14. Passenger liners are designed to move passengers from one land base to another. Cruise ships are designed for pleasure trips. Passenger liners have been severely affected by the growth of air service which is much faster and cheaper.

Hydrofoils, Fig. 5-15, are high-speed people movers that operate in inland waterways. With the aid of winglike structures on their hulls, they skim over the water at speeds of 50 mph (about 80 km/h). The foils are attached to the hull near the waterline. The forward motion of the craft causes the hull to be lifted

Fig. 5-14. This ferry operates between Victoria, British Columbia and Port Angeles, Washington. It is first shown approaching the dock and then unloading passenger vehicles.

Fig. 5-12. Tugboats are the workhorses of marine transportation.

Fig. 5-15. This is an artist's conception of a SWATH vessel-cruise ship. A hydrofoil, it is being built in Switzerland for an American hotel chain. (Radisson Hotels International)

above the water and supported on the foils. The foils are supported by the buoyancy of the water. Propulsion is provided by turbines that are connected to water jet pumps. However, such craft are experimental even though several have operated briefly in coastal and inland waterways.

Hovercraft® or air cushion vehicles are another class of experimental water craft. Hovercraft depend on a cushion of air for propulsion and to support them above the surface. Such craft are capable of operating both on water and on land. Other names for them are "surface effects" and "ground effects" machines. Air is sucked in from the top of the machine by way of large fans. The resulting cushion of air lifts the craft a safe height above the surface. Refer to Fig. 5-15.

WATERWAYS

Waterways are the routes used by marine craft. Like railways, these routes are more restrictive than roadways.

There are inland and transocean waterways. Inland waterways are the rivers, lakes, and other bodies of water within a continent. Examples are the Mississippi River and the St. Lawrence seaway. The Mississippi allows transport from Minnesota to the southernmost point of Louisiana. Materials and products are moved by barge up and down the river. The St. Lawrence seaway stretches inland from the Atlantic more than 2300 miles to ports on the Great Lakes. It gives cities like Toronto, Buffalo, Erie, Cleveland, Toledo, Detroit, Chicago, Gary, Milwaukee, and Duluth access to the Atlantic Ocean. Millions of tons of grain, coal, iron ore, and other cargo move through it. Inland routes have channels where the ships and boats may travel without danger of running aground. These chan-

nels are marked with buoys or, sometimes, with shore-bound beacons. Navigators in these waters must follow "Rules of the Road" to avoid collision with other craft. These are somewhat like the regulations for driving on highways and streets.

Although oceans are large expanses of water where ships might travel anywhere, transocean routes are laid out as sea lanes or shipping lanes. Ships are navigated along these lanes very carefully to avoid collision with other transocean vessels. Various instruments will be used to assist the navigator. These include a compass to keep the ship on course, charts for laying out courses, sextants to navigate by the stars, and radar to locate distant objects such as land or other ships. Other equipment used by marine navigators are loran, omni-VOR, Decca, omega, consolan/consol, and satellite navigation systems.

Occasionally there are barriers in waterways such as rapids or falls in a river or land masses between oceans. To allow ship passage, locks must be built to "lift" water craft over the barrier. The Panama Canal, which connects the Caribbean Sea and the Pacific Ocean, has several locks that lift ships from sea level to inland waters. See Fig. 5-16. Locks are huge watertight chambers which can be alternately flooded and drained to raise or lower ships. Fig. 5-17 shows ships traveling through the Miraflores Lock on the Caribbean side of the Canal.

AIR TRANSPORTATION

Air transportation systems are used for transport of both cargo and people. Their market share of the transportation industry has increased markedly in the last 40 years.

The aviation industry includes two basic areas: general aviation and commercial aviation.

General aviation has six classifications:
- Business flying. Many companies maintain one or more aircraft to transport company personnel who

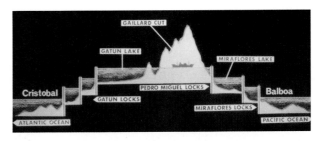

Fig. 5-16. This drawing profiles the Panama Canal which traverses the Isthmus of Panama for a distance of 50.7 miles (82 km). There are a total of 12 locks, each 1000 ft. long, 110 ft. wide, and 70 ft. deep. A lock chamber holds about 66 million gallons of water. (Panama Canal Commission)

Fig. 5-17. Ships are shown going through Miraflores lock at the Panama Canal. (Panama Canal Commission)

Fig. 5-19. A typical commercial airliner. This one is designed to carry air passengers over long distances. (McDonnell Douglas)

must confer with personnel at other offices or plants. The largest of the seven sectors, it comprises over 150,000 planes.

- Air taxi, rental, and commuter services. These are companies whose business is providing short-flight transport or charter service. See Fig. 5-18.
- Special purpose flying. Not organized to transport, these services perform a wide variety of activities such as aerial inspection, aerial photography, crop dusting, and traffic control.
- Personal flying. People in this category own or rent planes for personal business or enjoyment.
- Sportflying. In this category are stunt flying, skywriting, and sailplaning.
- Instructional flying. About 150,000 people take flying instruction in any year. This includes those in flight training at military bases.

Commercial airline service, Fig. 5-19, is more familiar to the general public than other types. It is the service that most people use when traveling by air.

Fig. 5-18. This plane provides air taxi service to reach remote fishing waters in Canada.

Commercial airlines carry the highest volume of travelers. They are capable of carrying several hundred passengers at one time. Commercial airlines are grouped into roughly three categories: domestic trunk carriers, international carriers, and regional carriers.

WHY AIRCRAFT FLY

The ability of aircraft to fly when they are heavier than air is based on a number of aerodynamic principles. While they are in flight, aircraft are subject to several forces: gravity, drag, lift, and thrust. **Gravity** and **drag** draw the craft earthward and tend to hold it back. **Thrust** is the force that moves the craft forward. **Lift** is an upward force that keeps the craft in the air. By changing these forces, the pilot can control the movement of the aircraft in flight.

Thrust is controlled by the speed of the engine. Lift is created by the rush of air past the wings. Because of the angle of the wing, air moving past the upper surface speeds up. This creates a vacuum along the top of the wing. A vacuum is a lowering of air pressure. Thus, there is a difference of air pressure between the top and the underside of the wings. This is one of the factors which produces lift.

Another factor producing lift is known as Newton's law of action and reaction. Newton said that for every action there is an opposite and equal reaction. As the fast-moving wing pushes against the air, the air pushes back, causing lift.

The thrust of the engine or engines controls the speed of the air moving past the wings. The pilot can cause the craft to lift or drop by controlling engine speed.

AIRCRAFT TYPES

Aircraft include several types: fixed wing, rotary wing, and lighter than air. Each has been developed to fit a need.

Fixed-wing aircraft have wings that do not move. They are attached to the fuselage (body) of the aircraft.

The aircraft is heavier than air.

The rotary wing type is known as a helicopter, Fig. 5-20. The craft's engine rotates the wing at high speed to produce the degree of lift required. A second rotor in the tail section overcomes the tendency of the helicopter to rotate with the wing and controls the direction of flight. Like the fixed wing craft, it is heavier than air.

Lighter-than-air (LTA) craft are more generally called **airships.** They are powered, maneuverable craft that operate in the air. Their buoyancy is the result of their being filled with a gas that is lighter than air. Both helium and hydrogen have been used.

Three different types of LTA have been used: nonrigid, semirigid, and rigid. There are important differences in these types.

Nonrigid airships, commonly called blimps, are small and have no framework. They are more than balloons, however, since they have an engine and some sort of housing that supports and carries the engine, fuel, and crew. See Fig. 5-21.

Semirigid airships have the housing and equipment of the nonrigid. In addition, they have a simple metal framework reinforcing the bow which gives the envelope (balloon) some shape. There is no other structural support.

Rigid airships, or **dirigibles,** are large and have an intricate framework consisting of ribs and stringers made of aluminum or high-strength plastics. The hull is divided into 15 or more compartments. Each compartment has its own gasbag.

Airships are patterned after balloons which were a French invention of 1782. The first powered balloon flight came in 1852 when Henri Gifford took a 40 ft.-diameter balloon aloft under steam power. Germany, Great Britain, and France engaged in development of airships from 1865 through the middle of the 20th century.

Being susceptible to high winds and attack by enemy craft, the military abandoned research in airships after World War II. Blimps and hot air balloons, however, continued to be more or less popular. The former are used in advertising and sports enthusiasts have claimed the hot air balloon.

Now, decades after they were abandoned, rigid airships are again under consideration and development. By the end of 1991, Westinghouse Airships had developed and delivered to the U.S. Navy a test craft of greatly improved design. Known as the Sentinel 1000, it is, at 222 ft., the largest airship in the world. The Sentinel is to be followed by another even larger airship, the YEZ-2A, which will measure 425 ft. from nose to tail. Its height will be roughly that of a 15- or 16-story building.

Space-age materials and technology are responsible for this renewed interest and development. Where older designs used relatively heavy and weak fabrics for the envelope, today's envelopes are made from lightweight, high-strength synthetic fabrics including dacron, mylar, and tedlar. Lightweight, miniaturized electronic units provide improved control and data-collecting ability. Improved propulsion is made possible with a vectored duct thruster and a sprint propeller.

Airship development is also underway in The Commonwealth of Independent States (Russia). One of their large projects is the Albatross, about the same size as the American YEZ-2A. It will have helicopter rotors to provide the extra lift needed to transport heavy cargo. Russia plans to use it to develop resources in vast regions where roads, airports, and railroads are scarce.

In theory, an airship's ability to stay aloft has no limit. The need for periodic refueling, however, brings them back to earth. What about a solar-powered airship? Eric Raymond, who built the first solar-powered airplane to fly across the United States, has developed plans for one that he hopes to fly around the world.

Central to his design are the new space-age materials. The keel and support framework are of small-diameter

Fig. 5-20. A rotary-winged aircraft is shown in flight. (Bell Helicopter)

Fig. 5-21. Blimps have no frame. The envelope (balloon) is filled with helium which is lighter than air. (Goodyear Tire Co.)

aluminum tubing wrapped with kevlar thread. Supporting the photovoltaic array is a thin foam sandwich of kevlar. The envelope is also of tedlar. This is a plastic similar to mylar film but stronger and resistant to ultraviolet rays.

Lift for the 100 ft. (30.5 m), 18 ft. (5.5 m)-diameter airship will be provided by six helium-filled bladders. Propulsion is provided by a 5 hp. (3.75 kW) electric motor that drives a 16 ft. (about 5 m) propeller. It should drive the airship along at an estimated 60 mph. Night flying at about 40 mph will be made possible by batteries which will store solar energy.

A gondola of a carbon fiber sandwich will have room for two passengers, equipment for navigation, communication, and survival equipment with room to spare for two electric motors. These are needed to provide emergency power and additional thrust for takeoffs and landings. In the event of bad weather, a rocket propelled anchor will be fired into the ground without the help of a ground crew.

SPACE TRANSPORTATION

Space is those regions that are beyond the atmosphere that surrounds earth. Near space begins at about 500 to 1000 miles (805 to 1600 km) from earth's surface and extends 10,000 or more miles (16,100 km) beyond. Outer space is the region beyond 10,000 miles away from earth.

Space travel began in 1957 when Russia launched Sputnik I, an artificial satellite, into earth orbit. On January 31, the United States also launched its first satellite, Explorer I. Russia achieved another first in 1961 by putting a person in orbit around the earth. The craft was the Vostok I.

Fueled by the determination to take a leading role in space exploration, the United States began the Apollo program. Its goals were:
• To land American astronauts on the moon and bring them back safely.
• To develop technology for other interests in space.
• To make the United States a leader in space exploration and technology.
• To carry out a series of scientific experiments on the lunar surface.
• To develop the human capability to work in the moon's environment.

A try for a lunar landing was successful in July, 1969, less than eight years after President John F. Kennedy's announcement of a national effort to place people on the moon. Before the end of the decade six more flights were made with five of them making successful moon landings.

The Apollo also had other tasks: to carry crews and equipment to an orbiting space station called Skylab and to take part in an international manned space flight. Skylab was a scientific laboratory in space. It was used on four space missions. The purpose of these missions was to:
• Make scientific investigations during earth orbit.
• Observe the earth from space.
• Test people and systems engaged in space flight.

Skylab prepared the way for the Apollo-Soyuz international space flight. In 1975 an Apollo spacecraft and a Russian craft were launched simultaneously and were joined together while both were in earth orbit.

The next significant development was the building and testing of a reusable spacecraft. The first space shuttle, Columbia, was designed for launch into space like a rocket. It can carry heavy payloads into space and serves as a space station while orbiting the earth. It can glide to a landing like an airplane.

Columbia's ability to be used over and over again significantly reduces the costs of a shuttle system.

Unlike other spacecraft, Columbia is a true transportation system. It involves several important pieces of space hardware. These include:
• Rocket boosters which assist the shuttle's own main propulsion engines during launching.
• The shuttle vehicle or orbiter, Fig. 5-22.

Fig. 5-22. The space shuttle is a true transportation vehicle. It will be used over and over to move people, materials, and supplies between earth and a space station. Here an artist shows the orbiter in space and docked with a beam builder. One day, products and/or structures may be produced in space. (NASA)

- A space lab with all the equipment for scientists to develop and conduct simple experiments.
- A space tug whose purpose is to carry out synchronous and high-altitude orbits during escape missions.

Columbia's purpose is to transport payloads between outer space and earth. Its large cargo bay is capable of containing large structures. Because of this cargo capacity, the shuttle has been used for carrying satellites into space and launching them. It has also been used for retrieving and repairing orbiting satellites, Fig. 5-23. Several shuttle flights carried a laboratory in the cargo bay. This lab is available to scientists, research institutes, commercial interests, government agencies, and individuals.

PROPULSION SYSTEMS

Every type of transportation requires a **propulsion system.** Such machines are known as energy converters. Their function is to change one form of energy into another form that the propelled vehicle can use. The types of propulsion units include:
- Human powered. The work or effort of a person moves the vehicle. An example is a bicycle. Experimental flights have been made in human-powered aircraft.
- Wind. The force of wind has been utilized for many centuries. Sails are the devices used to capture the wind. Early ocean-going craft used many sails. Today their use is recreational except in primitive cultures.
- Heat engines. These machines burn fuel to produce heat. Expanding gases of combustion or steam produce mechanical force to do work. Heat engines include internal combustion engines and external combustion engines. In the late 19th century steam engines powered some land vehicles such as locomotives, automobiles, and tractors. The most popular engine for land transportation is the internal combustion reciprocating engine. It is found in most automobiles, trucks, and diesel locomotives. Other types of heat engines are turbine and jet engines. Jet engines are used in air transportation.
- Rockets. A type of heat engine, rockets are used to power space vehicles in an environment where there is no oxygen to support combustion. The space vehicle must carry a supply of oxygen to burn the rocket fuel.

Regardless of the environment where it operates, every transportation system has common types of technology that are necessary to make the system usable as well as an efficient transport medium in its environment. The next chapter will be devoted to the various units and systems that convert energy so that it can perform work.

SUMMARY

Transportation has a profound effect on our society. Without it our lives would be drastically changed. Its purpose is to transport humans or cargo (goods) from one place to another.

Transportation environments are the different mediums in which transportation vehicles travel. They are: land, water, air, and space. Vehicles must be designed to suit the environment in which it will operate. The term for these different methods of travel is "modes."

The vehicles of transportation must have a propulsion system. Their function is to change one form of energy into another in such a way that the propulsion system is able to transport (or power) the vehicle and its passengers or cargo. These propulsion systems are:
- Human power.
- Wind power.
- Heat engines.
- Rockets (a special type of heat engine).

DO YOU KNOW THESE TERMS?

airship
bulk freighters
buoyant
container ships
dirigible
drag
fixed path

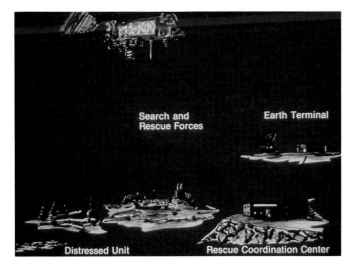

Fig. 5-23. Orbiters would also be used in space search and rescue missions. (NASA)

general cargo freighter
gravity
hovercraft
hydrofoils
lift
modes
on-site transportation
passenger liners
people movers
propulsion system
random path
space transportation
tankers
thrust
towboats
transportation environments
tugboats
vessels

SUGGESTED ACTIVITIES

1. Using a road map, select a route to a distant city. On a separate sheet of paper list the highways you will take. Then, using the scale of miles on the map, determine the distance you would drive to your destination.
2. Compute the amount of gasoline consumed by an automobile traveling to and from the destination selected in Activity 1. Assume that the automobile averages 25 miles per gallon at an average speed of 55 mph.
3. Construct a working model of a vehicle designed to travel by some form of propulsion in any of the transportation environments. Be prepared to name the type of energy converted to propel your vehicle.

TEST YOUR KNOWLEDGE

1. Define transportation system.
2. List the four transportation environments.
3. Bicycles, cable cars, elevators, escalators, scooters, and carts make up a major grouping of vehicles in land transportation called _____ _____.
4. Name three cargo transporters used to move cargo over long distances.
5. Name the principle that keeps water craft afloat.
6. Name the two types of waterways used in water transportation.
7. It is not necessary to lay out routes for ocean transport since the oceans are so large. True or False?
8. Which of the following are instruments that help a ship's navigator keep the ship on course and determine the ship's position while out of contact with land?
   a. Charts.   d. Loran.
   b. Compass.   e. All of the above.
   c. Sextant.
9. The two general areas of air transportation are _____ aviation and _____ aviation.
10. Name the three basic types of aircraft.
11. The limits of near space are from _____ to _____ miles (kilometers) from the earth's surface.
12. _____ are a type of heat engine used to power spacecraft in space where there is no oxygen.
13. Define the term, heat engines.
14. Every type of transportation vehicle needs a _____ _____ to make it move under its own power.

6 CHAPTER
ENERGY CONVERSION SYSTEMS

The information given in this chapter will enable you to:
- *Define energy conversion.*
- *Describe what happens in energy conversion.*
- *Explain the difference between indirect and direct energy conversion systems.*
- *List and describe the main indirect energy conversion systems.*
- *List and describe the main direct energy conversion systems.*

While you were studying how energy is harnessed in Chapter 2, you learned the term, **energy conversion**. As you may recall, it means the changing of energy from one form to another. This conversion is an important step in any use of energy. It accomplishes two things:

- It unlocks the energy so that it may perform work.
- It changes the energy into a form that is appropriate for the job to be done.

Sometimes a series of conversions are made before the energy is finally used, Fig. 6-1. Some energy is lost during each conversion.

TYPES OF CONVERSION

There are many different ways of converting energy. It is possible to separate the converters into different types. One way is to group them by the number of conversions that must take place before the energy can be used. If there are several conversions before the energy is in the right form to do work, the method is called **indirect conversion**. However, if the energy is used after one energy conversion, the method is called **direct conversion**.

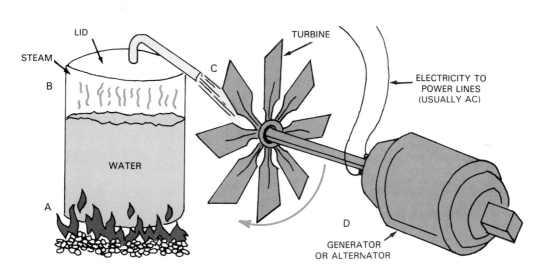

Fig. 6-1. Simplified diagram of an electrical power plant. Many conversions take place before the electric power is produced. A—Combustion changes chemical energy in coal to heat energy. B—Heat energy has created steam from water. C—The mechanical energy of the steam pressure drives the turbine. D—Finally, the generator converts the mechanical energy of the turbine to electrical energy.

INDIRECT CONVERSION METHODS

Because our society has become so dependent upon fossil fuels, most converters, up to now, have been designed to harness the energy of hydrocarbons (fossil fuels). The converters use combustion (burning) to produce heat from substances such as gasoline, coal, natural gas, diesel fuel, kerosene, and other materials containing carbon. The heat energy produced is used to produce mechanical energy. They are called combustion heat converters or heat engines because they convert the fuel to heat energy and then the heat energy to mechanical energy.

INTERNAL COMBUSTION HEAT ENGINES

One category of heat engine, Fig. 6-2, is designed to vaporize and burn a mixture of air and gasoline inside a closed chamber called a cylinder. This type is called an **internal combustion engine.** It changes chemical energy to heat energy and then converts the heat energy to mechanical energy. Being similar to the steam engine discussed in Chapter 2, it consists of one or more cylinders molded and then machined within a heavy metal casting called a block.

Four-stroke cycle engine

A four-stroke cycle engine is so called because four strokes of the piston are needed to produce one complete cycle. A cycle consists of two upward strokes and two downward strokes. Only one of the strokes does any work (produces power).

The top of the cylinder of the four-stroke cycle engine block is sealed except for two small openings

Fig. 6-2. An automotive engine is a heat engine. It burns fuel inside its cylinders to provide power for an automobile. (Chevrolet)

covered by valves. Valves are like a stopper on a bottle. They can be opened to allow passage of gases or liquids or closed to stop any leakage. On this engine, one valve controls the intake of fuel. The other opens to get rid of exhaust (spent fuel) and remains closed during the other strokes of the piston. Fig. 6-3 is a simplified cutaway drawing showing a single cylinder and the related engine parts.

Each cylinder in an engine has a piston. The piston seals the other end of the cylinder so that no gases can move past it and out the bottom of the cylinder. This piston is fastened to a crank by a shaft called a piston rod. The pistons are free to move up and down inside the cylinders.

How the four strokes produce power

A piston on its first downward stroke draws in an explosive vapor (mixture of fuel and air). The mixture is compressed on the first upward stroke of the piston. As the piston reaches the top of its stroke, a spark from

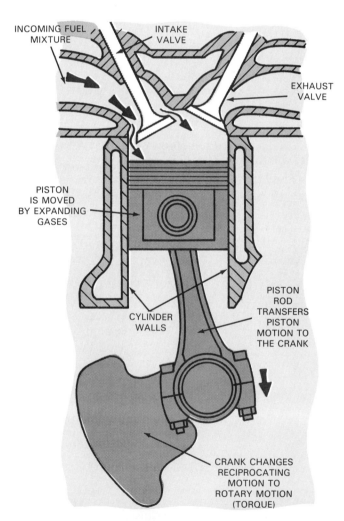

INCOMING FUEL MIXTURE

INTAKE VALVE

EXHAUST VALVE

PISTON IS MOVED BY EXPANDING GASES

CYLINDER WALLS

PISTON ROD TRANSFERS PISTON MOTION TO THE CRANK

CRANK CHANGES RECIPROCATING MOTION TO ROTARY MOTION (TORQUE)

Fig. 6-3. A cross section shows the parts of one cylinder of a four-stroke cycle engine. Note how the parts work together to produce energy.

a spark plug or the heat of the compressed gas ignites the vaporized fuel. The vaporized mixture burns rapidly, causing the gases to expand with explosive force. Since the top of the cylinder is sealed, the expanding gas pushes on the piston. The piston is driven downward in a power stroke. It is this stroke that provides the energy for the engine to do work. The piston rod attached to the piston and to the crankshaft transfers the power to the crankshaft and spins it rapidly. The crankshaft transfers the motion (called torque) to its load. The load may be the wheels of an automobile or the cutting blade of a lawn mower. The load could also be another machine.

On the second upward stroke, the piston pushes the spent fuel mixture out of the cylinder through the open exhaust valve at the top of the cylinder. When the piston reaches the top of its stroke, the exhaust valve closes and the intake valve opens once more. The cycle, consisting of intake, compression, power, and exhaust, begins all over again. Fig. 6-4 is a series of drawings illustrating the four events of the four-stroke cycle engine. Fig. 6-5 shows two engine designs.

Most automobiles use the four-stroke cycle engine. See Fig. 6-5.

Another heat engine design requires only two strokes to complete the combustion cycle. Those of this type are known as two-stroke cycle engines. See Fig. 6-6.

Operation of the two-stroke cycle engine

The two-cycle operation starts with a downstroke of the piston. As this action occurs, a vacuum occurs which pulls an air-fuel mixture into the combustion chamber. On the upward stroke of the piston the fuel entry port (opening) is closed. The fuel charge is compressed by the piston. As the piston reaches the top of its stroke, the compressed fuel charge ignites with an explosion that drives the piston down in a power stroke. As the piston nears the bottom of the downstroke, the spent gases leave the chamber and the next fuel charge rushes in. The cycle begins again on the next upward stroke.

Two-stroke cycle engines are used on lawn mowers, snowmobiles, and other applications where a small, light engine is needed.

Diesel engines

Diesel engines, Fig. 6-7, like the gasoline engine, have cylinders and pistons and burn a liquid fuel. They also may be either two or four-stroke cycle. The major differences are:

• The air and fuel are delivered to the combustion chamber separately and mix inside the cylinder rather than in the carburetor as in a carbureted gasoline engine. (Fuel injected gasoline systems are very similar to diesel injection systems.)
• The diesel fuel is not as explosive as gasoline.
• There is no ignition system. The diesel does not depend on an electrical spark to ignite the fuel. Ignition depends on temperature and pressure of the fuel vapors.

Since the diesel engine depends on high pressure on the fuel vapors to cause ignition, compression ratios

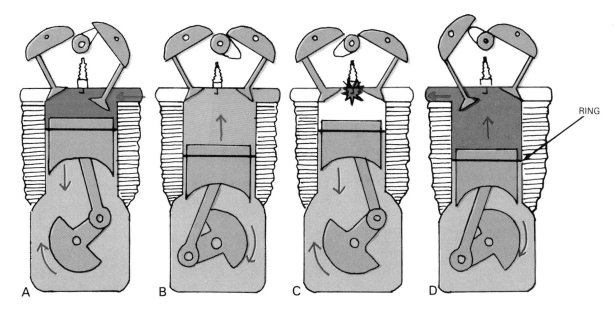

Fig. 6-4. Operation of a four-stroke cycle internal combustion engine. This type is also called a reciprocating engine because the pistons move up and down. A—The first stroke of the cycle draws in fuel vapor through open valve at upper right. B—The next stroke is upward to compress the gas vapor. Valve is closed. C—A spark ignites the fuel, driving the piston down in a power stroke. D—Exhaust stroke. The piston moves upward, driving out spent fuel through the open exhaust valve at top left.

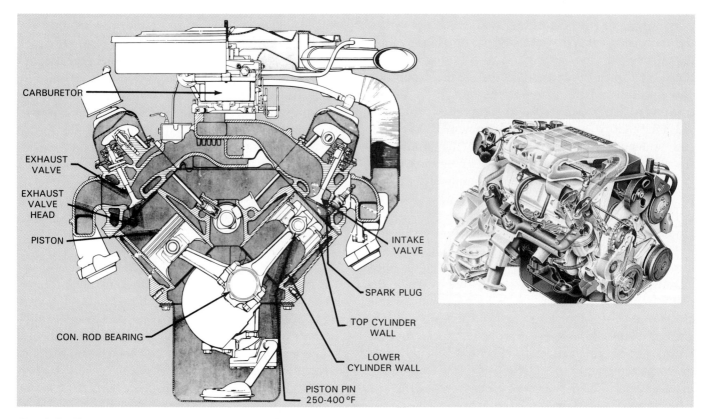

Fig. 6-5. Two automotive engine designs. A—Cross-sectional view of a V6 four-stroke cycle engine. B—Cutaway of a 3.1 liter V6, four-stroke cycle engine.   (Chevrolet)

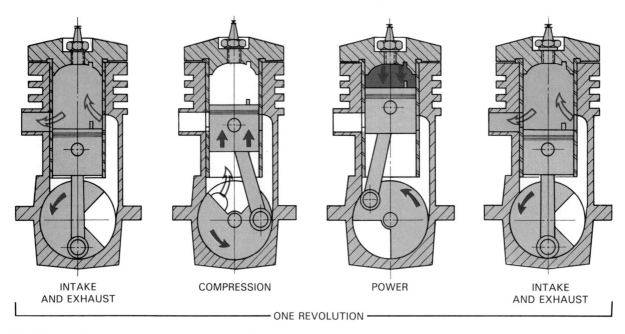

Fig. 6-6. Two-stroke cycle engine operation. A—At the bottom of the engine's first downstroke, spent fuel is exhausted and a new charge of fuel vapor enters the cylinder. B—On the upstroke, the fuel is compressed. C—Fuel ignites when the spark plug delivers a spark. Expanding gases move the piston downward in a power stroke. D—At the end of the downstroke, spent fuel is exhausted and a new charge of fuel enters.

of diesel engines are higher than those of gasoline engines. See Fig. 6-8. The air drawn into the diesel engine combustion chamber is heated up to 1000°F or more when compressed rapidly. Just before the piston reaches top dead center (the top of its stroke) the fuel injector sprays a measured amount of diesel fuel into the combustion chamber. The heated air ignites the fuel and produces the power stroke. Diesel engines, for this

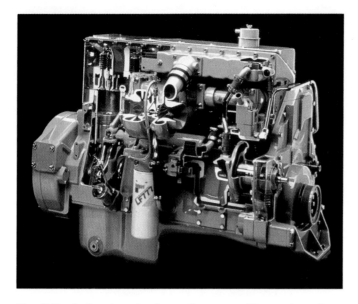

Fig. 6-7. A diesel engine is another type of internal combustion heat engine that ignites its fuel with the heat of compressed air. Note that inside parts are shown by the cutaway. (Cummins Engine Co., Inc.)

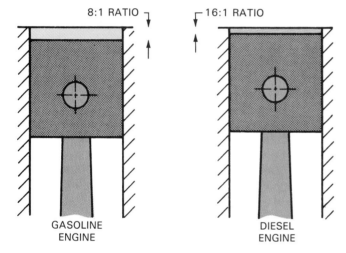

Fig. 6-8. To properly heat up the compressed air that ignites its fuel, the diesel engine must have a much higher compression ratio than the gasoline engine.

reason, are often called compression ignition engines.

Both two-stroke cycle and four-stroke cycle diesel engines are manufactured. The four-stroke cycle operates on the same principle as the four-stroke cycle gasoline engine. The two-stroke cycle diesel has a power stroke every revolution and can produce about twice the power of a four-stroke cycle diesel. Fig. 6-9 shows one revolution of a two-stroke cycle diesel.

Rotary internal combustion engine

Rotary combustion engines, Fig. 6-10, develop power through a triangular shaped rotor or rotors which spin inside a combustion chamber, Fig. 6-11. At three different points during its rotation, it compresses a vaporized fuel-air mixture which is then ignited. As with the conventional gasoline engine, the burning gases expand and drive the rotor.

Usually referred to as the Wankel engine, after its German inventor, Felix Wankel, the rotary engine has several advantages over the reciprocating internal combustion engine.

• It is smaller, quieter, and free of vibration.

Fig. 6-10. This rotary engine is also known as a "Wankel" engine after its inventor. Each of its rotors has three power thrusts during every revolution. (Mazda Motors of America)

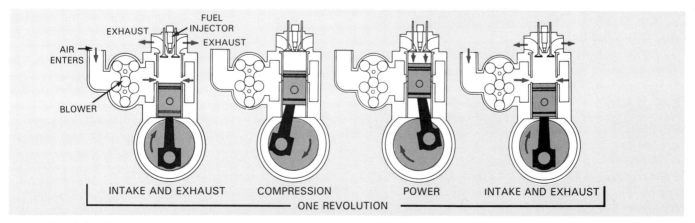

Fig. 6-9. How the two-stroke cycle diesel engine operates. This type engine is about two times as powerful as its four-stroke counterpart.

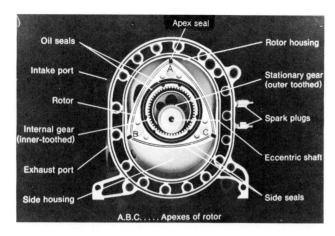

Fig. 6-11. These are the major parts of a rotary combustion engine. The rotor replaces the traditional piston while the housing replaces the traditional cylinder.
(Mazda Motors of America, Inc.)

- It is more efficient and develops more torque for its size.
- The Wankel has only about 70 basic parts. A six cylinder reciprocating engine of comparable power has about 230 basic parts. This should make it easier and cheaper to manufacture and assemble.
- A Wankel has but three moving parts. A reciprocating engine has 166.
- Easier to fit with pollution control systems. With its size and weight advantage, it is easier to add such controls.

The Wankel's efficiency and power are easy to understand. It has three power impulses during every revolution whereas the reciprocating engine has but one power stroke every two or four revolutions. Further, the reciprocating engine's piston must reverse direction at the end of every stroke. The rotor of the Wankel travels only in one direction.

For each power cycle, the Wankel goes through four phases:

1. Intake. A fuel-air mixture is sucked in as the combustion chamber expands.
2. Compression. The chamber contracts and compresses the mixture.
3. Power. As the fuel-air mixture is fully compressed, the leading spark plug fires. An instant later, the second plug fires to assure that combustion is complete.
4. Exhaust. The rotor pushes the burned gases through the exhaust port.

Fig. 6-12 shows the operation of one chamber of the Wankel engine as the rotor moves through each of these phases. The rotor has three faces and three working chambers. The chambers are separated by seals placed at the apexes (tips) of the rotor. The engine is timed so that each face produces one power impulse during

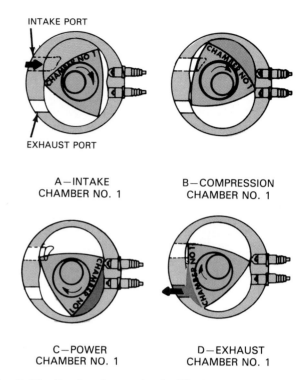

A—INTAKE CHAMBER NO. 1

B—COMPRESSION CHAMBER NO. 1

C—POWER CHAMBER NO. 1

D—EXHAUST CHAMBER NO. 1

Fig. 6-12. Combustion cycle of a Wankel engine shows one chamber during one rotation of the crankshaft. Each of the other chambers is going through the same cycle but at different times.

a single revolution of the rotor.

Fig. 6-13 shows a cutaway of a rotary engine coupled with a thermal reactor. The reactor reduces exhaust emissions to meet EPA standards.

Jet engines

Being powerful for their size and weight, **jet engines** are a natural for powering aircraft and missiles. They often move aircraft at speeds exceeding 750 miles (about 1207 km) an hour.

Jets and rockets work in about the same way as a balloon from which the air is escaping, Fig. 6-14.

The balloon is moved by the pressure of the air inside not by the air escaping. This is what happens:

- Air escaping from the balloon reduces the interior pressure on that end of the balloon.
- The high pressure continues to push out on the other end of the balloon.
- This unequal pressure propels the balloon in the direction of the greater pressure.

In the same way, the jet engine moves in the direction of the greater pressure inside of it, Fig. 6-15.

Jet propulsion is possible because of a natural principle known as Newton's Third Law of Motion. According to this law, for every action there is an equal and opposite reaction. In the balloon the action is the air rushing out of the neck. In the jet engine the action is the hot gases rushing out the rear of the engine

Fig. 6-13. Rotary engine fitted with a thermal reactor to control emissions. The reactor has no parts to wear out and uses no special chemicals.

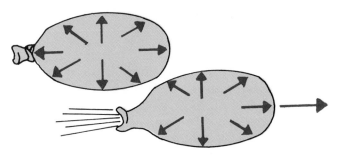

Fig. 6-14. Balloon demonstrates physical principle that ''for every action there is an opposite and equal reaction.'' Top. As long as air cannot escape from balloon, pressure inside is equal on all sides. Bottom. As soon as air is allowed to escape, pressure at front of balloon is greater and propels balloon in that direction.

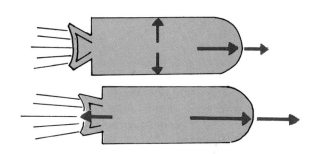

Fig. 6-15. How jet engine operates. Left. Like the balloon, jet moves in direction of the greatest pressure. Right. Action and reaction in a jet.

at a high pressure. The reaction in both cases is the unequal pressure inside.

The most widely used jet engine today is the gas turbine jet. Jets have replaced piston engines in all military and most commercial aircraft.

Ramjet engines

The **ramjet engine** is noted for its simplicity and has no moving parts. In fact, it is little more than an open-ended cylinder. Little wonder that it is usually referred to as a "flying stovepipe"! Its basic parts and types are shown in Fig. 6-16.

One limitation of the ramjet is that it cannot develop any thrust while it is standing still. It must be boosted

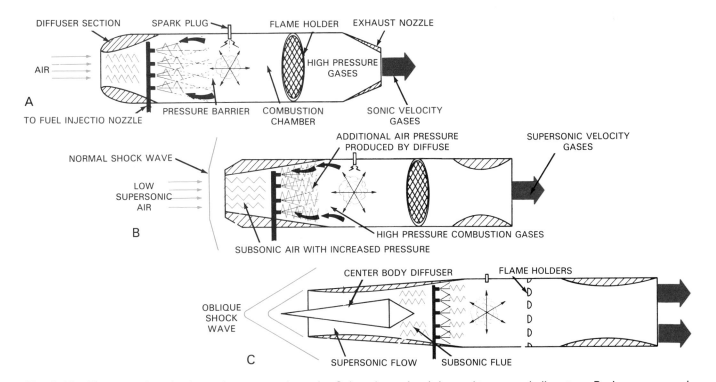

Fig. 6-16. There are three basic ramjet type engines. A—Subsonic ramjet. It is used to power helicopters. B—Low supersonic ramjet. One of its applications is to power target drones for the armed forces. C—High supersonic ramjets are used when speeds of Mach 2 (twice the speed of sound) are needed. They are known as scramjets (for **S**upersonic **C**ombustion **Ram**JETS). Other types of jet engines must be used to get ramjets up to thrust speed.

to a suitable speed before it will function. Once up to thrust speed its speed is theoretically unlimited; the faster it travels the better it operates and the more thrust it develops. For practical purposes its top speed is limited to Mach 5.0. In wind tunnel tests, higher speeds generated too much heat. Because of its characteristics, the ramjet engine is well suited as a propulsion unit for guided missiles.

Turbojet engine

The **turbojet** represents a separate class of engines that are designed for use in aircraft. That is still their main use today. Turbojets are most generally used in military aircraft. Like other types of internal combustion engines it has four actions: intake, compression, combustion, and exhaust. However, unlike the reciprocating engine, the actions take place continuously rather than in cycles.

Air is brought in to the engine by the compressor. A starter is used to bring the compressor up to speed. The compressed air is forced into the combustion chamber. Here, fuel (usually kerosene) is injected and ignited first by spark plugs. An instant later ignition takes place in the remaining chambers by way of flame crossover tubes. Burning is continuous, once started. Superheated gases expand with great force. They push toward the exhaust nozzle. Before leaving, they pass through a turbine. Gases strike the blades of the turbine causing it to spin at high speed. The turbine drives the air compressor to continue bringing in compressed air to the combustion chamber. A simplified sketch of a typical turbojet engine is shown in Fig. 6-17.

Output from a turbojet is measured in **pounds** or **newtons of thrust.** However, for comparison, a conversion can be made to horsepower by dividing pounds of thrust by 2.6.

A second class of turbojets is known as the **turbofan engine,** Fig. 6-18. A fan has been added in front of the compressor. The fan draws in extra air, partially compresses it before it enters the compressor. Additional airflow from the fan moves past the outside of

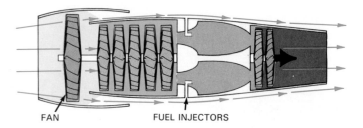

Fig. 6-18. Diagram of a turbofan jet engine. The fan is driven by a separate turbine wheel at the rear of the engine.

the engine. This arrangement gives the engine extra power.

Another variation of the turbojet engine is the **turboprop engine,** Fig. 6-19. Its main difference from the turbofan design is that a propeller replaces the fan. A reduction gear arrangement, driven by the compressor turbine connects the propeller to the engine. Almost all of the energy developed by the burning fuel goes to operate the compressor and propeller. Little energy is used directly to produce forward thrust.

Smaller aircraft are the main users of the turboprop engine. It is appropriate for flights of short or medium distance.

Gas turbine engine

The **gas turbine** is somewhat similar to a steam turbine. It has large, finned rotors which absorb the high energy of pressurized gas and spin at a high rate of speed.

Where the steam turbine is designed primarily for stationary operation, the gas turbine is found in mobile (moving) vehicles such as aircraft, boats, locomotives, and large trucks. Fig. 6-20 shows a simple sketch and a photograph of a gas turbine. It consists chiefly of a compressor, combustion chamber, a turbine, and an exhaust nozzle.

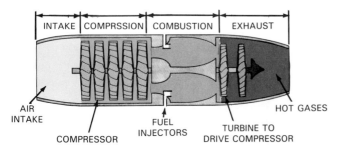

Fig. 6-17. A turbojet engine is like any other internal combustion engine with the same combustion cycle. However, the cycle is continuous.

Fig. 6-19. A turboprop jet engine. The power developed by the engine drives a propeller.

Fig. 6-20. Gas turbine engine. Top. Photo of an experimental engine designed for use in an automobile. It is compact and powerful for its size. Bottom. Schematic diagram of same turbine shows airflow and temperatures. At present, no manufacturer is producing this type of engine for automotive use.

The gas turbine is basically a turbojet engine that has been adapted to drive a free turbine. By using a speed reduction gear arrangement it easily adapts to its various uses. It is simple, reliable, easy to start in cold weather, and has a relatively clean exhaust. Marine gas turbines are huge, being in the 40,000 hp range. Industrial gas turbines are usually found generating electric power for utility companies during periods of peak demand.

Gas turbine operation

Compressed air is forced into the combustion chamber by a compressor impeller. This action heats the air to around 425°F (about 218°C). It picks up additional heat passing through a regenerator or heat exchanger. The regenerators, found on both sides of the engine, absorb heat from the hot exhaust gases for transfer to the incoming, compressed air. Fuel is added and ignited by the hot, compressed air. Expanding gases, at 1300 to 1700°F (1040 to 927°C) move through the turbine. There they expand to atmospheric pressure and transfer their energy to drive the compressor and power turbines. The power turbine drives either a propeller or drive wheels.

After the hot gases have passed through the turbine they are used in one of two ways. In the turbojet and turbofan engines, the gases provide jet propulsion. In the turboprop and turboshaft engines, they are used to drive rotors or propellers.

ROCKET ENGINES

Like jets, **rocket engines** are based on the Newton's third law of motion: every action has an equal and opposite reaction. Rocket engines are, for this reason, often called "reaction engines."

The rocket engine differs from the jet engine in these respects:
- The rocket can operate in outer space since it carries both its own fuel supply and the oxygen needed for combustion. The oxygen is contained in an oxidizer which is stored in a tank aboard the rocket. Jets are suitable only for flights within the earth's atmosphere since they draw their combustion air from outside the engine.
- Rockets have few, if any, moving parts.
- Operation of the rocket is very simple. See Fig. 6-21. Fuel and oxygen are released or mixed and burnt inside the combustion chamber. Unequal pressure inside the rocket drives the rocket forward.

There are two basic types of rockets. One uses a liquid fuel or propellant; the other uses a solid propellant.

Liquid propellant rockets

A liquid fuel rocket has five major parts:
- Propellant.
- Propellant feed system.
- Combustion chamber.
- Igniter for nonhypergolic (need external air source) fuels.
- Exhaust system.

Fuels are either monopropellants or bipropellants. Monopropellants contain both fuel and oxidizer in a single substance. When bipropellant is used, fuel and oxidizer are kept separate until injected into the combustion chamber. Fig. 6-22 shows a diagram of a rocket that uses a bipropellant system. Igniters are used to fire the bipropellant.

An igniter is not needed if fuel and oxidizer are **hypergolic**. This means that they ignite simultaneously when brought in contact with each other in the combustion chamber.

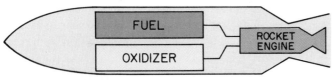

A

B

Fig. 6-21. Rocket engine is designed to propel vehicle in space. The air and fuel it consumes is carried in the spaceship itself. A—Simplified drawing of the self-contained propulsion system. B—Space shuttle's main engine is shown during a static (not moving) test in the manufacturer's laboratory. Propellants are liquid oxygen and liquid hydrogen. The fuel is clean-burning. Its only combustion by-product is steam.
(Rocketdyne Div., Rockwell International)

Solid propellant rockets

Propellants for solid rockets are made up of a hydrocarbon and an oxidizer. The oxidizer contains a high percentage of oxygen.

The materials are mixed in proportions to produce a solid. The finished solid fuel is called a grain or a stick. A charge is one or more grains or sticks. The mixed propellant is poured into the combustion chamber before it hardens.

Solid rockets are extremely simple in design. They have a short burn rate and are frequently used as booster units or for powering high-speed missiles.

EXTERNAL COMBUSTION CONVERTERS

External combustion converters burn their fuel outside of the cylinder or combustion chamber where the

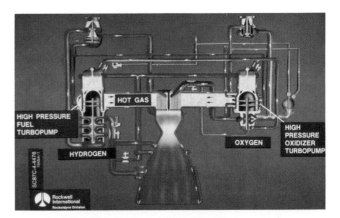

Fig. 6-22. Schematic view of how the Space Shuttle main engine operates. When the engine starts hydrogen and oxygen sides operate simultaneously. Igniters at the left and right domes start combustion that operate the turbopumps. A third igniter is located above the engine (center). It ignites a mixture of one part hydrogen and six parts oxygen.
(Rocketdyne Div., Rockwell International)

conversion from heat to motion takes place. There are several types:
- Steam engine. You may wish to review what was discussed in Chapter 2 about the use of steam.
- The Stirling cycle engine.
- The free piston engine.

Reciprocating steam engines at one time were popular as a propulsion unit or power source for tractors and as stationary power to operate factory machines. For a brief period in the history of automotive development, there were steam-powered engines in automobiles. Such engines failed to develop as a major power source for personal land transportation vehicles. The internal combustion gasoline engine was more convenient and soon displaced the "steamer." However, a great deal of experimentation was being done in the 1970s to perfect some type of external combustion heat engine for automotive use.

Until replaced by other power units, the steam engine was popular in rail and ship transportation as well as for powering farm tractors. See Fig. 6-23.

STIRLING CYCLE ENGINE

The **Stirling cycle engine**, Fig. 6-24, was invented in 1916 by Rev. Robert Stirling. Though fossil fuels are the usual combustion material used, any heat source will serve including solar-produced heat.

The combustion process takes place in a chamber called the burner. The heat is transferred to a fluid such as gas or air through an exchanger. This is a device that can collect the heat of combustion and transfer it without taking on any of the matter which is carrying the heat.

Fig. 6-23. Steam-powered tractors once provided reliable power for farm use. They are now ''museum pieces.''

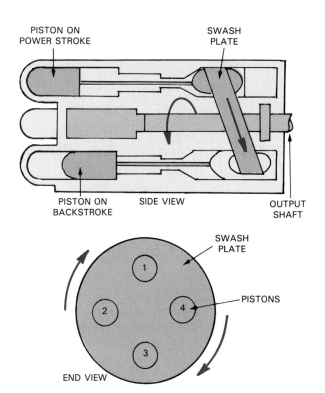

Fig. 6-25. Cutaway drawing of the pistons of a Stirling cycle engine. As the pistons move back and forth, they rotate the swash plate providing torque.

The exchanger heats up the compressed gas or air, which, in turn, expands. As pressure increases, the gas pushes on pistons to create mechanical energy. A crank or swash plate, Fig. 6-25, changes the linear (straight-line) motion to rotary (circular) motion.

Then the gas in the cylinder is cooled by an outside cooler. The gas contracts (takes up less space) and pressure drops.

Later development has added refinements to the Stirling cycle engine. However, the basic operation is the same as described. The engine has two power

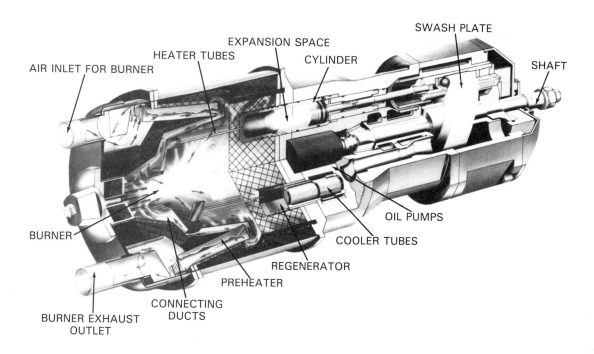

Fig. 6-24. Stirling cycle engine. More efficient than the reciprocating engine, it may be the automotive power plant of the future if some of its problems can be solved through engineering.   (Ford Motor Co.)

strokes of the piston for every revolution of the crank. This makes it more efficient than the two-stroke or four-stroke cycle internal combustion engines.

Efficiency of the most recent experimental models was about 30 percent. This is better than that of modern gas engines, Fig. 6-26.

PROBLEMS WITH THE STIRLING ENGINE

Further experimentation has ceased because it has several problems that engineers have not solved. One problem is the proper cooling of the engine. Since it runs at higher temperatures, it has to get rid of more heat than a conventional gasoline engine.

A second problem is the working gases that operate the pistons. Helium and hydrogen work best but are difficult to contain because they are under a high pressure of 200 atmospheres or 2940 psi (about 420 Pa). Hydrogen could be a safety hazard if a faulty seal leaked this highly flammable gas.

FREE PISTON ENGINE

Another energy converter engineers have tried to perfect is the **free piston engine**, Fig. 6-27. Somewhat of a hybrid, it combines elements of both reciprocating and gas turbine engines.

The pistons are called "free" because they are not connected to a rod or crankshaft and they have no ignition system to ignite fuel. See Fig. 6-28. The cylinders in which the pistons move are stepped so that there are two diameters to each piston. The large ends of the pistons are called the "bounce" pistons. As the engine starts, air pressure is forced into the bounce chamber

Fig. 6-26. An experimental Stirling engine developed for automotive use. (Ford Motor Co.)

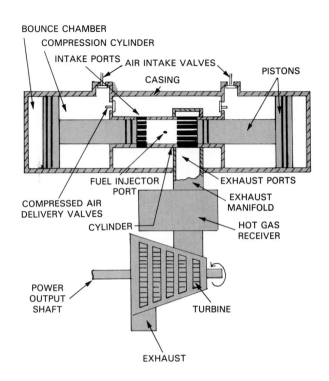

Fig. 6-27. Simple diagram of the free piston engine. It combines a turbine with parts of a reciprocating engine.

and two opposing pistons move toward each other. When the pistons are very close to each other, fuel is injected into the cylinder between them. The fuel is ignited by the high pressure and the pistons are forced apart violently. This causes air in the bounce chamber to be compressed providing energy for the return stroke. At a certain point, ports open in the cylinder and allow the hot, expanding gases to move into the turbine and cause it to rotate.

Even though the free piston engine will run on almost any fuel, it is not in commercial use. Various problems exist with the design. It is noisy, has low power, is hard to start, and violent forces of combustion soon cause breakdown.

ELECTRICAL CONVERTERS

Much of the energy consumed worldwide is converted to electrical power. The power is used everywhere. It lights streets, homes, factories and all manner of buildings and other structures. It also provides power to operate motors, appliances, furnaces, and heaters. All types of communication systems depend on electricity.

Most of the electricity we consume is produced by first converting the energy to heat. About 80 percent of the power stations use steam turbines to convert the heat to electrical power.

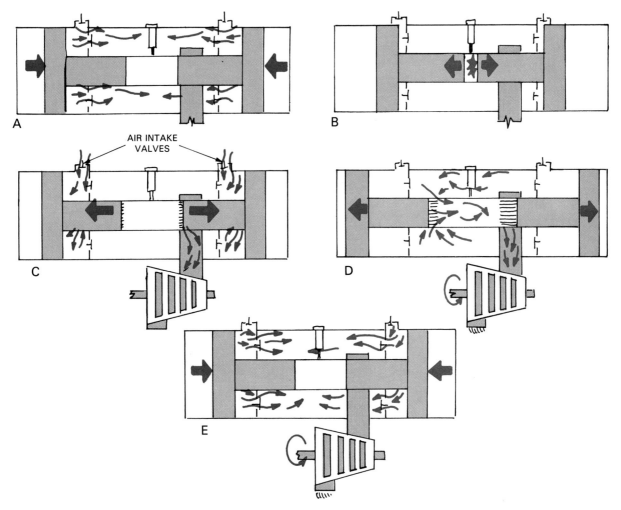

Fig. 6-28. Operating sequence of free piston engine driving a turbine. A—Air pressure from the starting system feeds into the bounce chambers at either end. B—When the ends of the pistons are close to each other, a spark ignites fuel fed into the chamber. C—Pistons, moving outward, compress air in the bounce chambers. This provides energy for the return stroke. Intake valves open to let air into the combustion chamber. Exhaust ports open and allow air into the turbine. D—Intake ports open at the power stroke. Then, compressed air in the enclosure around the cylinder rushes into the power chamber pushing the remaining exhaust gases out to the turbine. E— Compressed air in the bounce chambers move the pistons back together to start the next cycle.

THE NATURE OF ELECTRICITY

Electricity is thought to be the result of electrons traveling from one atom of certain materials to another atom in a continuous stream. These types of materials are known as conductors. Electrons traveling through conducting material are able to produce light as they travel through light bulbs. Their movement through electric motors makes the motors rotate. When the electrons travel through certain other materials, they cause the material to heat up.

The reason that electrons travel has to do with their attraction to other elements in an atom called protons. This attraction is the result of electrons being negatively charged and protons having a positive charge. This is similar to the attraction opposite poles of a magnet have for each other.

HOW ELECTRICITY IS PRODUCED

Electricity is also produced in nature. When electrons build up in the atmosphere during a storm, they move violently toward an excess of protons. The result is lightning. Running a comb through one's hair will produce what is called static electricity.

Electricity is generated by using magnetism. Through experiments, it was found that passing a wire conductor through a magnetic field (between the poles of a magnet) produced an electric current in the conducting wire, Fig. 6-29A. This principle is used in the design of the electric generator. A loop of conducting wire rotates continuously through this magnetic field producing a continuous flow of electrons. We call this flow electric current. Fig. 6-29B is a simplified drawing of a direct current generator and how it works.

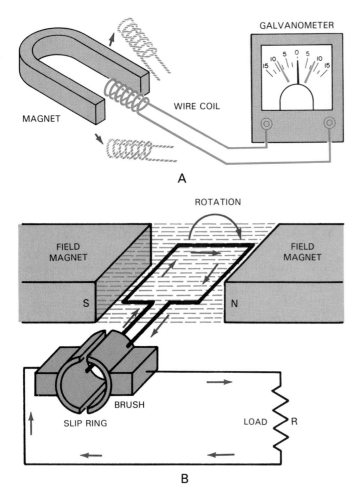

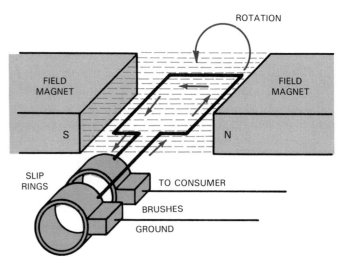

Fig. 6-30. Simple working diagram of an alternator. Current changes direction every half revolution of the conducting loop of wire.

Fig. 6-29. Electricity can be produced through use of magnetism. A—Passing a wire across a magnetic field between the poles of a magnet will produce a current in the wire. If the wire is connected to a galvanometer, a downward pass across the field will deflect the needle in one direction. An upward pass will deflect the needle in the opposite direction. B—A generator uses this same principle to generate direct current.

Generator types

There are two types of generators. In one the current always moves in the same direction. For this reason, the generator is called a **direct current generator.**

In the other, the electric current alternates, changing direction many times a second. In North America, this alternation takes place 60 times a second. Cycles per second are measured in hertz. Thus, a hertz is 1/60 second in duration.

Usually, the alternating current generator is called an **alternator.** Fig. 6-30 is a simplified drawing of an alternator. Note the names of its parts.

How generators/alternators work

Note that the armature of a generator or alternator is, in its simplest form, nothing more than a coil or loop of wire. In a working generator or alternator it

will be made up of hundreds of loops. To produce electricity, the loop must be rotated inside the magnetic field of a magnet. Large units actually use an electromagnet.

Current from the rotating armature can be passed through a conductor and attached to a load. The load for a power plant might be all the lights and power needs of a whole region. The load of a small generator may be the power needs of one home.

Power stations

Most generation of power is done at large installations called power plants. Energy sources used are fossil fuels, such as coal; hydropower (from a waterfall or water held behind a dam); and nuclear power. The power station in Fig. 6-31 gets its energy from coal.

Once generated, electricity must be delivered to the points of use. To be moved any distance, electricity must have some kind of pressure to move it along. Pressure that moves electricity through conductors is

Fig. 6-31. A coal-fired electric power station. A coal dock to the right allows coal delivery by boat.

called voltage. Another name for it is electromotive force. As it comes from the generator, **voltage** is high but not high enough to move through power cables without much loss of energy. Therefore, step-up transformers are used.

A **transformer** is an electrical device that is able, through use of a magnetic field, to change the voltage in an electrical current. See Fig. 6-32. However, current at a voltage of about 340,000 volts is too high to be used in motors, lights, and other devices. Other transformers, called step-down transformers are located in substations to reduce the voltage, Fig. 6-33. Then it is sent out to factories and other places where it will be used for power. Small transformers, located on poles or on the ground reduce voltage to 120 and 240 volts for use in homes and office buildings. See Fig. 6-34.

Operation of electric motors

An **electric motor**, Fig. 6-35, is very similar to a generator or alternator. The only difference is that the

Fig. 6-33. A substation reduces voltage to levels that can be utilized by industry. A protective fence always encloses substations because of the danger created by the high voltage.

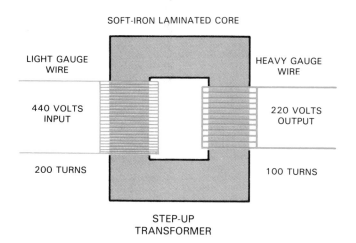

SOFT-IRON LAMINATED CORE

LIGHT GAUGE WIRE

HEAVY GAUGE WIRE

440 VOLTS INPUT

220 VOLTS OUTPUT

200 TURNS

100 TURNS

STEP-UP TRANSFORMER

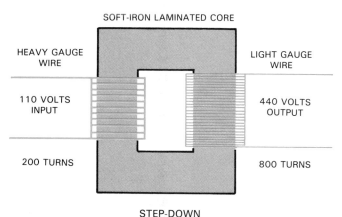

SOFT-IRON LAMINATED CORE

HEAVY GAUGE WIRE

LIGHT GAUGE WIRE

110 VOLTS INPUT

440 VOLTS OUTPUT

200 TURNS

800 TURNS

STEP-DOWN TRANSFORMER

Fig. 6-32. Transformers also make use of magnetism to change electrical voltage. Windings of wire set up a magnetic field (called magnetic flux). The magnetic field induces an electric current which is transferred from one coil to the other.

Fig. 6-34. Electrical service is provided to homes through power lines like those shown here. Transformers, located on the poles reduce voltage to 120 and 240 so the electricity can be used in homes.

Fig. 6-35. An electric motor uses electrical energy to create mechanical energy.

generator changes mechanical energy into electrical energy while the motor changes electrical energy back into mechanical energy.

Again, the operating principle of the motor is magnetism. This is how it works:

- Electric current from a generator passes through the armature.
- The current creates a magnetic field with one side of the coil having a positive charge and the other side a negative charge.
- The coil interacts with the motor's magnets. This causes the coil or armature to spin as one side of the coil is repelled and the other side is attracted to the magnet.
- The spinning action is strong enough so that the motor can drive machines.
- A shaft on the end of the armature transfers the rotating action to a pulley. A belt transfers the power to a machine.

Cells and batteries

Chemical energy can be stored and used later to produce electrical energy. Cells and batteries, Fig. 6-36, can be considered both as storage devices and as energy converters. The chemicals can be in the form of a paste (dry battery) or liquid (wet battery). A simple wet cell can be made by suspending a strip of copper and a strip of zinc in a chemical solution of water and sulphuric acid. If conductors are attached to the strips, and then to a light meter, electrons will flow to create an electrical current. A simple cell is shown in Fig. 6-37.

BOILERS AND FURNACES

Another type of indirect energy conversion device is the boilers and furnaces used to produce steam and hot air for space heating and industrial processes. These systems are a type of heat engine. They use a fossil fuel or solar energy to produce heat. See Fig. 6-38.

The Rankine engine

The **Rankine engine** is a reversible heat engine. It uses hot water to heat up a volatile (turns to vapor easily) fluid such as Freon.

The Rankine engine is generally used to drive air conditioners or generators. One of its recent uses is to produce electric power in ocean thermal energy conversion (OTEC). This is covered in Chapter 14 on Thermal Energy. See Fig. 6-39.

DIRECT ENERGY CONVERTERS

The battery is a familiar example of a device which makes a direct conversion of energy. It uses a paste or liquid chemical solution to store electricity. The stored electricity produces a chemical change in the solution. When the battery is connected to a circuit, electrons will move out of the chemical solution through the conductor. The battery produces the current without any moving parts or any intermediate steps.

Fig. 6-36. Batteries and cells come in various sizes and voltages.

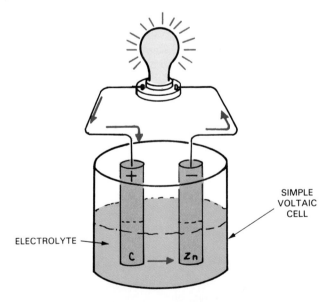

Fig. 6-37. The principle of a cell or battery can be demonstrated in this homemade cell.

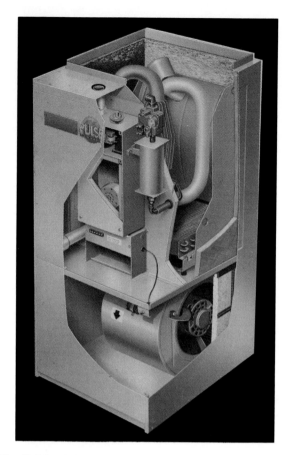

Fig. 6-38. A hot air furnace is a type of heat engine. (Lennox Industries, Inc.)

Scientists and engineers are working on other devices that will convert some form of energy directly. This is desirable for several reasons:
- For one thing, direct conversions are generally more efficient.

- Another reason is the belief that a direct conversion device will be more reliable and less prone to breakdown. This factor would make direct converters attractive for use where repairers would not be on hand to fix them. Small units of this type already operate the power systems in satellites and rockets used in the space program.

There are several emerging direct conversion systems. Some of them are in limited production. Others are just being developed.
- **Fuel cells.** Somewhat like a battery, they produce electricity from a fuel source and oxygen.
- The **MHD** (magnetoheterodyne) **generator** which turns hot gases directly into an electric current. It is an adaptation of a simpler device known as a thermocouple.
- **Solar cells** which turn sunlight directly into electricity.

These and other direct conversion devices that are still under development are discussed elsewhere.

SUMMARY

Energy conversion is the changing of energy from one form to another form. This unlocks the energy to do work and changes the energy into a form more appropriate for the work to be done.

Energy conversion can be made directly or indirectly. Indirect conversion involves more than one conversion to make the energy useful for the purpose intended. Direct conversion has the energy being used for work after one conversion.

Of the many energy converters, one of the most used is the internal combustion engine. It is so called because

RANKINE CYCLE

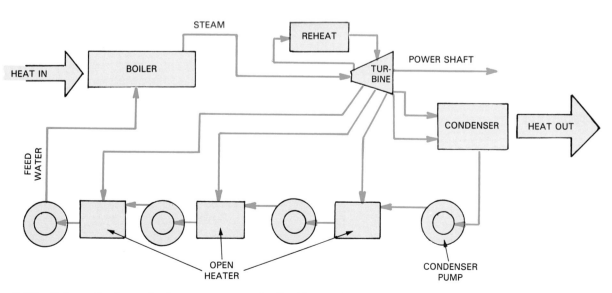

Fig. 6-39. Schematic shows basic arrangement of a Rankine engine. Its most common use is for driving air conditioners.

its fuel is burned inside the engine. The types are: four-stroke cycle engine, two-stroke cycle engine, diesel engine, rotary engine, jet engine, gas turbine engine, and rocket engine.

External combustion engines burn their fuel outside the engine. The oldest of this type is the steam engine, now little used. Other types are: Stirling cycle engine, and the free piston engine.

Because electrical power is very important to us, electrical energy converters are much used. The most common name for electrical converters is generator. The types are: direct current generator and alternating current generator. The latter is usually known as an alternator.

Generators are usually located at power stations and use steam or water power as an energy source. Generated electricity is sent out over power lines to its users. Voltage can be changed by transformers. Step-up transformers increase voltage; step-down transformers decrease voltage.

Electrical energy is changed to mechanical energy by electric motors. Electricity is also used to produce light and heat.

Direct energy converters include batteries, fuel cells, solar cells, and magnetoheterodyne generators. These are discussed in a later chapter.

When working with and repairing converters certain tools are needed. Tool users must follow certain rules when using tools to avoid injury.

DO YOU KNOW THESE TERMS?

alternator
direct conversion
direct current generator
electric motor
energy conversion
external combustion converters
free piston engine
fuel cell
gas turbine
hypergolic
indirect conversion
internal combustion engine
jet engines
MHD generator
newtons of thrust
pounds of thrust
ramjet engine
Rankine engine
rocket engine
solar cell
Stirling cycle engine

transformer
turbofan engine
turbojet engine
turboprop engine
voltage

SUGGESTED ACTIVITIES

1. Make a list of power systems in your community, name the type of energy converter each uses and, if you can, list the type of energy it converts.
2. Take the list of energy converters developed for Activity 1 and discuss the types of pollution each converter produces. Take part in a class discussion on possible methods of reducing or eliminating the pollution of the conversion process.
3. Again, considering the list of energy converters in Activity 1, take part in a class discussion on what you would do to replace these converters of energy if their energy supply was exhausted. List alternative fuels and converters.
4. With your instructor's help, build a simple model or toy demonstrating one of the types of energy converters discussed in this chapter.

TEST YOUR KNOWLEDGE

1. An energy converter which changes energy into several different forms before it is used to do work is called a (direct, indirect) converter.
2. A heat engine which produces the burning of fuel inside a cylinder is called:
   a. A reciprocating engine.
   b. An external combustion engine.
   c. A two-stroke cycle engine.
   d. An internal combustion engine.
3. A rotary engine has _____ power thrusts in every revolution.
4. State Newton's third law of motion and explain how it applies to jet propulsion.
5. The Stirling cycle engine has how many power strokes for every revolution?
6. Explain in what respect a free piston engine is like a diesel engine.
7. Electricity is thought to depend on the movement of _____ through a conductor.
8. Name the two types of machines that convert other forms of energy into electricity.
9. List the number of conversions for a fossil-fueled power station to produce electricity.
10. What safety precautions should be used when using a grinder?

Selecting tools

Energy conversion units require periodic maintenance and service. Your technology education laboratory may have tools for the service of small internal combustion engines such as those used in lawn mowers.

Among these tools there will be a complete set of wrenches, pliers and screwdrivers. Any engine work will also require socket wrench sets, punches, chisels, hammers, files, drills and drill bits, reamers, and tap and die sets. Refer to Fig. 6A.

Engine service and repair will also require certain measurement tools: inside and outside micrometers, feeler gauges, pressure gauges, and torque wrenches. See Fig. 6B.

Tools are safest when used properly:

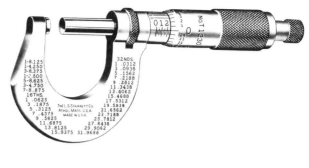

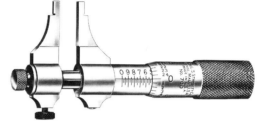

Fig. 6B. Micrometers are measuring instruments used to measure inside and outside diameters. (L. S. Starrett Co.)

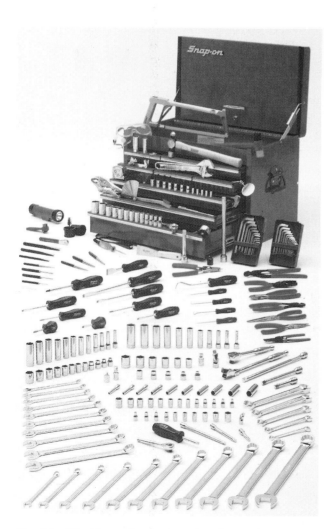

Fig. 6A. Working on internal combustion engines can require a large inventory of tools. Over time, engine repair technicians build up a large assortment of hand tools and specialty tools so they will always have the right tool for the service or repair task at hand. (Snap-On Tool Corp.)

- Never strike two hammerheads together. Their faces are very hard; chips of metal may fly off at high speed.
- Never use a striking tool with a loose or damaged handle.
- Avoid "choking up" too far on the handle of a striking tool. Knuckles can be injured when striking a blow.
- Do not use chisels or punches with mushroomed heads. Bits of metal can become dislodged, causing injury.
- Do not use screwdrivers as a pry or chisel.
- When using pliers, keep hands clear of cutting edges.
- Use pliers with insulated handles when working around electricity.
- Pull rather than push on a wrench. If a pushed wrench slips or the fastener loosens suddenly, you may injure your knuckles on the work. The movable jaw of an adjustable wrench should always face the direction the fastener is being turned.
- Remove metal chips with a brush, never with the hand.
- Keep fingers and hands away from the cutting edges of tools.
- When using a grinder, wear safety goggles and be sure the tool's safety shield is in place. Be sure that ALL guards are in place.

Integrated circuitry has made electronic control possible in the transportation industry.
(Harris Corp.)

7

POWER TRANSMISSION SYSTEMS

The information given in this chapter will enable you to:
- Define the term, power transmission systems.
- Discuss the essential elements of mechanical transmission systems, fluid transmission systems, and electrical transmission systems.
- Demonstrate an understanding of the three systems.
- Define terms used in all three systems.

The force and work produced by energy converters would be useless without a method of moving the force applied by the converter to do useful work. Power derived from a form of energy can be transmitted from one place to another by a power transmission system.

A **power transmission system** is a method of transferring, controlling, and adapting the output of an energy converter. By transmitting and controlling power we adapt it to the work that needs to be done.

Various systems have been invented or developed to provide for power transfer. This chapter will discuss:
- Mechanical systems.
- Fluid systems.
- Electrical systems.

A transmission system might be as simple as a mechanical or hydraulic jack that lifts an automobile to change a flat tire or as complicated as the electrical system in a factory. See Fig. 7-1.

MECHANICAL POWER SYSTEMS

A mechanical power system transmits and controls power through a solid, mechanical linkage between the energy converter and the machine it powers and controls. Various linkage and control devices are used. The most common are:

Fig. 7-1. A car jack transmits human force and multiplies it to lift the heavy car. The threads on the jack are an adaptation of the inclined plane.

- Shafts.
- Pulleys and belts.
- Gears.
- Sprockets and chains.
- Clutches.
- Cams.

KINDS OF CHANGE

Recalling for a moment what was said in Chapter 1, systems have three parts: input, process, and output. The energy converter provides the input. Mechanical devices provide process or change in several ways:
- Moving the power over some distance.
- Connecting power when needed, disconnecting power when not needed.
- Changing direction of power transmission.

• Changing force and/or speed.

Output of the system can be linear motion, like a car traveling down a highway; rotary, like the spinning action of a drill press; reciprocating, like the hammering action of a forge. It can also be the many operations of an industrial machine.

Transmitting power to the load

A converter or propulsion unit, as they are also called, is usually some distance from its load. A mechanical link must be made so that power can be carried to the load. The transfer device may continue the twisting motion of an engine, for example, or convert it to straight-line motion.

Without a means of turning off power, the power source would be producing power when it is not wanted. There must be a means of disconnect without stopping the production of power by the converter.

Connecting and disconnecting

A **clutch** is a simple type of connect/disconnect device. It is used with the manual transmission in a tractor or a car. A hand lever or a foot-operated pedal disconnects the engine from the drive wheels when the operator wants to stop, change direction, or change to a different gear. See Fig. 7-2.

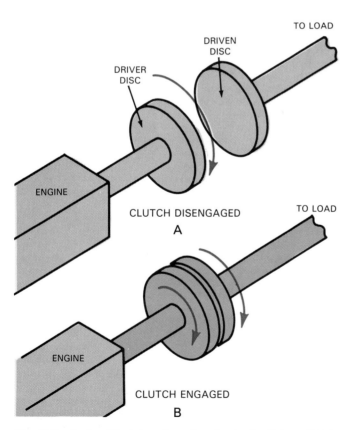

Fig. 7-2. A simplified drawing of a clutch. A—Driver disc is not in contact with the driven disc; driven shaft is motionless. B—Drive disc contacts driven disc. Driven disc and shaft rotate.

Changing direction

Many of the converters or propulsion units in use today have a rotating input of power. (The technical name for rotating force is **torque**.) Often the direction of the force must be changed to get the motion needed by the load. Most often, the change is:
• Reversing of direction. This happens when you want an automobile to go from forward to reverse.
• Turning. Power rotating on one plane must be delivered in another plane. This happens when a rear end assembly of a car redirects power 90° to deliver the power from the engine to the rear wheels.

Changing force or speed

Force and speed are always present in mechanical power. There is a close relationship between them. When one increases the other decreases by the same amount. This relationship is explained in the principle called mechanical advantage. How this works will be explained later.

In our use of mechanical power, our needs change with conditions. Sometimes we need more power; sometimes more speed is required. If you have ever ridden a three-, 10-, or 12-speed bicycle you have experienced this changing need. Look at Fig. 7-3. As you first started to move you needed extra power. Then, as the bicycle began to move faster, you did not have to pump so hard but wanted to increase your speed. By changing gears, you were able to produce the right amount of force and speed as they were needed.

Mechanical advantage

There are times when more power or more speed is desired than the propulsion unit is designed to provide.

Fig. 7-3. Gear cluster on the rear wheel of a 10-speed bicycle. When more power is needed, the rider shifts the chain to a larger gear.

Then a means of increasing power or distance must be found. When such a means is employed, it gives the applied force an advantage. This advantage, either greater force or greater speed than that applied, is called **mechanical advantage.**

The principle of mechanical advantage is closely tied to the lever. The lever is a simple machine consisting of an arm and a fulcrum (pivot point). See Fig. 7-4. The lever rotates on the pivot point. Depending on how the fulcrum and lever arm are arranged, the lever can be either a force multiplier or a distance multiplier. Fig. 7-5 shows an arrangement that makes the lever a force multiplier. A small amount of force will multiply at the load end. For example, if the load is 1 ft. from the fulcrum and the force is 4 ft. away from the fulcrum, the actual applied force is multiplied four times. A 25 lb. force will lift a 100 lb. load. (Force delivered at the load is always measured in foot-pounds since it is a combination of force and distance.)

Problem: Given the lever arrangement in Fig. 7-5, what force would be needed to lift a 280 lb. rock? Solution: Since the distance from fulcrum to force is 4 times the distance from fulcrum to the rock, divide the load by 4 to determine the force needed: 280/4 = 70. Thus force = 70 ft./lb.

To multiply distance, the force must be nearer the fulcrum than the load. See Fig. 7-6. Thus, a force moving a short distance can cause a load to be moved a

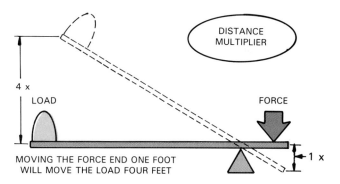

Fig. 7-6. When the fulcrum is nearer the force end of the lever, the lever becomes a distance multiplier.

much greater distance. A much larger force must be applied, of course.

Problem: A force to be applied to a lever is 1 ft. from the fulcrum and the load to be moved is 4 ft. from the fulcrum. How far will the load move if the force travels 6 in.?

Solution: Since the load is 4 ft. from the fulcrum and the force 1 ft., the distance traveled by the load will be four times the distance traveled by the force. Thus, .5 ft. x 4 ft. = 2 ft.

Problem: What is the mechanical advantage of a lever if a force of 30 lb. will lift a load of 90 lb.? Solution: The formula for mechanical advantage is:

$$\text{Mechanical advantage (MA)} = \frac{\text{Resistance}}{\text{Effort}}$$

Thus: a resistance of 90 lb. divided by the effort, 30 lb., has a mechanical advantage of 30.

Mechanical power transfer is often made using gears, belts and pulleys, shafts, sprockets, clutches, and cams. Gears and pulleys are, in reality, an application of the principle of the lever. The center of the gear, called the axis, is the fulcrum. The distance from the center to the outside rim of the gear or pulley is the lever arm. If a force is applied at the rim, the force at the axis is multiplied. If the force were applied at the axis, force would be reduced at the axis but speed would be multiplied at the rim.

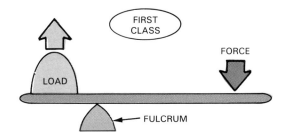

Fig. 7-4. A lever, when used as shown, multiplies force. A small force applied at the right end will have greater force at the left end. This increase in force is called "mechanical advantage."

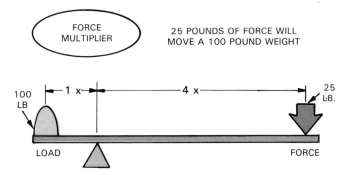

Fig. 7-5. An example of force multiplication.

GEARS

A **gear** is like a wheel with notches called teeth around its rim. Gears are designed to be used with other gears. The teeth of one mesh with (fit between) the teeth of another gear. See Fig. 7-7. This meshing allows one gear to transmit power to another gear. The gear connected to the power source is called the **drive gear.** The gear that receives the power from the drive gear is called the **driven gear.** An advantage of gears is that there is no slippage. Transfer of power is positive. Fig. 7-8

Fig. 7-7. Gears are a method of transmitting power as the teeth of a drive gear engage the teeth of a driven gear.

shows the different ways that gears transfer and control power.

Gear ratio

A gear ratio is a comparison of the number of teeth on a drive gear to the number of teeth on a driven gear. Ratio is important to know because it allows you to determine changes in torque and speed by comparing the difference in the number of teeth. If the tooth count is the same, there is no difference in torque or speed. When the drive gear has less teeth than the driven gear, speed of the driven gear is less but torque is increased. If the drive gear has more teeth, speed of the driven gear is increased while torque of the drive gear is decreased.

Computing gear ratio

Using division, you can determine the gear ratio of two gears. Suppose that a drive gear has 40 teeth and a driven gear has 20 teeth. By dividing the numbers 40 and 20, by a common number to the lowest number possible you get a gear ratio:

$40/20 = 20$ and $20/20 = 1$; therefore, the ratio is 20:1

If you want to know how many revolutions the driven gear will have, compared to the drive gear, you would divide the driven gear's tooth count into the drive gear's tooth count:

$40/20 = 2$; therefore, for every revolution of the drive gear, the driven gear will have two revolutions

When the drive gear is smaller and has fewer teeth, it has an increase of torque. This means that it has a mechanical advantage. You can find the mechanical advantage by dividing the number of teeth in the driven gear by the number of teeth in the drive gear.

Example: A driven gear has 40 teeth and the drive gear 10 teeth.

$40/10 = 4$

Thus, the gear arrangement has a mechanical advantage of 4. The driven gear will provide three times the torque but will turn only 1/4 as fast.

BELTS AND PULLEYS

Belts and pulleys are an often-used method of transferring and controlling power in modern factories and in many transportation applications. Accessories to heat engines, such as alternators, water pumps, and

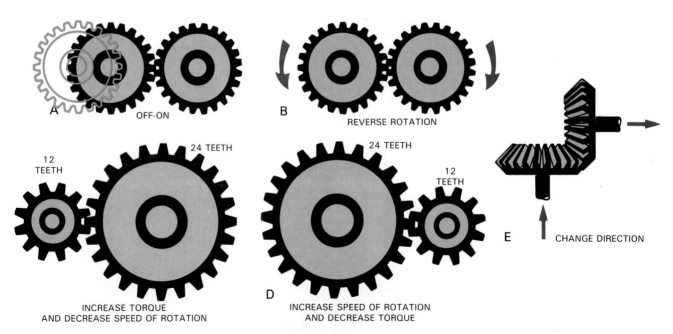

A — OFF-ON

B — REVERSE ROTATION

C — 12 TEETH — 24 TEETH — INCREASE TORQUE AND DECREASE SPEED OF ROTATION

D — 24 TEETH — 12 TEETH — INCREASE SPEED OF ROTATION AND DECREASE TORQUE

E — CHANGE DIRECTION

Fig. 7-8. Gears control transfer of power in several different ways.

hydraulic pumps are powered by belt arrangements between the engine crankshaft and the accessory, Figs. 7-9 and 7-10.

Like gears, belts and pulleys can:
- Connect and disconnect power.
- Reverse direction of rotation.
- Change direction or plane of motion.
- Change torque.
- Change speed of rotation.

One advantage of the belt and pulley is its flexibility. Because of this feature, the system works well in spite of vibration or slight misalignment. Belt and pulley systems are also cheaper. Fig. 7-11 shows different belt-and-pulley arrangements designed to control transfer of power.

CHAINS AND SPROCKETS

Chains and sprockets are similar to belts and pulleys. They combine features of gears. The **sprocket** is a thin, flat gear with teeth. The teeth mesh with gaps in a chain. The chain transmits the torque of the sprocket over some distance to another sprocket. In all other respects the operating principle is the same as for gears and belts and pulleys. A bicycle is a good example of a machine transferring power with a sprocket and chain arrangement.

SHAFTS

A **shaft**, Fig. 7-12, is a metal rod or pipe sometimes used to transmit power. Because of the design, shafts cannot change the direction, speed, torque, or connect and disconnect power. Their main advantage is that they do not slip like belt-driven systems. Because of this advantage they are said to offer positive power

Fig. 7-9. Belts and pulleys. Belts transfer rotary motion from one pulley to another.

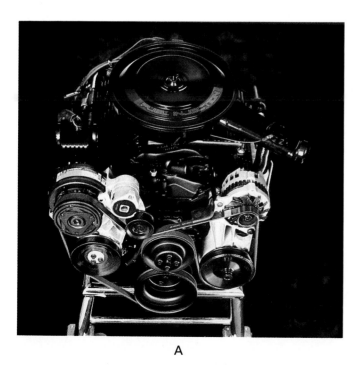

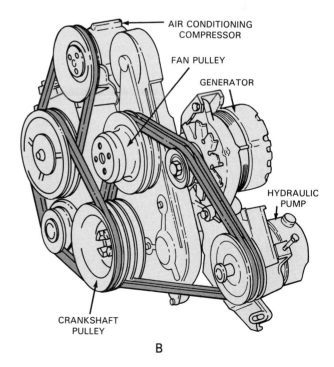

AIR CONDITIONING COMPRESSOR

FAN PULLEY

GENERATOR

HYDRAULIC PUMP

CRANKSHAFT PULLEY

A

B

Fig. 7-10. Belts and pulleys are used to transfer engine torque to operate engine accessories. A—This arrangement powers the alternator of a passenger vehicle. (General Motors) B—Drawing of belt and pulley arrangement in a typical automotive application.

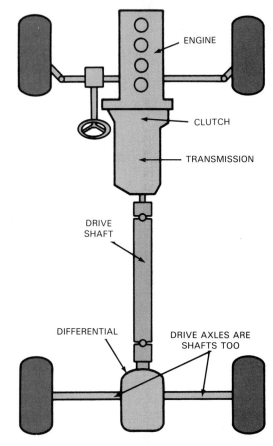

Fig. 7-11. Various methods of controlling power with belts and pulleys. A—Switching power on and off. B—Multiplying power and multiplying speed. C—Reversing direction. D—Changing direction.

transfer. Fig. 7-13 shows a common application of the shaft.

CAMS

It is sometimes an advantage to convert rotary motion to reciprocating (up and down) motion. Cams may

Fig. 7-13. A shaft is often used in transportation vehicles to transmit engine torque to the drive wheels.

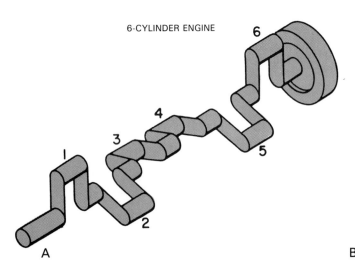

Fig. 7-12. Shafts transmit power in and away from engines. A—A crank shaft. It converts reciprocating force to rotating force (torque) in an engine. B—A drive shaft transmits torque in a straight line to a distant load.

be used for this purpose. The **cam** is a device that employs the principle of the inclined plane to achieve a lifting motion. As the cam rotates its sloping surface exerts pressure on rod. Automobile engines have cams that open and close valves. See Fig. 7-14.

FLUID POWER

Like mechanical systems, **fluid power** transfer systems are not able to produce power. They use fluids in transmitting power from one place to another and for altering power to do certain types of work. Fluid power systems do this by compressing the fluids which may be either liquids or gases.

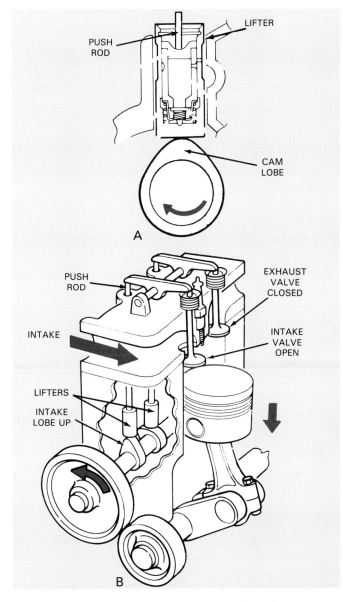

Fig. 7-14. Automotive engines have cams that change rotary motion to reciprocating motion. A—Cam first pushes on lifter which transfers the motion to the push rod. B—Push rods rock the arms which convert the upward motion to downward motion that opens the valves.   (John Walker)

Fluid power goes by two names and includes two separate systems: hydraulics and pneumatics. Both names are derived from Greek words, hydraulics for liquids and pneumatics for air. Both find applications in many areas of technology. See Fig. 7-15. Fluid power is also often used to move liquids and solid materials long distances through pipelines.

One of the major advantages of fluid systems over mechanical systems is that they do away with the need for a complex arrangement of gears, belts, pulleys, and levers. Another advantage is more accurate control of the applied power. There is no slack or play because of loose-fitting parts. Forces applied can be transmitted up, down, around corners, and over relatively long distances. In the process, there is little loss of power. Motion is smooth and constant. Forces remain constant even during a change of direction.

HYDRAULIC SYSTEMS

Hydraulic transfer systems, as stated before, use liquids as a transfer media. Several different liquids are suitable. These are: water-oil emulsions, water-glycol mixtures, and phosphate esters. In a few applications, for transfer of materials, water is the preferred fluid. Refined petroleum oil is the best fluid for most applications.

Liquids have certain properties that make them suitable for transfer of power:
- Shapelessness. Liquids have no shape of their own. They readily take the shape of their container. This allows them to transfer power easily through pipes, tubing, cylinders, and hoses.
- Resistance to compression. Liquids decrease in volume very little under pressure. When force is applied to a confined liquid, it quickly acts like a solid.
- Liquids transfer applied force in all directions at the same time, Fig. 7-16. Pascal's Law relates to this characteristic: Pressure set up in a confined liquid acts equally in all directions. The Law is the basis for modern hydraulics.

Multiplying force

A hydraulic system consists basically of two cylinders with movable pistons connected by a tube or pipe. By varying the sizes of the pistons, the hydraulic system can be used to multiply either force or speed. When the cylinders are of equal size, the piston travel is the same. When force is applied to one the other will move the same distance. When one cylinder is smaller, however, piston travel is no longer the same. See Fig. 7-17.

Certain rules apply to the use of hydraulic power. These are based on Pascal's Law:

A

B

C

D

Fig. 7-15. Fluid power has many uses. A—Hydraulic hoist lifts front of a dump truck box. B—Hydraulic cylinders control a back-hoe. C—Hydraulic pump lifts concrete to a construction form. D—Bucket of a bulldozer is lifted by hydraulic cylinders.

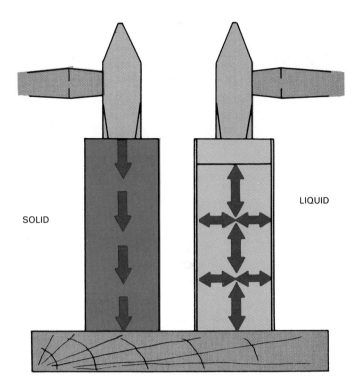

SOLID

LIQUID

Fig. 7-16. Solids transmit force in only the direction of the force. Liquids, when contained, transmit force in all directions.

- The force of each piston is in direct proportion to the areas of the pistons.
- The magnitude (amount) of each force is the product of pressure and piston area.
- Distances pistons will travel are inversely proportional to their respective areas.

Simply put, if pistons in a hydraulic system are the same size, the force at the output cylinder or piston will be the same as the force at the input piston or cylinder. On the other hand, if the pistons have different areas, force applied to a smaller piston will be multiplied at an output piston in direct relation to its size. Also, the smaller the input piston, the farther it must travel. The larger the output piston, the shorter will be the distance it travels.

Pressure

Pressure is measurement used most commonly in working with confined fluids. As you know, confined fluids exert force in all directions. Fluids press equally on the insides of their containers.

To calculate the force of fluids, we multiply the pressure or force by the area. Suppose that the surface (inside) of a tank is 200 sq. in. and the pressure of air pumped into it is 40 lb. per square in. (psi). Calculate the force.

Force = Pressure times area
Force = 40 x 200
Force = 8000 lb.

In fluid power, a small amount of pressure on a small piston acting against a large piston can create a larger force on the large piston. Consider the pistons shown in Fig. 7-17. A force of 20 lb. on the smaller piston becomes a force of 200 lb. at the larger piston.

Parts of a system

A hydraulic system, in addition to the hydraulic cylinders, will be made up of these parts:
- A pump. Driven by a power source like an electric motor or some other energy converter, it applies pressure to the hydraulic fluid in the system.
- A relief valve. Set to a predetermined value, the valve limits the pressure that is applied to the system. Should pressure exceed the set value, the valve will allow fluid to bleed off and return to the fluid reservoir.
- A reservoir. Since a hydraulic system needs more fluid than might be held in the pressurized part of the system, the reservoir contains the reserve supply. The pump draws from the reservoir as needed. The fluid returns to the reservoir when the system is at rest.
- **Accumulator.** This is a second storage place for liquid that is under pressure in the system. It acts as a damper or "shock absorber" as pressure changes or movement stops in the system.
- Directional control valves. These are installed in the pressure side of the system to control movement of fluid to and away from the output piston. It governs the direction in which the output piston will move.
- Filters. Every system has one; large systems may have several.

Fig. 7-18 shows a simple hydraulic system. Note the movement of fluid through the system and how its parts control the system.

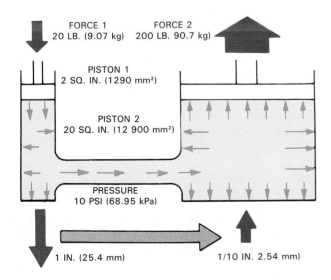

FORCE 1
20 LB. (9.07 kg)

FORCE 2
200 LB. 90.7 kg)

PISTON 1
2 SQ. IN. (1290 mm²)

PISTON 2
20 SQ. IN. (12 900 mm²)

PRESSURE
10 PSI (68.95 kPa)

1 IN. (25.4 mm)

1/10 IN. 2.54 mm)

Fig. 7-17. Multiplying force with hydraulics. A small force on the left-hand piston produces a large force at the larger piston. Which piston moves farthest?

PNEUMATIC SYSTEMS

Pneumatic systems are power transfer systems that use air as the transfer medium. In other respects, they are similar to hydraulic systems. Unlike liquids which do not compress very much, gases can be compressed to occupy a much smaller space than when in their free state. This property makes pneumatic energy transfer systems suited to many tasks in industry and transportation.

Pneumatic systems have other advantages. No expensive fluids must be purchased since they use atmospheric air. Also, there are no problems with leakage creating housekeeping problems in a factory environment where cleanliness may be critical to a manufacturing process.

Basic pneumatic system

A typical pneumatic system will have the following:
- A source of pressurized air. This source will include an air compressor driven by an engine or an electric motor.
- A storage tank, called a **receiver**, to hold the compressed air.
- Filters to take impurities out of the air.
- Pressure gauges.
- A heat exchanger to cool the compressor.
- Control valves to stop airflow or change its direction.
- An actuator. This is an air-powered device that converts the compressed air's energy to mechanical energy.

Uses of pneumatics

Industries are heavy users of pneumatics. They have largely replaced manual labor in factories. Garages use it to hoist vehicles into the air when they are being serviced. The hoist is raised by an air-over-oil system. Both air and oil are used.

Many tools are powered by air motors. They are more rugged and trouble-free than electric motors. See Fig. 7-19.

ELECTRICAL POWER

An **electrical power system** transfers and controls electric current. It does this through the use of conductors to carry the current and through devices that switch the current on and off, change its value, or change its direction.

The production of electrical power was explained in a previous chapter on energy converters. You learned in Chapter 6 that electrical power is produced by power stations from the chemical energy in fossil fuels or the kinetic energy in flowing or falling water. Electricity

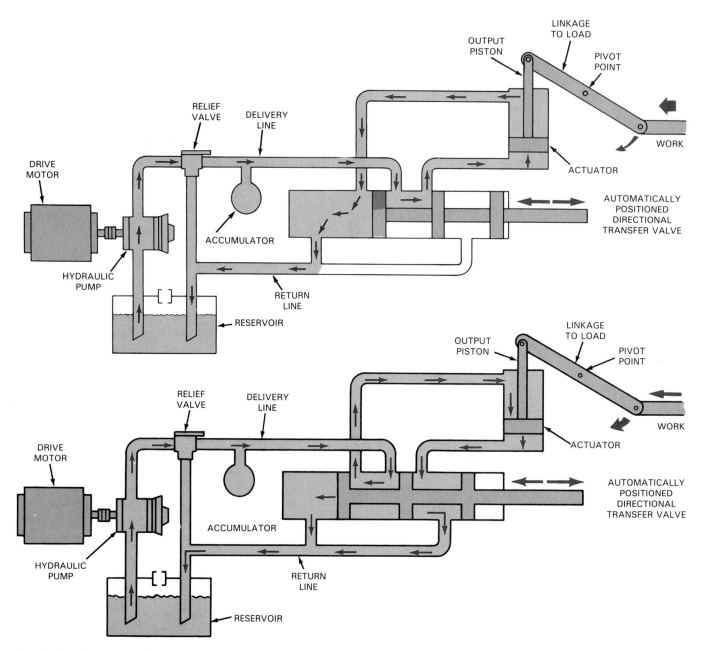

Fig. 7-18. Simple hydraulic system designed to move output piston in either direction. A—With transfer valve in this position, output piston will move upward. B—With transfer valve moved to the left, hydraulic pressure is applied to the other side of the piston and it will move downward.

Fig. 7-19. Pneumatic systems, besides being used as a transfer and control medium for power, operates door closers, jack-hammers, and other power tools especially in garages.

can also be generated by wind generators and solar cells. These units will be discussed in later chapters. You also learned that after electrical power was produced from other energy sources, conductors moved it to users and transformers changed its power so it fit the needs of various users.

A system of routing and controlling electrical power is called a **circuit.** The circuit is a path that directs electrical power to lights, machinery and appliances and other loads. Then the circuit directs the electric current back to its source. The word circuit comes from a word meaning a circle. The current does travel in a circle, doesn't it?

Electrical current travel is possible because electrons, attracted by protons, can move from one atom to another like water moves through a pipe. Fig. 7-20 shows how this is possible.

Electrons will move through a circuit as long as the power source provides electromotive force and the circuit is not interrupted by a switch. **Electromotive force** is the electrical pressure that causes electrons to move from one atom to another in a conductor. Another name for electromotive force is voltage.

TYPES OF CIRCUITS

Circuits are specially designed for the work that they do. There are three types:
- Series.
- Parallel.
- Series-parallel.

A **series circuit** allows only one path for current. A current traveling in such a circuit will go through every device in the circuit. Fig. 7-21 is an example of a series circuit. If a switch in the circuit should interrupt the electron flow, current would stop. The entire circuit would go "dead." There would be no current.

A **parallel circuit** is one that provides more than one path for the current to take. Look at Fig. 7-22. Three different paths are available for current. If any one of

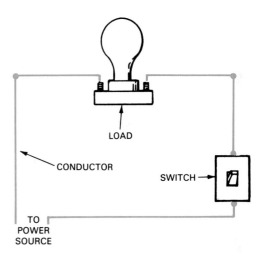

Fig. 7-21. A simple series circuit. It has but one path for electron flow. If the switch is in the "off" position, electron flow will stop.

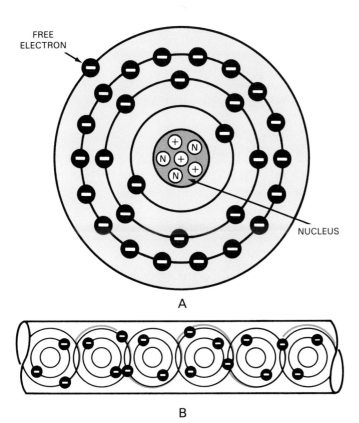

Fig. 7-20. How electrons are arranged and travel in a copper conductor. A—Each copper molecule nucleus is made up of neutrons and protons. A number of electrons equal to the number of protons orbit in four shells around the nucleus. B—The outer free electrons of each molecule can hop from one molecule to another. In this way, they are able to flow through the conductor.

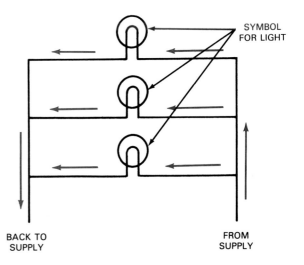

Fig. 7-22. A parallel circuit. If the path to any of the lights is interrupted, current will stop and that light will go out. However, the other lights will continue to burn since electrons still have a path.

Power Transmission Systems   99

the paths is interrupted, current would continue to flow in the other paths.

Circuits in a home are examples of parallel circuits. A load such as a light or an appliance can be switched off without affecting current to other loads.

A **series-parallel** circuit combines series and parallel elements. Part of the circuit has parallel conductors and part of it has but one conductor with devices in series with it. See Fig. 7-23. If one of the lamps, at part A, for example, burns out or is switched off, the remaining lamps and the rest of the circuit will remain on. If the switch at B is switched off, current will stop.

ELECTRICAL CONTROLS

In addition to a power source and conductors, an electrical circuit must have controls and a load. Controls are electrical devices like switches, relays, and protective devices.

Switches

Switches are disconnect devices that stop and start current in a circuit. Fig. 7-24 shows a common single pole toggle switch. It acts like the knife switch shown in Fig. 7-25. The toggle switch should be familiar to everyone. It is the type of wall switch commonly used to control lights. The toggle works on spring tension. When the toggle is flipped one way or the other, it causes throws to flip from one electrical contact to another with a snapping noise, Fig. 7-26. One direction turns the light on; the other turns it off.

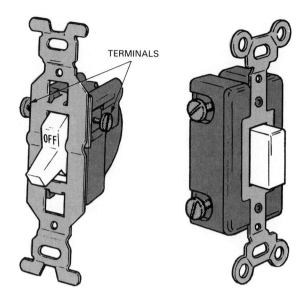

Fig. 7-24. Single-pole switches. A—This one is known as a toggle switch because of its handle. It has two terminals and will control a light or an appliance. B—Single-pole switch with touch control.

Fig. 7-25. A knife switch is the simplest example of a single-pole switch. This one is shown in the "open" position.

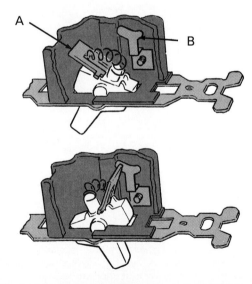

Fig. 7-26. Inside view of a single-pole switch. Spring causes the "snapping" noise as it moves contact "A" onto or away from contact "B."

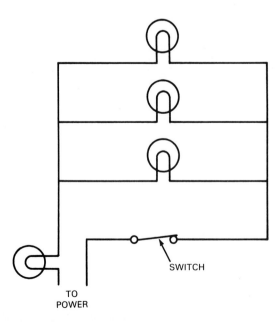

Fig. 7-23. A series-parallel circuit combines both types of circuit. What would happen if the switch shown were put in the "off" position? Would part of the circuit remain lit?

A **three-way switch**, Fig. 7-27, is like a single pole switch but with three terminals instead of two. It is used along with a second three-way switch when current to a light must be controlled from two different locations in a building. One terminal, usually a darker color than the other two, has the "hot" (current-carrying conductor) connected to it. A set of conductors called "travelers" conduct current between the switches. This arrangement between the switches allows the light to be turned on or off from either location. Fig. 7-28 shows how a three-way switch works in a circuit.

A **four-way switch** is used when a light must be controlled from three locations. It has four terminals and is used between two three-way switches, Fig. 7-29. A four-way switch always allows current to flow through it. Fig. 7-30 shows a circuit with a four-way switch in it.

RELAYS

A **relay** is a switch that is operated by an electric current. It controls on and off switching with an electromagnet. The electromagnet is a small iron cylinder that becomes a magnet when energized by electricity. The relay consists of a conducting coil of wire wrapped around the iron cylinder several times. When an operator closes the switch, a current travels through

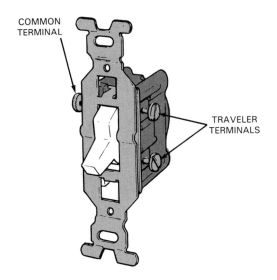

Fig. 7-27. A three-way switch has three terminals.

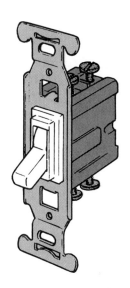

Fig. 7-29. A four-way switch has four terminals.

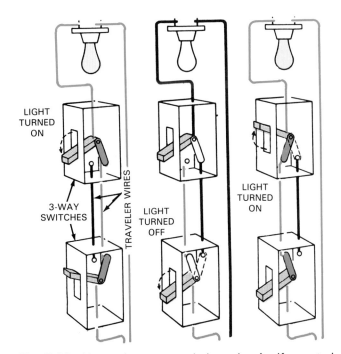

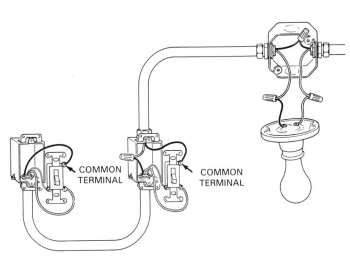

Fig. 7-28. How a three-way switch works. A—If you study the switch positions, you can see how a three-way switch works. B—This is how an electrician might wire up a three-way switch.

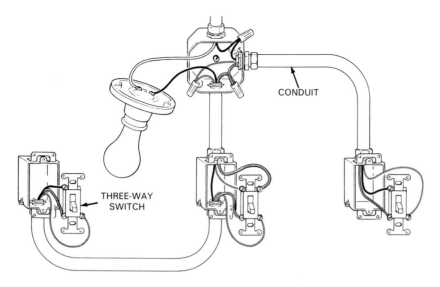

Fig. 7-30. Installation of a four-way switch as an electrician might do it.

the coil. The iron cylinder, being magnetized, pulls on two contacts and allows current to flow through a circuit.

Relays can be used to operate such devices as locks, lights, and motors. The relay can be operated by other devices such as clocks, thermostats, light meters, and other sensing devices. Fig. 7-31 is a schematic of a relay.

FUSES AND CIRCUIT BREAKERS

Fuses and **circuit breakers** are electrical devices placed in circuits to protect them from overcurrents. If an abnormal condition exists in the circuit or if current demand becomes too high, a fuse or circuit breaker will stop current in the circuit.

One of the common faults that cause an overcurrent is a short circuit. A **short circuit** is an accidental, overly strong current in the circuit. It occurs when current, due to a fault in the conductors, returns to its source too soon. If the resulting overcurrent is allowed to continue, it could overheat the conductors, creating damage or even a fire. The overcurrent is a natural result of a sharp drop in the circuit's resistance to current. This follows Ohm's Law which states that current is inversely proportional to resistance. Thus, if resistance drops, current becomes stronger.

Short circuits can occur when insulation breaks down and allows the conductors to touch. Insulation failure can be caused by repeated bending of a lamp cord, for example.

An **overload** is an overcurrent caused by an unusually high demand put on a circuit for electric power. The current in a overload never leaves its normal electrical path through the circuit.

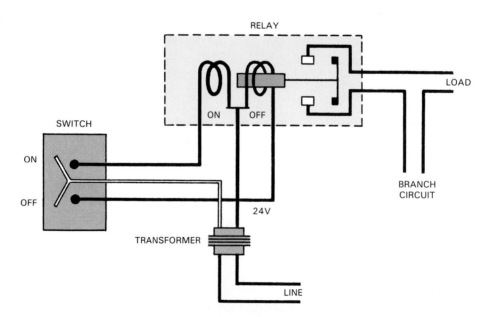

Fig. 7-31. Schematic for a circuit that is controlled by a relay.

A short-circuit overcurrent can exceed the normal current by hundreds of times. An overload causes overcurrent only in a range from 2 times to 10 times more than normal.

Fuses

The simplest kind of fuse is simply a length of thin wire. If current becomes too high it melts the wire and breaks the circuit. See Fig. 7-32.

A circuit breaker is more complex than a fuse. Overcurrent will trip contacts in the circuit breaker's housing. Once the problem has been corrected, the circuit breaker can be reset. Fig. 7-33 shows a cutaway of a circuit breaker. Fig. 7-34 shows symbols used in house wiring.

NAMES AND SYMBOLS FOR ELECTRICAL SYSTEMS

Electricity has force, flow rate, resistance, and power. The terms or units of measure and symbols for these are:

- **Amperes** — the measure of the rate of electron flow (current). The ampere refers to the number of electrons moving through a conductor in a period of time. The ampere is 6240 electrons per second past a given point in a conductor. It is equal to an electromotive force of 1 volt acting through a resistance

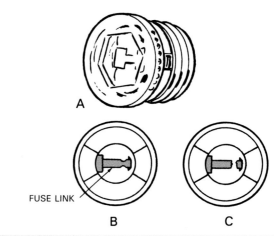

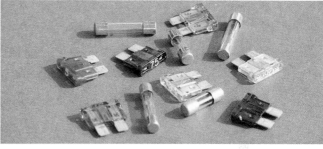

Fig. 7-32. Fuses are designed to ''self-destruct'' and burn out to stop overcurrents from damaging an electrical circuit. A— A plug fuse. B—The current must pass through the fusible link, a strip of thin metal. C—An overload caused this fuse to blow. Note that the fusible link is partly melted away stopping electron flow. D—Fuses of this type protect automotive circuits.

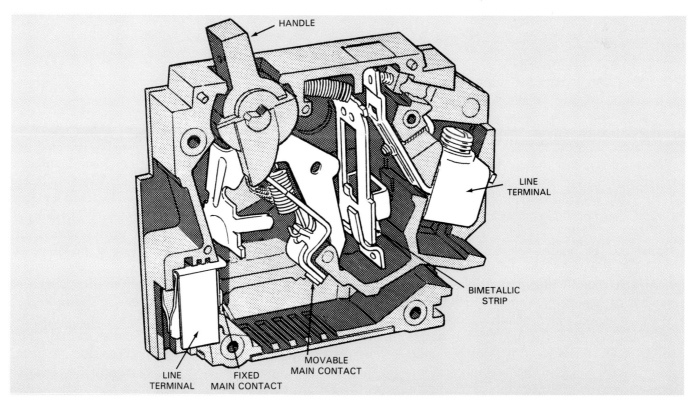

Fig. 7-33. A circuit breaker is an automatic switch in the hot wire of a circuit. It will open automatically, cutting off current when it becomes too high. (General Electric)

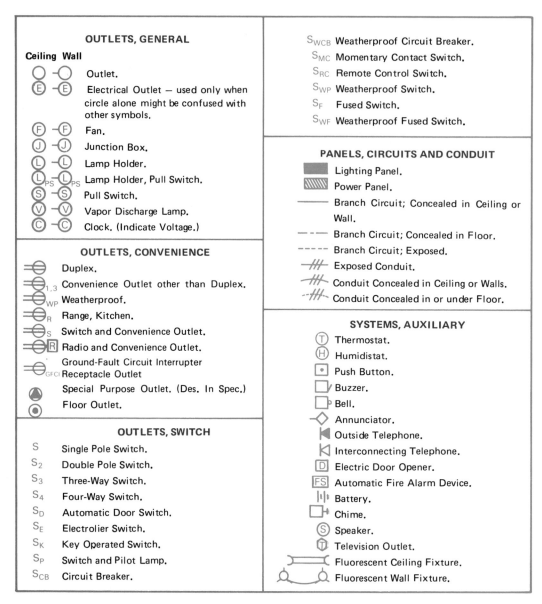

OUTLETS, GENERAL

Ceiling Wall

○ ─○	Outlet.
Ⓔ ─Ⓔ	Electrical Outlet — used only when circle alone might be confused with other symbols.
Ⓕ ─Ⓕ	Fan.
Ⓙ ─Ⓙ	Junction Box.
Ⓛ ─Ⓛ	Lamp Holder.
Ⓛ₍ₚₛ₎ ─Ⓛ₍ₚₛ₎	Lamp Holder, Pull Switch.
Ⓢ ─Ⓢ	Pull Switch.
Ⓥ ─Ⓥ	Vapor Discharge Lamp.
Ⓒ ─Ⓒ	Clock. (Indicate Voltage.)

OUTLETS, CONVENIENCE

Duplex.	
Convenience Outlet other than Duplex.	
Weatherproof.	
Range, Kitchen.	
Switch and Convenience Outlet.	
Radio and Convenience Outlet.	
Ground-Fault Circuit Interrupter Receptacle Outlet	
Special Purpose Outlet. (Des. In Spec.)	
Floor Outlet.	

OUTLETS, SWITCH

S	Single Pole Switch.
S₂	Double Pole Switch.
S₃	Three-Way Switch.
S₄	Four-Way Switch.
S_D	Automatic Door Switch.
S_E	Electrolier Switch.
S_K	Key Operated Switch.
S_P	Switch and Pilot Lamp.
S_CB	Circuit Breaker.

S_WCB	Weatherproof Circuit Breaker.
S_MC	Momentary Contact Switch.
S_RC	Remote Control Switch.
S_WP	Weatherproof Switch.
S_F	Fused Switch.
S_WF	Weatherproof Fused Switch.

PANELS, CIRCUITS AND CONDUIT

	Lighting Panel.
	Power Panel.
	Branch Circuit; Concealed in Ceiling or Wall.
	Branch Circuit; Concealed in Floor.
	Branch Circuit; Exposed.
	Exposed Conduit.
	Conduit Concealed in Ceiling or Walls.
	Conduit Concealed in or under Floor.

SYSTEMS, AUXILIARY

Ⓣ	Thermostat.
Ⓗ	Humidistat.
	Push Button.
	Buzzer.
	Bell.
	Annunciator.
	Outside Telephone.
	Interconnecting Telephone.
Ⓓ	Electric Door Opener.
FS	Automatic Fire Alarm Device.
	Battery.
	Chime.
Ⓢ	Speaker.
Ⓣ	Television Outlet.
	Fluorescent Ceiling Fixture.
	Fluorescent Wall Fixture.

Fig. 7-34. Every electrician must know these symbols.

of 1 ohm. Its symbol is A.

- **Volts**—the pressure that moves electrons through a conductor. This pressure is also called electromotive force. One volt, moving through a resistance of 1 ohm (Ω), will produce 1 ampere of current. Its symbol is V.
- **Ohm**—the resistance to electron flow in a circuit. One ohm is equal to the resistance of a conductor to an electromotive force of 1 volt, maintaining a current of 1 ampere. Its symbol is Ω.
- **Watt**—a measure of the electrical power produced from an electrical device or system. A watt (symbol, W) is equal to a current of 1 ampere under 1 volt of pressure, or 1 joule per second, or about 1/746 horsepower. Watts are found by multiplying volts times current ($W = V \times A$).

To solve equations in electrical power, you need to follow Ohm's Law. This Law states that the intensity of a constant electrical current in a circuit is directly proportional to the electromotive force and inversely proportional to the resistance. In simpler terms it means that whenever current increases, voltage increases at the same rate and vice versa; also, when resistance increases, voltage and current decrease at the same rate and vice versa.

It is easier to remember this and to solve electrical equations by using the memory device shown in Fig. 7-35. You can tell whether to divide or multiply by covering up the electrical value you wish to find. Then observe the position of the two other values. See Fig. 7-36. A sample problem will show how this works.

Problems:

If the voltage through a circuit is 120 volts and the current is 10 amperes, what is the resistance?

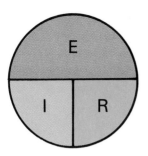

Fig. 7-35. This memory device will help you in knowing when to multiply and when to divide in working electrical math problems.

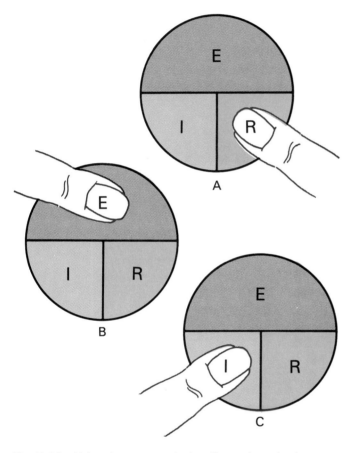

Fig. 7-36. Using the memory device. Cover the value (current, voltage, or resistance) that you wish to determine. Multiply when the remaining symbols are side by side. Divide by the lower when one symbol is above the other.

First, cover up the "R" in the memory device, Fig. 7-36A. Since the other two symbols, E and I, are shown with the E over the I, this means you must divide to the answer:

R = E/I
R = 120/10
R = 12 ohms

If the current in a circuit is 20 amperes and the resistance is 20 ohms, what is the voltage?
Covering up the voltage (E), as in Fig. 7-36B, shows the I alongside the R. This indicates that the values are to be multiplied:

E = I × R
E = 20 × 20
E = 400 V

If the voltage in a circuit is 240 volts, and the resistance is 10 ohms, what is the current?
Covering up the I, as in Fig. 7-36C, will show one of the other two values above the other. Thus the formula must be:

I = E/R
I = 240/10
I = 24 A

ELECTRICAL SAFETY

Working with and around electricity requires great care to avoid accidental contact with electrical current. Electrical shock is always unpleasant, dangerous, and potentially fatal. Current, rather than voltage, is the main concern with shock. Voltage is no more deadly at 10,000 volts than at 120. A current approaching as little as 10 milliamperes (0.01 ampere), however, is painful and may have severe effects on the body.

Because of the grave danger of injury or death, not to mention property damage, several safety agencies have set down rules that apply to the design, installation, and use of electricity. The National Fire Protection Association sponsors The National Electrical Code. This is a set of rules, regulations, and standards for the installation of electrical power systems. It is drafted by a team of experts and revised every three years. While it does not have the force of law, it is given legal status by many states and communities. Essentially, it is a guide for those who install electrical systems in structures.

Another body of electrical safety standards and regulations are set by the Occupational Safety and Health Act (OSHA). These standards are contained in the publication, "Design Safety Standards for Electrical System." The purpose of these standards is to ensure the safe use of electrical equipment by workers.

ELECTRONIC SYSTEMS

One area of electrical circuitry depends on the use of electronic devices to perform certain functions. Electronics is a technology involving the application of electricity through these special devices. At one time, these electronic functions were performed by vacuum tubes. Today, solid materials called semiconductors perform the same functions. For that reason, semiconductor devices are commonly known as **solid-state** devices. To understand why and how they work we need to know more about semiconductor material.

Semiconductor material

So far, our discussion of electricity has considered only materials that conduct electricity well. Insulators, such as the material covering electrical conductors do not conduct at all. Like their name suggests, semiconductors resist current. They do not stop it.

A commonly used semiconductor material is silicon. Pure silicon is an insulator. To make it somewhat conducting, impurities such as arsenic or aluminum are added. Arsenic frees up some of the electrons in the silicon. This makes the silicon negatively charged. On the other hand, when aluminum is added to the silicon, the silicon will accept free electrons. This makes the silicon positively charged. Fig. 7-37 shows a solid-state device made up of these two types of semiconducting materials.

Hole flow

Electrical theory is based on electron flow through metal conductors (wires). To understand how solid-state devices work, we must learn the theory of **hole flow**.

Hole flow is always the opposite of electron flow. It is best explained with the marbles shown in Fig. 7-38. Each marble represents an electron in a circuit. Each of the empty spaces is a hole. Each time a marble is moved from the left to the right, the hole moves to the left. This movement continues as the marbles are moved from left to right. When a current exists in a

Fig. 7-37. Semiconductor devices come in many sizes and shapes. (General Electric)

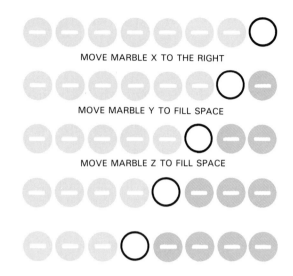

MOVE MARBLE X TO THE RIGHT

MOVE MARBLE Y TO FILL SPACE

MOVE MARBLE Z TO FILL SPACE

Fig. 7-38. Hole flow is the opposite of electron flow. Think of the colored balls in the illustration as electrons and the white ball as a hole. As electrons move to the right, the hole moves to the left.

circuit which has a solid-state device, the movement of electrons and holes is continuous as long as the device is allowing electron flow.

Diodes

A diode is a solid state device whose purpose is usually to allow electron flow in one direction, Fig. 7-39. If current is applied in the opposite direction, the diode will block it. To understand this, we need to understand the PN junction.

PN junction

The PN junction is the line of separation between the P material and the N material in the diode. See Fig. 7-40. This junction is important in the biasing of a transistor or any other semiconductor device. This will be seen clearly later.

To get electron flow across a semiconductor device, the electrons must travel through the P and N material. The electron flow can be likened to water attempting to flow from behind a dam. The water will back up behind the dam until it reaches a certain height. Then it will flow through gates installed in the dam for that

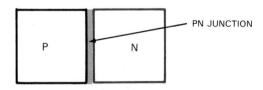

Fig. 7-39. A diode is like a check valve in a water pipe. It allows electron flow in one direction but not in the opposite direction.

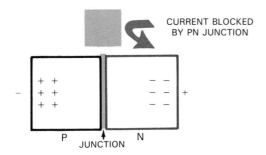

Fig. 7-40. Drawing of a PN junction. It is the barrier formed at the joint between the positive P material and the negative N material.

purpose. Once through the gates, the water cannot return.

A diode acts like a dam. It blocks any flow of electrons until they reach a certain level. When that level is reached, the electrons flow through the diode to perform their function in the circuit. If they try to move in the opposite direction, however, the diode will stop them.

Biasing

As described earlier, P material is fused with N material. When the N material side of the device is connected to the negative side of the power supply and the P material side is connected to the positive terminal of the power supply, the device begins to act like a dam. Since like charges repel each other, the electrons in the N material are driven toward the junction. On the other side of the junction, the holes (+ particles) are also repelled toward the junction. The electrons and the holes meet at the junction, Fig. 7-41. Current will then flow through the diode. It is said to be **forward biased.**

What happens when a diode is hooked up to the power supply positive terminal to N side and negative terminal to P side? This hookup will draw electrons and holes away from the junction. When this occurs,

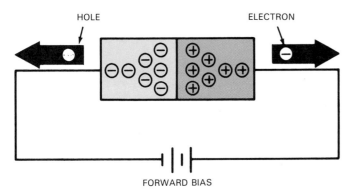

Fig. 7-41. Drawing of a forward-biased diode. Current causes the positive holes and the negative electrons to ''pile up'' at the junction. Current then is allowed to pass easily through the diode.

few, if any, electrons will flow through the device, Fig. 7-42.

The symbol of a diode is shown in Fig. 7-43. The tip of the "arrow" formed by the symbol always indicates the direction of "hole flow." Electron flow is in the opposite direction.

Diodes have a number of uses. Since they can convert alternating current to direct current, an important function is to convert current from an automobile's alternator to direct current so the gasoline engine can use it. LEDs (light-emitting diodes) perform a special function. They give off light when connected in forward bias. LEDs are used in digital instrument panels for automobiles, in digital clocks, and digital watches, as well as other places where digital displays are desired.

Transistors

Transistors are key semiconductors in electronic circuits. They can:
- Amplify current in a circuit.
- Create alternating current signals at desired frequencies.
- Switch current on and off.

Their switching function makes transistors important to the operation of computers.

The bipolar transistor is made up of three layers of doped (impure) silicon. There are two types: NPN and PNP. A PNP bipolar transistor has a thin layer of N-type crystals separating two N-type crystals. The NPN type bipolar transistor has a thin layer of P-type crystals separating two layers of N-type crystals. Drawings

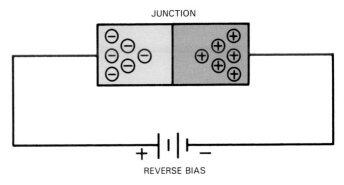

Fig. 7-42. Reverse-biased diode. If current tries to pass through, holes and electrons move away from the junction and little, if any, current can pass through the diode.

Fig. 7-43. Symbol for a diode.

showing their makeup and symbols are shown in Fig. 7-44. In both types, the first crystal is called the **emitter**, the second is called the **base**, and the third, the **collector.**

In the symbols, note the direction of the arrow inside the circle. By looking at the symbol in a circuit you can tell if it is an NPN or a PNP type by the way the arrow is pointing. A PNP type has the arrow pointing into the circle. An NPN type has the arrow pointing out of the circle. (You can remember them with this memory device: PNP type has the arrow **P**ointing i**N**; NPN type has the arrow **N**ot **P**ointing i**N**.)

Integrated circuits

An **integrated circuit** is a circuit containing resistors, transistors, and capacitors. It is a complete electronic circuit contained in one package. The IC, as it is generally called, is very small and compact, a circuit in miniature. Fig. 7-45 is a photo of an IC. Its smallness is its major advantage. Circuitry for an entire AM radio can be contained in a single IC no larger than a book of matches.

Making an IC

There are many types of ICs and there are different methods of making them. One method starts with a

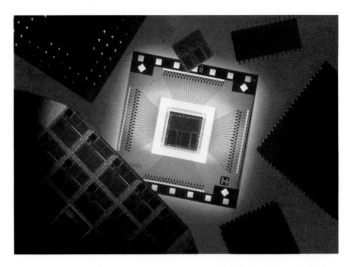

Fig. 7-45. Integrated circuits are the basis of much modern electronic technology such as signal processing. Integrated circuits contain many miniaturized circuits and circuit devices. (Harris Corp.)

block or wafer, of silicon. Then a drawing of the circuit is made much larger than the final circuit. Next, a special camera, called a reducing camera, reduces the drawing to 1/400 its original size. This allows the placing of several hundred complete circuits on a piece of silicon less than 1/2 in. in diameter. Special furnaces add the impurities to the silicon chips. This type of IC is called "monolithic." See Fig. 7-46.

ICs are used where smallness is important or necessary. Space communication equipment aboard satellites and space vehicles make extensive use of them. Limitations on weight and size are understandably severe in space operations. Closer to home and more common, are the limitations of space in a hearing aid. Can you think of any other place where smallness would be an advantage?

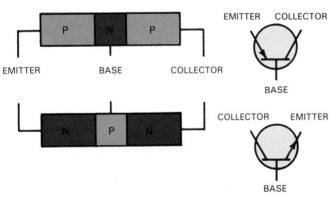

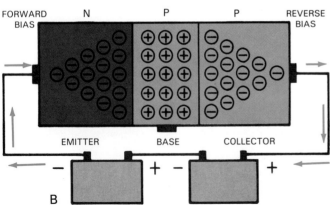

Fig. 7-44. Bipolar transistor. It has three layers of silicon crystals. A—Block diagrams and symbols for NPN and PNP transistors. B—How current flows in an NPN transistor.

Fig. 7-46. An advanced-technology wafer etching system in a modern production facility. The etching system provides microscopic circuit detail not previously possible in semiconductor manufacturing. (Harris Corp.)

Computers

Development of solid-state electronics and integrated circuits has made possible the development of microprocessors and computers. A **microprocessor** is a single-purpose computer with integrated memory units and controller units. It is used as a controller in transportation, homes, and factories. A **computer** is an electronic device that accepts data, processes it according to a stored program of instructions, and produces various outputs. Its output may produce visual or print messages, control machines that produce parts or products, or control various functions of vehicular systems or whole transportation systems. Fig. 7-47 shows a computer. Fig. 7-48 shows some of its inputs and outputs. Computers have become indispensable tools for various technologies.

Transportation. An on-board microprocessor receives information from sensors placed at various locations on an automobile engine and its drive train. Using information stored in its memory, the microprocessor uses incoming information to control such things as fuel mixture, idle speed of the engine, timing of ignition (spark), engine emissions, shifting of automatic transmissions, and braking. Refer to Fig. 7-49.

When a component malfunctions, the computer can also produce a code to tell a technician what the problem is. Fig. 7-50 shows an electrical schematic for an on-board computer.

Communication. The advent of the computer has touched off a new era of information exchange. Computers take in, process, store, and pass along information at amazing speeds. Whole encyclopedias, for example, can be placed on a few compact disks.

Manufacturing. Computers with special machining programs can accurately operate machines to machine,

Fig. 7-47. A computer being used as a communication tool. The material on the screen can be output to a printer.

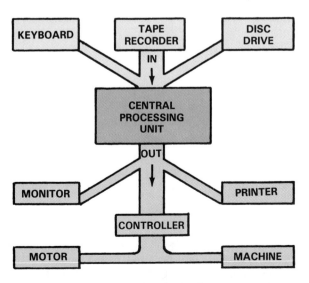

Fig. 7-48. This shows the parts of a computer system.

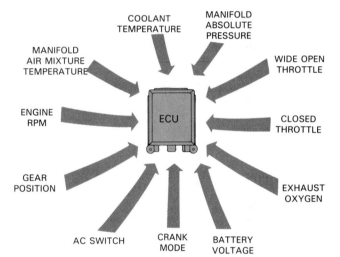

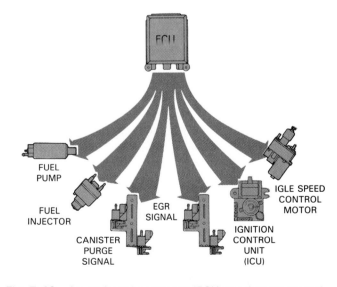

Fig. 7-49. An on-board computer (ECU or microprocessor). A—Sensors send it signals about the various operations of the vehicle. B—In response to signals, the computer controls various parts of the system.   (Renault)

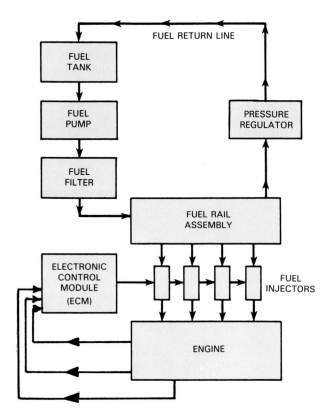

Fig. 7-50. How a computer operates an automobile's fuel delivery. Engine sensors feed information to the computer (ECM) as small electric currents. The ECM feeds current to operate injectors which feed fuel to the engine.

cast, and form products. They also control robots that perform such tasks as welding, transporting parts and materials, and control assembly lines.

SUMMARY

Power systems, sometimes referred to as power transmission systems, transfer, change, and control the power output of various types of energy converters. There are three common types of power systems: mechanical, fluid, and electrical.

Mechanical systems involve a number of devices. The most common are: shafts, pulleys and belts, gears, sprockets and chains. Clutches and couplings are devices often found in mechanical power systems.

Four kinds of control are provided by mechanical power systems. These are:

- Transmitting power over distance from the converter to the load (driven device or machine).
- Connecting and disconnecting power as needed.
- Changing the direction that power is transmitted.
- Changing the force and speed of power.

Power can be transmitted by direct mechanical linkage as a twisting (torque) or straight-line motion. In mechanical systems a clutch is a means of momen-

tarily disconnecting a load from its power source to execute a change in speed or to stop. Change of direction involves a reversal of direction or a turn to left or right.

There is a close relationship between force and speed. When force is increased there is a proportionate loss of speed; an increase in speed always involves a loss of force. This relationship is known as mechanical advantage. It is found in both mechanical and fluid power systems.

A mechanism that provides mechanical advantage is the lever. Gears and pulleys have arrangements that not only transfer power but can provide a mechanical advantage of force or speed. The relative size of the pistons in fluid power systems also provide either an advantage in speed or force.

Fluid power systems use liquids and gases to transmit power. When liquids are the transfer medium, the system is known as hydraulic. When air is the transfer medium, the system is known as pneumatic.

A major advantage of fluid systems over mechanical systems is their freedom from complex arrangements of gears, belts, pulleys, and clutches. Another advantage is more accurate control of the applied power.

Hydraulic and pneumatic power systems consist of a pump, movable pistons, reservoirs or pressure tanks, pipes, tubes, and valves. Hydraulic fluids include water-oil emulsions, water-glycol mixtures, and phosphate esters. Refined petroleum oil is the best for most applications.

Electrical power depends on the movement of electrons through paths called circuits. The circuits are wires called conductors along with devices that control the current between its source and its load. A series circuit allows only one path for electron flow. If something in the circuit interrupts the path, electron flow stops. A parallel circuit has more than one path for electron flow. If one is blocked, electron flow will continue through one or more alternate paths. A series-parallel circuit combines both single-path segments and multiple path segments.

As with other power transfer systems, electrical power uses control devices. Switches connect and disconnect electric power. Relays are switches that are operated by an electrical current. Fuses and circuit breakers are always placed in the electrical circuit to protect it from overcurrents caused by overloads and short circuits.

Electronic systems involve the movement of electron circuits that are controlled by solid-state devices. These devices switch current on and off, limit amount of current, amplify current, and control direction of electron flow. Electronic solid-state devices are also known as semiconductor devices. They include diodes, transistors, and variations of these devices. Integrated circuits include one or many circuits in one miniaturized package. Modern communication systems are the main beneficiaries of solid-state devices. Modern transportation vehicles are another major user.

Computers are also a result of the development of the semiconductor devices. They have become an important technological tool, not only in transportation, but other areas as well.

DO YOU KNOW THESE TERMS?

accumulator
amperes
base
cam
circuit
circuit breakers
clutch
collector
computer
drive gear
driven gear
electrical power system
electromotive force
emitter
fluid power
forward biased
four-way switch
fuses
gear
hole flow
integrated circuit
mechanical advantage
microprocessor
ohm
overload
parallel circuit
pneumatic power systems
power transmission system
receiver
relay
series circuit
series-parallel circuit
shaft
short circuit
solid-state
sprocket
switches
three-way switch
torque
transistors
volts
watt

SUGGESTED ACTIVITIES

1. Observe the mechanical transfer of power in the power lab at your school or on a field trip.
2. Construct a model vehicle that uses a power transmission device to propel the vehicle.
3. Construct a robot or a fluid power device using syringes, tubing, and other materials available.
4. Visit a computer store to learn about personal computers. Collect literature on various models. Learn about different uses of computers.
5. Visit a factory to learn how computers are used to control machines and robots.
6. Use a computer to write a report on a field trip.

TEST YOUR KNOWLEDGE

1. Define a power transmission system.
2. A transmission system is always very complex. True or False?
3. Name the devices that mechanically transfer and control power output.
4. A _____ is a simple type of mechanical device for quickly connecting and disconnecting power.
5. List the four ways that mechanical transmission systems control and change power output.
6. Where there is a mechanical advantage in transmission of power, an increase in force causes a _____ in speed.
7. State the formula for mechanical advantage.
8. _____ is a comparison of the number of teeth on a drive gear to the number of teeth on the gear it drives.
9. Name the type of power transmission device shown in the following illustration.

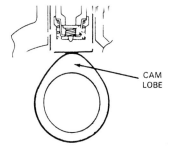

10. Fluid power includes only the use of liquids as transfer media. True or False?
11. Name three properties of liquids that make them suitable for transfer of power.
12. A transfer system based on the use of compressed air is called a _____ system.
13. Name one advantage of a transmission system using air.
14. Name the parts of a pneumatic system.

15. _____ power transmission systems use electron flow to transfer, control, and change power.

MATCHING TEST: Match the definitions or statements at the right with the terms at the left.

16. __ Series circuit.
17. __ Parallel circuit.
18. __ Series-parallel circuit.
19. __ Relay.
20. __ Short circuit.
21. __ Overload.
22. __ Voltage.

A. Electromotive force.
B. Accidental overcurrent caused by circuit fault.
C. Only one path for current.
D. Overcurrent resulting from too-high current demand.
E. Offers more than one path for current.
F. Switch operated by an electric current.
G. Part of the circuit has many paths; the other part, only one.

23. What is a semiconductor?
24. A _____ is a semiconductor device that will allow current in only one direction.
25. Name four functions that an on-board computer performs in an automobile.

FOSSIL FUELS—COAL

CHAPTER

The information given in this chapter will enable you to:
- Describe the process which turned dead plant matter into coal and list the stages in the coal forming process.
- List and describe the types of coal mining in the United States.
- Explain methods used to transport coal from mine to where it is used.
- List the processes for converting coal to gaseous and liquid fuels.
- Discuss the future of coal as an energy source.
- Explain the impact of coal on the environment and describe efforts to reduce pollution from burning of coal.

Coal is a fossil fuel, as you learned in Chapter 1. The coal is thought to be the remains of plant life which decayed and became buried millions of years ago. It is a rocklike material though much softer. Coal is usually black or brown, depending on its age.

HOW COAL WAS FORMED

Heat, pressure, and bacteria acting on the dead matter made new substances that are rich in carbon (82 percent) and hydrogen (12 percent). The substances are called **hydrocarbons** because they are a mixture of these two elements. The mixture burns well and makes an excellent fuel.

Coal is thought to have begun forming in the earth about 300-500 million years ago. The process took about 85 million years and was completed over 235 million years ago.

The world's great beds of coal are the remains of great near-tropical forests. See Fig. 8-1. The leaves and

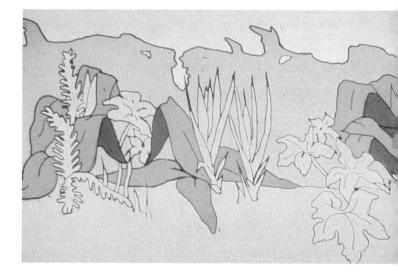

Fig. 8-1. Coal is a sedimentary organic rock. It formed from plants that died anywhere from a million to 500 million years ago. (American Coal Foundation)

the trees, themselves, fell into swampy water and decayed. First it was a slimy, watery mass. Later, it lost most of its water and the rotting vegetation went through various stages which are explained later. Scientists estimate that it required from 3 to 7 ft. (about 1 to 2 m) of compacted plant matter to form a foot of bituminous coal.

Coal was formed as seas advanced and receded in cycles all over the earth, Fig. 8-2. The waters deposited heavy layers of sandstone, shale, and other rock on top of the decomposing plant life. The successive deposits caused increasing pressure on the decaying matter and it became more and more dense.

STAGES OF COAL

Coal is found in several ranks (stages), depending on its age. See Fig. 8-3. The first stage is a product

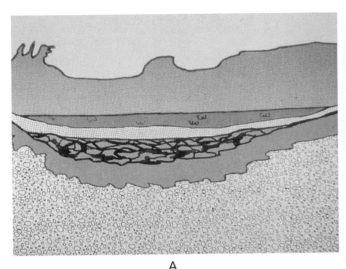

Fig. 8-2. Formation of coal. A—Remains of vegetation, in time, became buried under sediment from seas and rivers. The sediment formed, over time, into rock and pressed ever harder on the decomposing organic matter. Under constant pressure and heat, it turned into coal. (American Coal Foundation) B—A chunk of coal. (National Coal Foundation) These are the four ranks (stages) of coal. Development from one rank to another depends on the amount of heat and pressure on the coal bed.

RANKS OF COAL

- Peat
- Lignite
- Sub-bituminous
- Anthracite

Fig. 8-3. Coal has various ranks (stages) depending upon its age.   (American Coal Foundation)

that looks like decayed wood. It is called **peat**. As pressure continued to be placed on it, the peat turned slowly into a soft, brown fibrous material known as **lignite**. This is young coal. It burns well but does not give off much heat. The heat range of lignite is between 4000 and 8300 Btu/lb. Its carbon content is 25-35 percent. Sometimes referred to as brown coal, it is mainly used for the generation of electric power. At 8300 to 13,000 Btu/lb., the heat content of **sub bituminous coal** is somewhat higher than that of lignite. Its carbon content is between 35 and 45 percent.

Older coal becomes more dense and turns black. This type is called **bituminous** or soft coal. Its carbon content is much higher (45-86 percent) and it produces a hotter fire (10,500 - 15,500 Btu/lb). It is the most plen-

tiful and most used. **Anthracite coal** is older and much harder than bituminous. It burns with a hot, blue, smokeless flame. However, it is not very plentiful. In the United States, Pennsylvania is the only major producer.

COAL SUPPLIES IN THE UNITED STATES

Coal is abundant in North America. Altogether, Canada, the United States and Mexico have estimated recoverable reserves in excess of 277 billion short tons. U.S. reserves make up 90 percent of all its known fossil-fuel reserves. These reserves contain 12 times the energy in all of Saudi Arabia's petroleum.

Most of the United States coal reserves are found in two regions:
- Eastern and midwestern.
- Western.

Fig. 8-4 shows the location and types of coal deposits found in the United States.

MINING COAL

There are several types of mining.
- If the coal is near the surface, the coal is recovered through **surface** or **open-pit mining**, Fig. 8-5.
- If the coal is located in hilly country or if it lies too deep beneath the surface for surface mining, the mining company will use an **underground mine**, Fig. 8-6.

The type of mining to be used depends upon how deep the coal is buried in the ground. Most western and midwestern coal can be strip mined because it is near the surface. Eastern coal, being located deeper

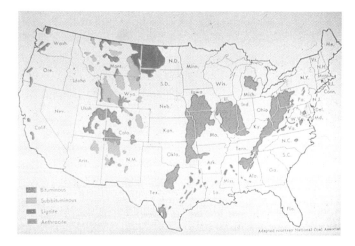

Fig. 8-4. Large coal deposits are found in the east and midwest.

Fig. 8-5. In open pit mining, huge shovels scoop the coal from shallow pits and load it onto trucks. (National Coal Assn.)

Fig. 8-6. An underground mine. This type is necessary where coal is located beneath rugged terrain or when it is buried too deep for open-pit mining.

underground, must be extracted through underground mining.

PROSPECTING THE COAL SEAM

To find the size of the coal bed, the engineers will make test borings. These are the earth samples taken out of deep holes. They examine the materials brought up by the drill. This is called "prospecting."

The engineers will also make a **seismic analysis** (study) to determine the hardness, depth, and nature of the overburden. **Overburden** is the soil and rock covering the coal seam. Explosives are set off in holes drilled into the ground. Shock waves travel into the ground and "bounce" back. Instruments measure the time lapse. An operator studies a **seismograph** (chart measuring the shock waves) and determines the type and thickness of underground materials. These data help engineers decide whether mining the coal seam is worthwhile.

If surface mining is being considered, the data collected from drilling and blasting will also help the engineers determine whether the overburden can be easily removed with a stripping machine or whether it must be drilled and blasted first. See Fig. 8-7.

SURFACE MINING

Before an open-pit mine is developed, engineers must study the **topography** of the land. (This is the way the

Fig. 8-7. A huge vertical drill bores foot-wide holes in the overburden. Explosive charges are set off in the holes to loosen the overburden so it can be removed by power shovels or draglines. (National Coal Assn.)

land lays.) It might be mountainous, rolling hills, or flat land. The engineers also study the thickness of the coal bed and the amount of overburden.

Obtaining this information may require aerial survey maps. These are photographs over which map lines are laid. The engineers will study the maps to:
- Locate coal outcroppings.
- Estimate the coal reserves.
- Determine property lines.
- Select areas where the excavated earth can be dumped. These are called the **spoil areas**. See Figs. 8-8 and 8-9.

STRIPPING

Removal of the overburden above a coal seam is called **stripping**. A bulldozer, Fig. 8-10, may be used first to level off the surface so that other, larger machines can work there.

Usually, specially designed machines will do the actual stripping to lay bare the coal seam, Fig. 8-11. One type is the **giant excavator**. It has a cutting wheel at the end of a long boom. The wheel rotates, cutting and

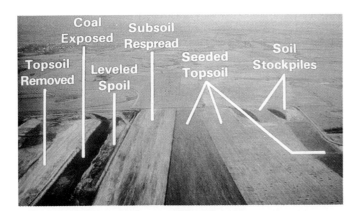

Fig. 8-8. This is a surface mine in North Dakota. The entire process of strip mining is shown and labeled. (American Coal Foundation)

Fig. 8-9. Before surface mining can commence, the mining company must show how the land will be mined and the land reclaimed later.

Fig. 8-10. Bulldozers are put to work clearing the land. Topsoil is removed and stored. It will be reused during the reclamation process.

A

B

Fig. 8-11. Besides bulldozers, giant excavators, such as draglines or power shovels clear away the overburden until a large area of the coal bed is uncovered. A—This shovel has a capacity of 245 cubic yards and could hold three school buses. (American Coal Foundation) B—Sometimes a huge wheel excavator is used to strip away overburden. A conveyor carries it to a spoil area in the background. (Consolidation Coal Co.)

scooping up the overburden with bucketlike claws. Another machine used for stripping is the **walking dragline**. Shown in Fig. 8-12, it is basically a large bucket suspended from the end of a boom.

A giant **surface mining coal shovel** may also be used. The dipper of one, shown in Fig. 8-13, will hold 210 tons of earth and rock or coal.

STRIP MINING METHODS

Two popular methods of strip mining are:
• Scraper haulback method, Fig. 8-14.

Fig. 8-12. Top. A walking dragline has a bucket suspended from a boom. Bottom. It is used to clear away overburden as well as to scoop up loosened coal. (Chevron Corp.)

Fig. 8-13. The dipper of a surface coal mining shovel removes about three carloads of earth and rock at one bite. (National Coal Assn.)

• Truck haulback method, Fig. 8-15.
    In either method, overburden is moved to a portion of the pit already mined.

LOADING AND TRANSPORTING

Coal from surface mining operations is often loaded by electric power shovels into huge trucks. The vehicles move the coal to a preparation plant. A conveyor belt may be used for the same purpose, Fig. 8-16.

UNDERGROUND OR DEEP MINING

Coal deep beneath the earth must be extracted through underground or deep mining. There are three types of underground mines. They are named for the kind of entry shaft used to reach the coal.
• Drift mine. The shaft is driven horizontally into a hill, Fig. 8-17.
• Slope mine. The mine shaft slopes gently downward. This type of mine shaft is designed for use where coal is too deep in the ground for pit mining but not deep enough for a shaft mine. See Fig. 8-18.
• Shaft mine. Its opening is sunk straight down, as shown in Fig. 8-19. An elevator is needed for access to the mine.
The mine entry serves two purposes. It takes miners and equipment into the earth to reach the coal. It also

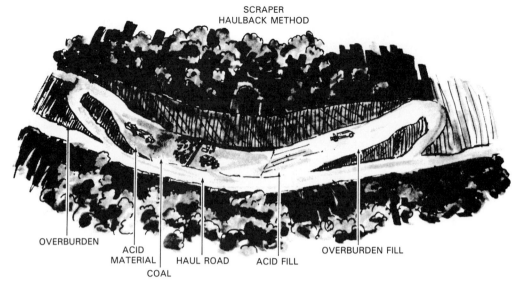

SCRAPER
HAULBACK METHOD

OVERBURDEN
ACID
MATERIAL
COAL
HAUL ROAD
ACID FILL
OVERBURDEN FILL

Fig. 8-14. Scraper haulback method of strip mining. Overburden is moved to a mined-out area by scrapers. Acid fill is laid down first. Rock and clay are put on top.   (Caterpillar Inc.)

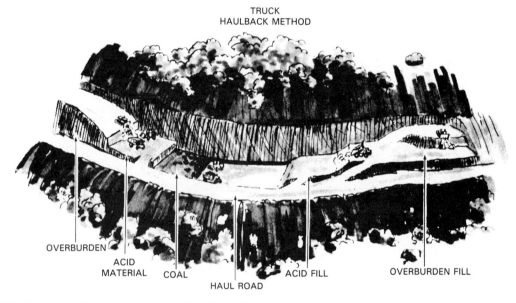

TRUCK
HAULBACK METHOD

OVERBURDEN
ACID
MATERIAL
COAL
HAUL ROAD
ACID FILL
OVERBURDEN FILL

Fig. 8-15. Truck haulback method of strip mining. Trucks replace scrapers in hauling acid fill and overburden. (Caterpillar Inc.)

Fig. 8-16. A conveyor belt is being used to move coal away from the mine. (Chevron Corp.)

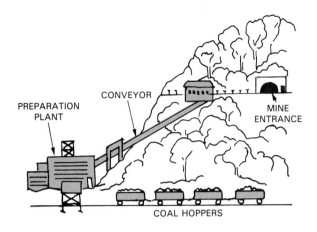

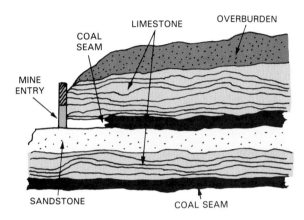

Fig. 8-17. A drift mine has a level shaft that is driven into the side of a hill. Top. Front view. Bottom. Side cutaway view.

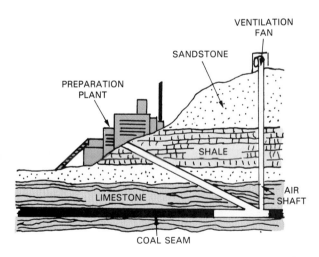

Fig. 8-18. Cutaway of a slope mine. This type of mine shaft is used where the coal is too deep for strip mining.

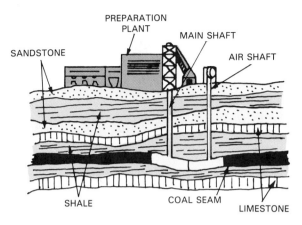

Fig. 8-19. A shaft mine in cutaway. This type is the most expensive. Two shafts are necessary. One provides access; the other provides ventilation to the mine and an escape route should the main shaft become blocked.

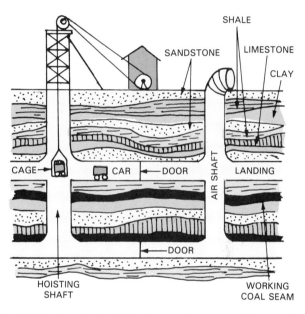

Fig. 8-20. A vertical shaft mine may have more than one level.

DEEP MINING METHODS

There are two methods for removing the coal from a deep mine:
• Conventional mining.
• Continuous mining.

DEEP MINING SYSTEMS

Coal must be mined in such a way that all the coal can be removed without having workers and machines trapped by cave-ins. The most used system in the United States is the **room and pillar** or **bond and pillar mining** system. In this method, coal is removed by making a series of "rooms" in the coal seam, Fig. 8-21. Large pillars of coal are left at regular intervals to act

provides a way to transport the mined coal to the surface.

In Fig. 8-20, the shaft on the right is designed for ventilation and has a blower that forces fresh air into the mine. The same shaft can also be used for raising and lowering workers and materials, if need be. This mine has two levels. The lower one is provided for drainage and/or to get at another coal seam.

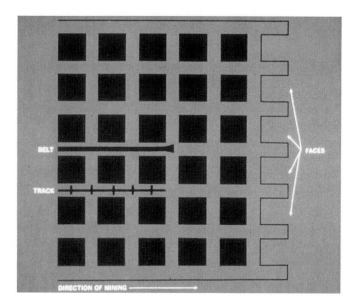

Fig. 8-21. Room and pillar method of mining. Coal is extracted by cutting a series of rooms into the coal seam. Pillars of coal remain to help support the roof. (Pillars are shown in black.) (American Coal Foundation)

Fig. 8-22. A continuous mining machine in operation. It is most often used in room and pillar mining. A rotating cylinder, studded with tungsten carbide tips, strips the coal from the seam.

as roof supports. The rooms, themselves, are 20 to 30 ft. (6 to 9 m) wide. Pillars are 20 to 90 ft. (6 to 27 m) on a side.

Usually, the pillars are removed after the coal seam has been worked out. The roof is allowed to cave in as the mining operation works back to the entry shaft.

CONVENTIONAL MINING

Mining by conventional methods involves four steps: cutting, drilling, blasting, and loading. The cutter looks like a giant chain saw. It can make horizontal or vertical cuts. Portable electric drills bore holes for the explosive charges.

Loosened coal is scooped up by a **loader**. This is an electrically powered machine with clawlike arms. An attached conveyor carries the coal to a shuttle car behind the loader. The shuttle car moves the coal to a conveyor or to a spot where it can be reloaded onto rail cars.

CONTINUOUS MINING

In **continuous mining**, special mobile (moving) machines mine and load coal in a nonstop operation. No explosives are used.

Machines used in continuous mining include:
- A continuous miner which tears the coal from the seam with horizontal rotating teeth, Fig. 8-22. Mined coal is loaded automatically into a shuttle car.
- A boring miner which tears coal from the seam by boring into the vertical working face.

With the addition of special equipment, continuous mining machines can bolt the roof at the same time that coal is being mined, Fig. 8-23.

A

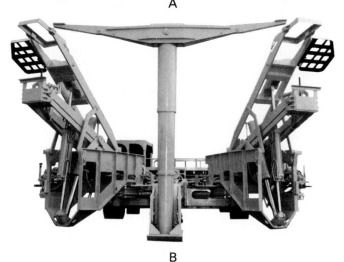

B

Fig. 8-23. After advancing a certain distance, the continuous miner backs out and a roof bolting machine moves in. A—Holes are drilled into the rock. Long metal rods, along with resin, are forced into the roof. (American Coal Foundation) B—A bolting machine has hydraulically operated temporary supports. (Lee-Norse Co.)

Bolting is a system for supporting mine roofs. Holes are drilled in the roof to accommodate long bolts. The bolts are anchored at their upper end by an expansion shell. As the bolts are tightened, they bind the rocks together to act like supporting beams, Fig. 8-24.

LONGWALL MINING

A second method of underground mining is called **longwall mining**. In this method, two parallel **entries** are driven through the coal seam. (Entries are like hallways. They give mining machinery access to the coal.) The entries may be from 500 to 800 ft. (152 to 244 m) apart and 10,000 ft. (3048 m) long.

The entries are joined at the far end by another entry called a **crosscut**. The coal face formed by the crosscut is the longwall.

A revolving cylinder with tungsten carbide bits moves back and forth across the face of the wall cutting away the coal. A water spray keeps down the dust. Mined coal falls onto a conveyor system that delivers it to the surface or to underground rail cars. Movable steel supports, shown at the right side of Fig. 8-25, support the roof over the work area. As the machine moves for-

Fig. 8-25. A longwall coal mining machine moves across a wall of coal that may be 500 to 800 ft. across and nearly two miles long. (American Coal Foundation.)

ward, the steel supports also advance. The roof over the mined out area is allowed to fall. This method recovers nearly 80 percent of the coal.

MINE SAFETY

Mining safety is a major concern in underground mining. Federal and state governments establish standards. Coal companies must follow these regulations which include proper ventilation and proper clothing. Fig. 8-26 shows a miner properly equipped and clothed for safety.

PROCESSING OF COAL

Unprocessed coal is called **run-of-the-mine coal**. It contains rock and other impurities, such as iron sulfide,

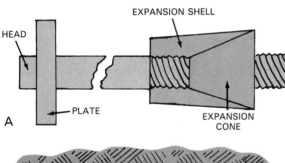

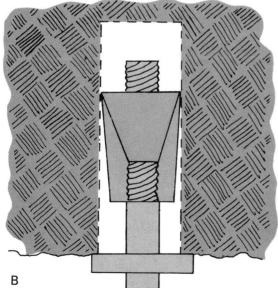

Fig. 8-24. A—A typical roof bolt designed to strengthen the mine roof against cave-ins. B—An installed roof bolt. Expansion shell grips the sides of the drilled hole.

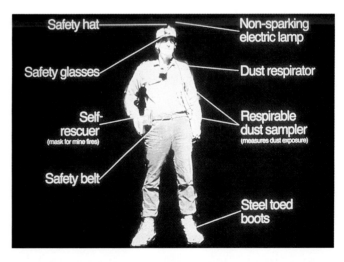
Fig. 8-26. Safety is important for today's mining operations. Some of the equipment and clothing worn by the miner are shown. In case of a fire or explosion, the self rescuer filters out carbon dioxide. (American Coal Foundation)

which must be removed.

Mined coal is hauled to a coal preparation plant, Fig. 8-27. At the same time, the coal is crushed, sized, and blended, Fig. 8-28.

Fig. 8-27. Most coal has impurities that must be removed before delivery to a user. In this preparation plant, devices using pulsating water currents and different liquids, separate impurities such as pyritic sulfur, rock, clay, and other ash-producing materials.

The plant where bituminous coal is prepared for use is called a **tipple**. Anthracite is prepared in a building called a **breaker**.

COAL STORAGE

Coal is stockpiled against the time when it will be used. Storage areas are usually located at shipping terminals or at electric power generating stations. Huge traveling stackers, such as the one pictured in Fig. 8-29, unload the coal from unit trains. Then the stackers move along steel rails and store the coal in gigantic piles. Some units are capable of unloading 10,000 tons of coal in less than six hours.

When the coal is reclaimed from the storage piles, a bucketwheel on the stacker transfers the coal at rates up to 6000 tons an hour. It can be moved by conveyor to dockside or to a power plant. The bucketwheel is also shown in Fig. 8-29.

MOVING COAL

Coal can be moved by rail, barge or ship, trucks or by pipeline. See Fig. 8-30. Much coal is moved by unit train. The most useful type of rail car for coal is the gondola car. It has an open top for easy loading. Some have hopper bottoms. The floor of the car opens up like double doors to unload the car quickly without need for additional equipment.

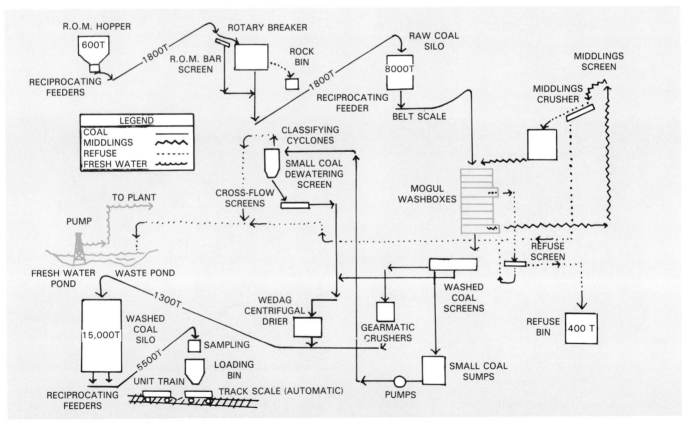

Fig. 8-28. Flow diagram shows how coal is processed in a coal preparation plant.

Fig. 8-29. A traveling stacker and a bucketwheel reclaimer. Constructed at Thunder Bay, Ontario, it stockpiles lignite being shipped by boat to eastern Canada. Left. Stacker at work. Right. Bucketwheel retrieves stored lignite.   (Allis Chalmers)

A

B

Fig. 8-30. A—Barges are being loaded with coal for shipment on an inland waterway such as the Ohio or Mississippi rivers. Each barge can hold about 1500 tons of coal. (American Coal Foundation) B—Unit trains such as this, made up of roughly 100 cars, transport between 10,000 and 11,000 tons of coal at one time. The cars can be loaded in about three hours and unloaded in four or five.   (Chevron Corp.)

PIPELINE TRANSPORT

It is also possible to transport coal through a pipeline. See Fig. 8-31. The moving process requires the use of several pieces of equipment and takes several steps:

• The coal is ground to the consistency of granulated sugar. See Fig. 8-32.
• A second crushing in a rod mill reduces the coal to even finer particles. Water or some other liquid is added.
• Pumps, Fig. 8-33, move the slurry through a pipeline

Fig. 8-31. In operation for more than 18 years, The Black Mesa Pipeline moves coal from the Black Mesa coal field in northern Arizona to an electric power generating station in southern Nevada, about 275 miles away.
(Williams Technologies, Inc.)

Fig. 8-32. Each of three process lines at Black Mesa consists of an elevated coal bin, a 1500 horsepower cage mill for dry crushing, a 1700 horsepower rod mill for wet grinding and a rod mill sump pump for transferring the slurry to a 63,000-gallon agitated tank.

Fig. 8-34. Main control station. A staff of 55 manage, maintain, and operate the preparation plant and its pumping stations.

Fig. 8-33. Slurry pumping equipment. The line also has four other reciprocating pumping stations along its 273-mile route. It can transport 5 million tons of coal a year. (Williams Technologies, Inc.)

at speeds of 5 to 6 ft. per sec. Pumping stations are located at intervals of 40 to 50 miles.

Fig. 8-34 is a view of the control room and Fig. 8-35 is a flow diagram of a coal slurry system. It is said that pipeline transport is cheaper. Unit train delivery costs 2.5 cents per ton/mile; pipeline delivery, 1.5 cents.

COAL'S POSITION AS AN ENERGY SOURCE

Even though petroleum became the most popular fuel source in the 1960s, the use of coal has continued to increase. In 1971 the United States used about 502 million tons of it. Then an international shortage of oil caused users to become concerned about future petroleum supplies. Ever since, use of coal has grown

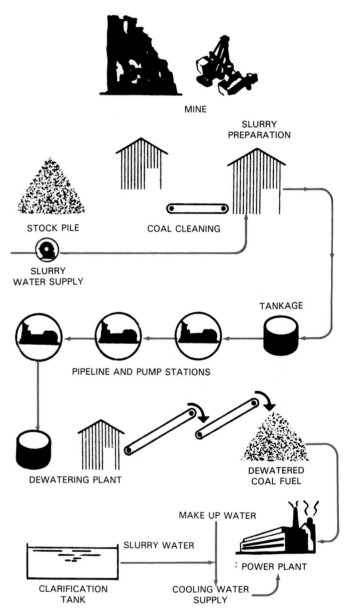

Fig. 8-35. This flow diagram shows the preparation and movement of coal slurry from mine to user. (Williams Technologies, Inc.)

an average of 20.5 million tons a year. See Fig. 8-36.

Coal's major market, the electric utility industry, accounts for nearly 85 percent of all the coal used. Coal-fired plants supply more than half of all the electricity generated in the United States, Fig. 8-37.

The second largest market is industrial/retail. Coal is a raw material for chemical manufacturers, paper mills, the primary metals industry, and the food industry. See Fig. 8-38.

Bituminous coal accounted for 94 percent of the coal mined in the United States in 1990. Electric utilities (power plants) are the largest coal consumers. In 1990 they burned 772 million short tons, 86 percent of the total produced. Before being fed into power station boilers, the coal is crushed to improve combustion.

More coal is mined east of the Mississippi than in the West. However, due to higher demand for low-sulfur coal, the West's share of production has been increasing every year.

HOW MUCH IS THERE?

The United States is second only to Russia in coal reserves. The supplies of unmined coal can be roughly estimated by geologists using seismographic soundings and borings.

The amount of these reserves for the United States in about 1.5 trillion tons. How long this will last is not certain.

Estimates are that, at 1975 levels of usage the U.S. would have enough coal to last 600 years. However, as other sources dwindle, and as technology finds more

Fig. 8-37. Electric utilities use about 85 percent of the coal burned in the U.S. (American Coal Foundation)

Fig. 8-38. Top. Coal is used in the manufacture of paper. Bottom. Coal is used to make coke, an ingredient in the making of steel. (American Coal Foundation)

ways to use coal, usage will increase. Assuming that usage would grow at 2 percent a year, coal reserves might be expected to last for 150 years.

COAL AND THE ENVIRONMENT

Coal mining is an important activity and contributes heavily to our need for large amounts of energy. However, extracting coal and using it as a fuel affects

YEAR	CONSUMPTION IN MILLION SHORT TONS	YEAR	CONSUMPTION IN MILLION SHORT TONS
1949	483.2	1970	523.2
1950	494.1	1971	501.6
1951	505.9	1972	524.3
1952	454.1	1973	562.6
1953	454.8	1974	558.4
1954	389.9	1975	562.6
1955	447.0	1976	603.8
1956	456.9	1977	625.3
1957	434.5	1978	625.2
1958	385.7	1979	680.5
1959	385.1	1980	702.7
1960	398.1	1981	732.6
1961	390.4	1982	706.9
1962	402.3	1983	736.7
1963	423.5	1984	791.3
1964	445.7	1985	818.0
1965	472.0	1986	804.3
1966	497.7	1987	836.9
1967	491.4	1988	883.7
1968	509.8	1989	890.6
1969	516.4	1990	892.5

Fig. 8-36. Consumption of coal in the United States from 1949 through 1990. Use has nearly doubled in 40 years. (Energy Information Administration)

the environment in a number of ways. We need to consider the following and decide how best to mine and utilize coal:

- If not properly handled, open pit mining could leave deep holes in the ground and unsightly piles of overburden. The overburden could pollute streams and be blown by the wind. Little vegetation would grow on it. About 2 million acres of land in 26 states have been affected by surface mining operations.
- Shaft mines can cause land subsidence (sinking). This has happened in the past in the Pennsylvania coal region and in southern Illinois.
- Coal, used as power plant fuel, is an air pollutant. Scrubbers and other new technology must be employed to reduce emissions. Coal also produces more carbon dioxide than other fossil fuels as well as a small amount of radiation.
- Open pit mining creates noise pollution. On the other hand, shaft mining uses very little land and creates very little noise.
- Open pit or strip mining uses up farmland in some regions such as southern Illinois but not in western regions because of mountainous terrain.
- Large quantities of water are required to mine and wash coal.

LAND RECLAMATION

On the positive side, most of the land disturbed by surface mining can be reclaimed. It can be made productive again and so continue to be a benefit to humankind. Mined sites can be redeveloped for many different purposes:
- Production of agricultural or horticultural crops.
- Recreation and wildlife habitat.
- Industrial and residential building sites.
- Planted to trees which will, in time, yield lumber and paper products.

The forestry service of the U.S. Department of Agriculture is concerned about the proper reclamation of such lands. Much study has gone into the development of proper reclamation methods. Mining companies are also concerned about the effects of mining on the environment and are successful with current reclamation practices. See Figs. 8-39 and 8-40.

NEW COMBUSTION PROCESSES

Different methods of burning and processing coal are lowering the pollution levels which have always made it a dirty fuel. **Fluidized bed combustion**, Fig. 8-41, is one process that has proven successful in cutting down sulfur and nitrogen oxide (NO_x) emissions

Fig. 8-39. Once the mine owners have removed the coal from an open-pit mine, they return the land to its former condition. After backfilling, grading, and replacing topsoil, the area is reseeded.

Fig. 8-40. At this midwestern mine, trees are being planted on reclaimed land. Other uses may be considered such as cropland, grazing land, fish habitat, or wildlife and outdoor recreation areas. (American Coal Foundation)

and fly ash from burning coal. The coal, reduced to small lumps, is fed into a heated bed of sand or ash while air is bubbled up through the sand.

At the same time, calcined (powdered) limestone is fed into the combustion chamber to absorb the sulfur dioxide (SO_2). Heat exchangers under the fire pot draw off the heat so that the combustion temperature stays at about 1550°F (843°C). At this temperature the lime captures the sulfur, and nitrogen oxide gases are kept below required pollution control standards.

It is also possible to break down coal and convert it into gaseous and liquid fuels.

NEW FUELS FROM COAL

Producing gaseous and liquid fuels from coal are not recent processes. Very early in the industrial revolution, kerosene was made from coal and is still called

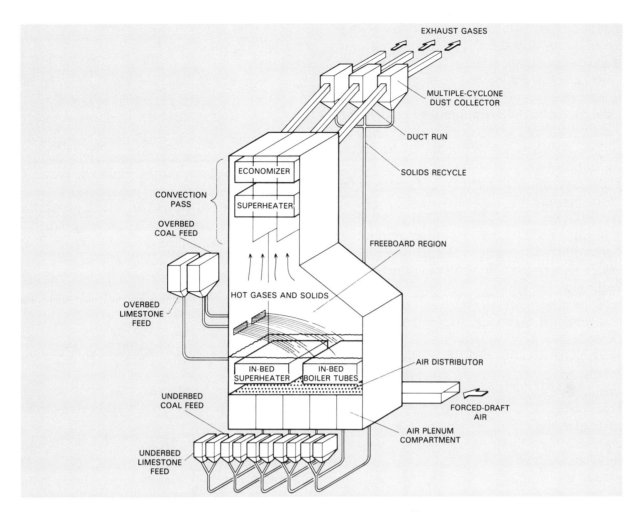

Fig. 8-41. An atmospheric fluidized bed combustion boiler. This method of burning coal without the heavy air pollution of standard boilers has been under experimentation in a pilot plant. (EPRI Journal)

coal oil in some rural areas. Another product of coal was the **town gas** or **water gas** used in the cities for cooking and light in the 19th and early 20th century.

It was only the discovery of large quantities of crude oil and natural gas that caused these coal conversion industries to be abandoned. During World War II, Germany improved the technology of conversion and demonstrated that large quantities of oil and gas could be produced this way.

Two processes can be used to convert coal into other fuel forms:
- Gasification. In this process, extra hydrogen atoms are added to turn the coal into a gaseous fuel.
- Liquification. By reducing the amount of the carbon in the coal to that found in fuel oil, liquification is achieved.

These processes are explained in Chapter 16.

SUMMARY

An important energy source, coal is a fossil fuel found in several stages: peat, lignite, sub bituminous coal, bituminous coal, and anthracite coal. Being 82 percent carbon and 12 percent hydrogen it burns well and is among substances known generically as hydrocarbons.

North America has abundant reserves of coal estimated at 277 billion short tons. This represents 12 times the energy of all of Saudi Arabia's petroleum. U.S. coal beds are most abundant in the western half of the country.

There are two chief types of coal mining: surface or open-pit mining and underground mining. Prospecting for coal involves test borings to sample the rock in deep holes, seismic testing to locate underground rock formations, and examination of the overburden for depth and hardness.

In open-pit mining, the overburden (soil and rock covering the coal) is removed. Then various machines may be used to harvest the coal. Overburden and topsoil is restored to mined out areas and the land is returned to other uses.

Underground mining requires sinking shafts into the ground to reach the coal seam. There are two methods of deep or underground mining. In conventional mining the coal is cut, drilled, blasted, and loaded onto rail cars and brought to the surface. In continuous mining, mobile machines mine and load the coal in a nonstop operation.

Once mined, coal must be transported to its destination. Trains, ships, and trucks are a common method of transport. Another, less common, is the use of pipelines. For the latter, the coal must be crushed and water added to produce a slurry that can be pumped through the pipeline.

Mining and using coal produces pollution. Much of this can be mitigated (reduced) by proper restoration of the overburden and topsoil and by use of scrubbers or fluidized bed combustion in burning the coal.

While the major use of coal is to fuel electric power stations, it has other markets. One is as a raw material for manufacture of chemicals, paper, and steel.

DO YOU KNOW THESE TERMS?

anthracite coal
bituminous coal
bond and pillar mining
breaker
continuous mining
crosscut
entries
fluidized bed combustion
giant excavator
hydrocarbons
lignite
loader
longwall mining
open-pit mining
overburden
peat
room and pillar mining
run-of-the-mine coal
seismic analysis
seismograph
spoil areas
stripping
sub bituminous coal
surface mining
surface mining coal shovel
tipple
topography
town gas
underground mine
walking dragline
water gas

SUGGESTED ACTIVITIES

1. Research what the coal mining industry is doing to restore mined out areas. Use a computer to prepare a report on the results of the research.
2. Build a model of any type of coal mine.
3. Research the topic and draw up a list of the products which use coal as a raw material.
4. Have a coal company representative speak on coal's future as an energy source.
5. With your instructor's help devise an experiment with coal, conduct the experiment, and report the results to your class.

TEST YOUR KNOWLEDGE

1. List the stages of coal.
2. Coal makes up _____ percent of U.S. fossil fuel reserves.
3. Coal that is found near the surface is mined by the following method (check correct answers):
   a. Open pit mining.
   b. Surface mining.
   c. Shaft mining.
   d. Drift mining.
4. In the _____ method of underground mining, machines rip out the coal along the working surface of the coal seam and haul it away in a single nonstop operation.
5. Why must run-of-the-mine coal be cleaned?
6. List the methods of transporting coal.
7. U.S. coal reserves amount to:
   a. A million tons.
   b. 500 million tons.
   c. 1.5 billion tons.
   d. 1.5 trillion tons.
8. Coal is our most plentiful fossil fuel and is expected to last somewhere between _____ and _____ years.
9. Coal is a pollutant as are other fossil fuels but technology is available to reduce its polluting effects. True or False?
10. What is fluidized bed combustion and how does it reduce pollution caused by burning coal?

9
CHAPTER
FOSSIL FUELS—
PETROLEUM AND NATURAL GAS

The information given in this chapter will enable you to:
• Define petroleum and natural gas and explain briefly their chemical makeup.
• Discuss the historical development of the oil and gas industry.
• Describe methods of extraction and refining.
• List the products that come from petroleum and natural gas.

Like coal, petroleum or **crude oil** is a fossil fuel. It is the natural state of liquid hydrocarbons. As it comes from the well, crude oil is a mixture of semisolids, liquids, and gases. The gaseous portion of the petroleum is called **natural gas.**

Natural gas is always found wherever oil deposits are discovered, Fig. 9-1. Often, however, gas is found in underground pockets that contain no oil. Natural gas is made up chiefly of methane, ethanes, propane, and butane. All are combustible (burnable) gases. Small quantities of inorganic gases such as carbon dioxide, hydrogen, and nitrogen are also mixed with the volatile gases.

HISTORY OF PETROLEUM

Petroleum was first used as a medicine. Semisolid forms, such as **asphalt**, were used for waterproofing as early as 3000 B.C. (Asphalt is a brownish or blackish coal-like natural state of petroleum.) Egyptians and Chinese drank petroleum as medicine and applied it to wounds. It was not used as a fuel until near the end of the 1700s. Then, the need for a cheaper fuel for heating homes and factories and for powering machinery caused rapid development of the petroleum industry.

FIRST OIL COMPANY

The modern petroleum industry really began with the Pennsylvania Rock Oil Company. In 1859 the company drilled the first oil well at Titusville, Pennsylvania. See Fig. 9-2. It was accomplished with redesigned salt well boring equipment. Two years later the first gusher (well) was drilled nearby at Oil City. From then on, the oil industry boomed. By 1900, wells in the United States were producing 64 million barrels a year. Demand led to deeper drilling and more scientific methods. The first off-shore drilling was attempted off the California coast in 1896. In 1938 a well was drilled from a platform built off the Louisiana coast.

Need for gasoline to fuel the automobile led to development of the refinery industry. The first successful cracking (separating) process was introduced in the United States by two chemists, William M. Burton and Robert E. Humphreys.

LOCATING PETROLEUM AND GAS DEPOSITS

Geologists are specialists in the study of rocks. In the search for petroleum, they examine surface rock formations that may give them an idea of the earth structure below. For instance, outcroppings of shale, sandstone, dolomite, or limestone may indicate the presence of oil or gas far below. Aerial photographs are sometimes taken because they show rock formations not noticeable from ground level.

Sites where rock formations seem promising are studied further. Geologists look for three natural formations that are necessary to the existence of commercially valuable petroleum deposits.
• **Source beds.** This is where petroleum first formed.
• **Reservoir beds.** These are porous rock structures where oil spreads throughout the porous rock.

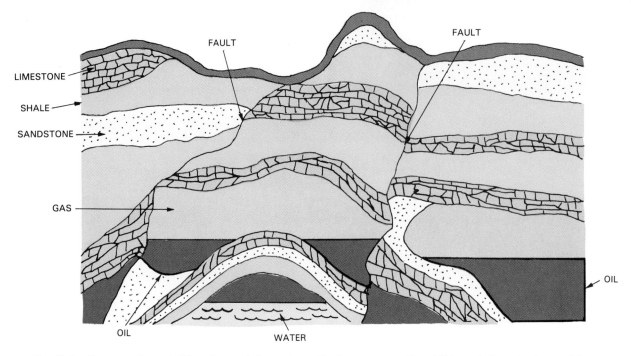

Fig. 9-1. Cross section of oil bearing rock formations. Faults are caused by shifting of plates in the earth's crust.

Fig. 9-2. The first oil well was drilled in Pennsylvania in 1859. (Shell Oil Co.)

- **Traps.** These are nonporous rock strata that lie next to the reservoir rock and prevent the petroleum from moving farther.

Traps are pinpointed through surface and subsurface geology. This data is compared with previously-known facts from nearby drilling and testing with geological instruments.

GEOLOGICAL TEST INSTRUMENTS

Tests will be made with sensitive instruments to determine the underground rock structure. The rock struc-

ture does not prove the existence of oil, but it tells the geologist if the subsurface rock formations are the type in which oil occurs.

Three instruments are used for these tests:

- **Gravimeter.** Dense rock exerts more force of gravity than less dense rock. The gravimeter, Fig. 9-3, measures gravity and will indicate the density of rock deep in the ground.
- **Magnetometer.** This device, Fig. 9-4, measures the strength of a magnetic field. It is used to locate rocks

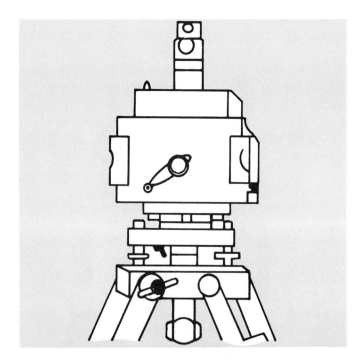

Fig. 9-3. A gravimeter is used by geologists to detect density of rock deep beneath the earth.

Fig. 9-4. Magnetometer locates rock formations likely to contain petroleum deposits. (Shell Oil Co.)

likely to carry oil. Oil itself has no magnetism, but rocks do. The magnetometer is usually towed behind an airplane, but it can be moved about by truck, as well.

• **Seismograph.** A modification of the instrument used for detecting and measuring the intensity of earthquakes, the seismograph measures shock waves indicating the presence of rock formations underground. Dynamite is placed in small holes bored in the ground. When exploded, the dynamite causes shock waves that bounce off the rock formations. These are picked up by the seismograph, Fig. 9-5. Recordings of the shock wave are "read" by the geologist. They tell him or her what kind of rock is below, how large the formation is, and its depth.

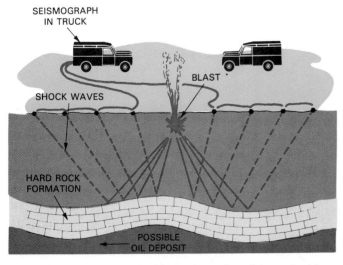

Fig. 9-5. A seismic study. Measuring echo of dynamite blast tells geologist where large rock formations are to be found underground.

DRILLING FOR OIL

Testing for rock formations can tell geologists where there is likely to be oil. However, the only way to be sure is to drill a well. There are two types of wells:

• The **exploratory well** that is drilled to find a new pool, to extend an existing oil field, or to find an entirely new oil field. In other words, it is drilled in search of more oil. Often, these are called **wildcat wells**.

• The **development well** which is drilled in an area that is already producing oil.

Well drilling is a big operation. More than 45,000 are drilled every year for exploration or development.

TEST CORES

During drilling, **test** cores are examined. These are cylindrical chunks of rock and other subsoil taken from the drill hole. Geologists study them looking for fossils and the type of rock that might suggest the presence of oil.

DRILLING METHODS

Before drilling can begin, workers erect an open steel tower called a **derrick**, Fig. 9-6. The derrick supports the drilling tools.

Drilling is done by either of two methods:

• Cable tool drilling.
• Rotary drilling.

Cable tool drilling

The cable tool method, adopted during the early days of the petroleum industry, is used for drilling shallow wells. The **cable tool string** (all of the tools that go into the well) includes a drill bit, a stem, connecting links called jars, sinkers for weight, and a cable for getting the tools down the hole to the drilling face. The string may weigh 2 tons.

The hole is punched into the ground by the alternate raising and lowering of a heavy chisel-like cutting tool. The tool is suspended on a cable.

During the drilling operation small amounts of water are fed into the hole. The water mixes with the rock dust and chips produced by the pounding of the bit. The resulting sludge can more easily be removed.

When the sludge becomes so great in volume and so thick that it interferes with the drilling, the tool is pulled out and the sludge is removed. More clean water is fed into the well and drilling continues. Fig. 9-7 is a sketch of a rig set up for cable tool drilling.

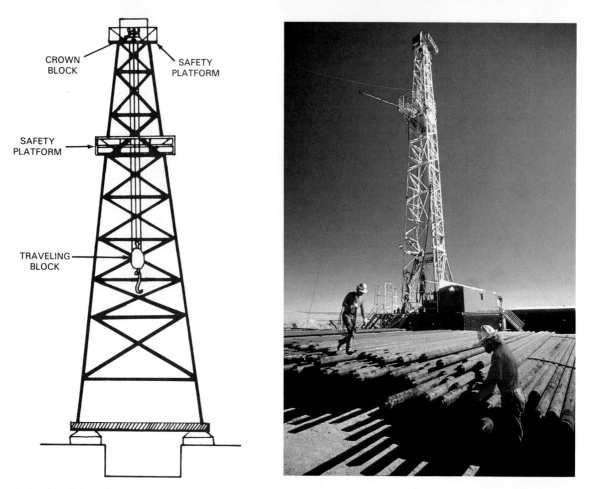

Fig. 9-6. An oil derrick supports tools used in well drilling. Left. Sketch of derrick. Note block and tackle for hoisting pipe and drilling bits. Right. Derrick on well drilling site.
(Union Pacific)

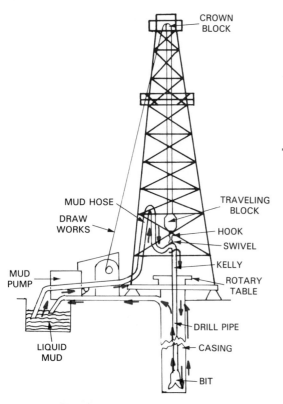

Fig. 9-7. Drilling operation. Mud is pumped down the well casing to cool the drill bit and bring up broken rock loosened by the drill.

Rotary drilling

In rotary drilling the cutting tool is a heavy steel bit. This is connected to a length of steel pipe called a **drill pipe**. A turntable supported on a platform at the base of the derrick turns the drill. The turntable is powered by diesel engines. As the drilling goes deeper into the earth, additional lengths of drill pipe must be added to the bit.

A special mixture of water, clay, and chemicals is continuously fed into the well during drilling. It cools the bit and carries away the rock pieces broken loose by the drilling. Then, this mixture, called **mud** is forced back up the hole. On the way up, it coats the walls of the hole to prevent cave-in or seepage. Later, a steel casing will be installed to line the hole. Fig. 9-8 shows a drill crew at work.

When the recirculated mud reaches the top, the rock particles are removed and the mixture is reused.

DRILL BITS

Many kinds of drill bits have been designed, Fig. 9-9. Selecting the right one is important. It has a direct bearing on the speed and efficiency of drilling. For exam-

Fig. 9-8. Drilling crews work around the clock finding new petroleum reserves. (Texaco, Inc.)

Fig. 9-9. Bits are also called rock drills. Toothed wheels turn against each other and crush rock broken loose at bottom of hole.

ple, a **fishtail bit** is best for soft clay and sand. **Cone bits** and **roller bits** are designed to cut through certain kinds of rock.

Since abrasive rock would wear out an ordinary bit, **tungsten carbide** (a hard metal) is used or the die bit may be studded with tiny industrial diamonds.

Even so, drill bits quickly become dull and must be exchanged frequently for sharp ones. This is the reason for the tall derricks. To change bits, all of the pipe must be pulled from the hole. If each 30 ft. length had to be disconnected, this method of drilling would be slow and inefficient. The great height of the derrick makes it possible to pull three or four sections from the well and stack them on end inside or near the tower.

OTHER WELL DRILLING METHODS

While cable tool and rotary drilling are the most common, there are other drill methods designed for special needs. These include:

- **Turbodrilling.** In this method, the force of the mud being pumped into the well powers a small turbine just above the bit. Its main advantage is the nearness of the power to the bit. There is no need for the heavy lengths of cable called "string."
- **Electrodrilling.** This type takes its power from an electric motor located just above the bit.
- **Sonic drilling.** The bit vibrates at the same level as a high-frequency sound generated at the surface. This makes the bit cut faster.
- **Flame drilling.** Jets of burning liquid rocket fuel burn a hole through rock and soil.

TESTING THE WELL

As drilling progresses, the drill cuttings are examined under ultraviolet light to see if oil is present. When the cuttings indicate that the drill has struck oil, a testing tool is lowered into the well to check for **bottom hole pressure**. This is the pressure of fluids and gases at the bottom of the well. The tool records these pressures as the fluids and gases enter it. Then the tool is sealed and brought back to the surface. Its contents are inspected.

SEALING

If oil is discovered, the well must be sealed. Steel pipe casing is lowered into the well. Sections are telescoped so the joints seal out water. Cement is pumped down through the casing where it rises up the outside of the pipe. This completes the seal that separates the oil from other formations. It keeps out sand, mud, and ground water that might contaminate (mix with) the oil.

PIERCING THE CASING

At this point the newly installed casing prevents oil from entering the well. Piercing of the well casing opens up the bottom of the sealed well. This job is done with an electric perforating gun. The gun is lowered into the well where it fires "bullets" or shaped charges through the casing into the oil-bearing rock. The rock fractures. Oil can now flow into the well through the openings created by the perforating gun. See Fig. 9-10.

Two-inch diameter tubing is now lowered into the well. Oil, gas, and, sometimes, salt water flow up through the tube.

A set of valves is installed on the top to control the flow from the well. Known as a **Christmas tree**, the valves control distribution of the oil and gas and prevent unwanted flow.

Some of the valves are **blowout preventers**. They prevent the violent gushers which once were a common occurrence during drilling of new wells.

Other valves direct the flow from the well head to gathering lines, collecting tanks, or separators. Separators remove the water and gas from the crude oil.

RAISING THE OIL

Bringing the oil to the surface from several thousand feet down, takes a great deal of force. This force is called **drive** or **lift**. Several methods are used:
• Natural or reservoir drive.
• Artificial drive.

NATURAL DRIVE

Natural drive is the pressure within the oil pool itself. Most new wells have this pressure. It brings the oil to the surface without the aid of pumps or any other type of artificial force. There are three types of natural drive:
• Water drive, the most efficient. Often, a large pool of salt water has been trapped around the oil pool. As shown in Fig. 9-11, the salt water, being under tremendous pressure, tries to expand. It pushes the oil through the fractures in the rock into the well casing and up the tube to the surface.
• Gas-cap drive. Pockets of natural gas, under tremendous pressure, drive the oil up the tubing. See Fig. 9-12.
• Dissolved gas drive. There is no gas cap in this type of drive. The gas is dissolved in the oil, Fig. 9-13. The low pressure in the well casing causes the trapped gas to move into the well, dragging oil with it. It is not as efficient as the water or gas-cap drive. Some of the oil gets left behind.

ARTIFICIAL DRIVE

As wells give up their oil, natural pressure drops and eventually the flow stops. Then the remaining oil must

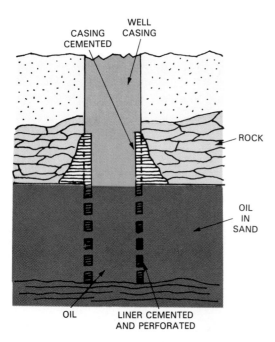

Fig. 9-10. Sketch of the bottom of an oil well. Oil flows into the casing through holes and is pumped to the surface.

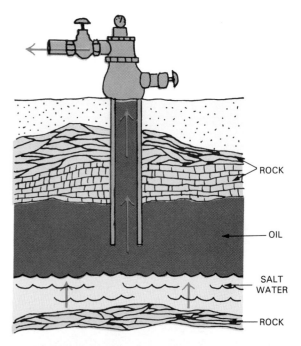

Fig. 9-11. Sketch of water drive. Salt water, occurring naturally, drives oil into well, up the pipe to the surface.

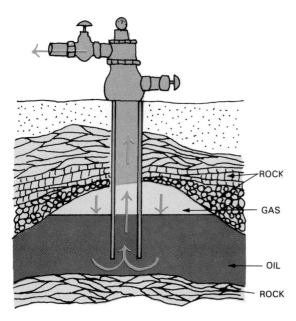

Fig. 9-12. Gas-cap drive. Trapped underground, natural gas in dome exerts pressure on oil to drive it to surface.

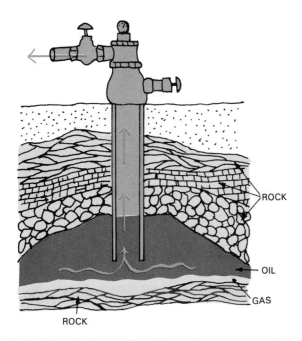

Fig. 9-13. Dissolved gas drive. Trapped gas mixes with oil, dragging mixture into well and to surface.

Fig. 9-14. Pumping with a lift pump is the most common method of bringing oil to surface. Top. This rocking beam pump is a familiar sight in oil fields. Bottom. Pumped petroleum is stored temporarily in tanks nearby.

be brought up by some other means. One of several artificial methods is used:

• Pumping. This is the most common. Some wells never flow naturally and must be pumped from the start, Fig. 9-14.

• Gas-cycling. This method uses gas previously removed from the well to maintain the pressure. The mixture of oil and gas is separated as it comes from the well. When oil goes to petroleum products such as propane, it is known as **wet gas**. These by-products are removed and some of the gas is returned to the

oil pool by way of an **input well**, Fig. 9-15. It helps maintain the pressure and keeps the oil moving into the well.

• Repressuring. In this method the oil reservoir is

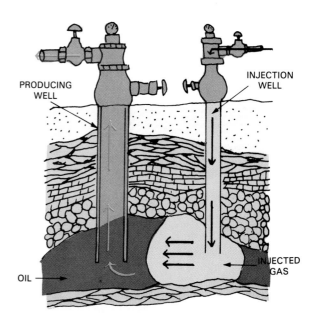

Fig. 9-15. Artificial drive. Gas pumped into an injection well may be used to force oil to the surface.

flooded with water or injected with more gas to raise the pressure.

- Front burning. This method is used in wells that contain heavy oil. A fire is lighted in the reservoir while air is pumped into it. The pressure of hot expanding gases and the heat make the oil flow better. Another name for this method is "in situ combustion." (In situ means "in place" or "on location.")
- Steam injection. Steam is generated at the well head and fed down the well into the oil-bearing rock. Steam causes the oil to expand and move toward the well. Some of the lighter oils vaporize and these gases provide additional pressure to move the oil.
- Other methods being considered for oil recovery are: high pressure gases, nuclear explosives, and solar and electrical heat.

INCREASING OIL FLOW

Low pressure and loss of pressure are not the only problems faced in extracting oil. Sometimes low porosity (too dense) of the rock formation keeps oil from the well. Then the porosity of the rock must be improved by:
- Fracturing treatment.
- Acid treatment.

FRACTURING

Fracturing actually produces cracks in the rock formations. High-pressure pumps force a mixture of sand and fluid down the hole and into the oil-producing area. The fluids produce cracks in the rock and the grains of sand keep the cracks open as the oil and fluid flow back to the well. See Fig. 9-16.

Fig. 9-16. Fracturing of dense rock formation at bottom of an oil well often is done to increase flow of oil in the well.

ACID TREATMENT

Treating the rock formation with acid etches tiny channels that allow the oil to move toward the well. Usually, hydrochloric acid is used. It is pumped into the well along with other chemicals at high pressure. The acid enters the rock formation at the point where the well casing has been pierced. Several treatments are usually needed.

OFFSHORE DRILLING FOR OIL

A great deal of oil is buried underwater in coastal areas. Drill rigs are mounted on platforms, Fig. 9-17.

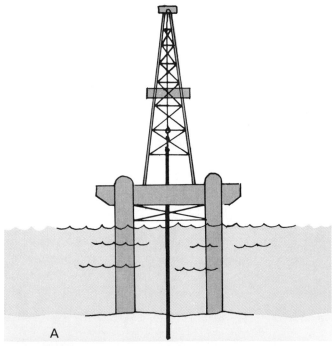

A

B

Fig. 9-17. Offshore oil rig. A—Rig is supported by piles sunk deep in the ocean floor. B—View of operating offshore wells. (Amoco Corp.)

The platforms are of three different types:
- The permanent platform which is supported by piles embedded deep in the floor of the ocean. The platform is large enough to house the drilling crew as well as the drill machinery and supplies.
- The small permanent platform. This structure, like the larger permanent platform, is supported by pilings. However, the crew and their supplies are housed in an attached floating tender.
- A mobile platform. This type is supported by a submersible barge. It is the most economical because it can be moved and used over and over again to drill many wells in different locations. In deep water the mobile rig is self-propelling.

UNDERWATER DRILLING RIGS

A newer type of drilling rig can be operated underwater. It is lowered to the ocean floor and anchored. Operation is controlled from the surface.

Another new system is the robot with a mechanical arm. It can perform the same tasks as a deep-diver. The robot is controlled from a boat. Equipped with sonar and television camera, it can locate objects underwater.

CLASSES OF CRUDE OIL

Petroleum is not uniform in its makeup. There are many complex mixtures of hydrocarbons. Also present in the petroleum are small amounts of sulfur, nitrogen, and oxygen compounds.

The most common petroleum mixtures are:
- Paraffinic hydrocarbons.
- Aromatic hydrocarbons.
- Asphaltic base hydrocarbons.

Also, petroleum may be sweet or sour. Sour crudes contain sulfur and have an unpleasant, rotten-egg odor. Sweet crudes contain very little sulfur and have more pleasant odors.

Fig. 9-18 shows the major products that are made from crude oil. As received from the well, the crude oil is not usable. It must be processed. This process is called **refining**.

REFINING OF OIL

Refining breaks down crude oil into various usable liquids and semisolids. In its original state, crude oil's carbon atoms are arranged in many different ways. Crude with many carbon atoms forms a thick, heavy petroleum such as asphalt. Crude with fewer carbon atoms makes up the lighter elements such as gasoline.

The products of crude are called **fractions**. They are separated by several refining processes which include:
- Separation.
- Conversion.
- Treating.

SEPARATION

The main separating processes are:
- Distillation.

REFINED PRODUCT	PROCESSES USED TO REFINE	USES OF THE PRODUCT
Light Gases	Pipe still distillation of crude oil	Manufacture of chemicals; Manufacture of gasoline; Gas fuels (liquid petroleum gas)
Gasoline	Distilling of crude in pipe still; Thermal and catalytic cracking; Catalytic and thermal reforming; Hydrocracking and coking of pitch; Polymerization of light olefins; Isomerization of paraffins; Alkylation of olefins with isoparaffins	Fuels for automobiles, trucks, and aircraft; Solvents; Fuels for cooking and lighting
Kerosene	Distilling of crude oil; Cracking of heavy fractions	Jet fuel; Fuel for cooking, illuminating, and heating; Solvents
Gas Oils	Distilling of crude oil; Cracking of heavy fractions; Hydrocracking of residual oils; Vacuum distilling	Residential and light industry fuel; Manufacture of chemicals; Manufacture of gasoline; Solvents; Asphaltic materials for roads
Lubricating Oils	Distilling of crude oil; Hydrocracking of heavy fraction; Solvent refining of residual oils	Many kinds of lubricants and oils; Pharmaceutical white oils; Waxes and petrolatums; Petroleum jelly; Asphalt; Greases
Residual	Vacuum distillation of petroleum; Distilling of synthetic petroleum	Manufacture of gasoline and fuels; Chemicals; Source of coke; Asphalt
Coke	Thermal cracking of residuals and pitches	Fuel; Manufacture of iron and steel; Industrial electrodes

Fig. 9-18. Major products listed above are all derived from crude petroleum.

- Solvent extraction.
- Adsorption.
- Refrigeration.

DISTILLATION

Distillation separates different parts of the petroleum through their different boiling points. Refineries today use a process called fractional distillation. First the petroleum is heated to 700°F (371°C). Heat is provided by a **pipe still**. This is a special type of furnace with many coils of tubing running through it. The oil picks up heat as it flows through the tubing.

Hot liquids and vapors resulting from the heating are released into a **fractionating** or **bubble tower**. The tower is about 100 ft. tall and has an arrangement of trays at different levels to collect the products of the distillation. See Fig. 9-19.

The rising vapors pass through holes in the trays and are caught and condensed by **bubble caps** placed above the holes, Fig. 9-20.

The various petroleum products condense at different temperature levels in the tower. The heaviest fractions (products) collect at the bottom. They are

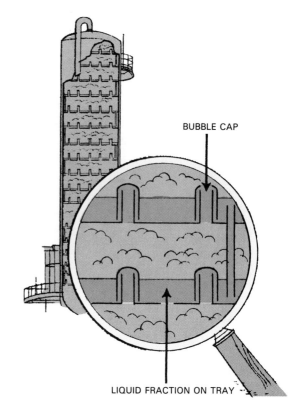

Fig. 9-20. Bubble cap catches rising vapors and deposits droplets into trays.

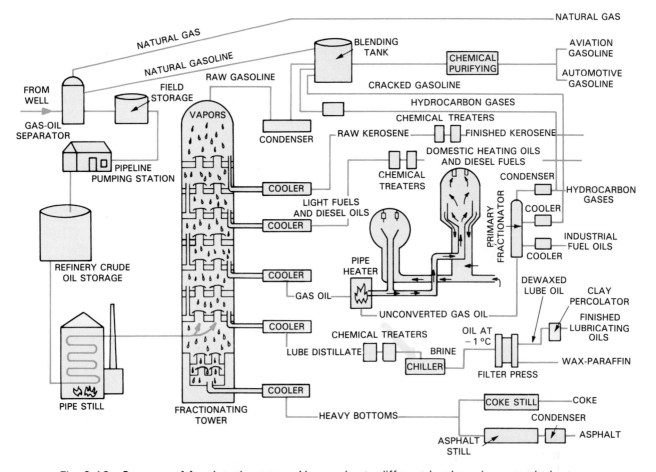

Fig. 9-19. Cutaway of fractionating tower. Vapors rise to different levels and are caught by trays.

either processed into asphalt, used as fuel oil, or are reprocessed into lighter products. The residue is sold as coke. This is a pure carbon that is used to make steel.

Gas oils condense at the next level, then kerosene and naptha. Straight-run gasoline condenses near the top of the tower.

Some gases do not condense at all. These are piped off at the top of the tower.

SOLVENT EXTRACTION

Solvent extraction is used mainly for refining different fractions produced in the fractionating tower. The solvents take into solution (mix with) the undesirable substances and can be drawn off with the solvent. One product produced this way is lubricating oil.

ADSORPTION

In the adsorption process a very porous solid material is used to collect the unwanted substances from a petroleum fraction. Silica gel is a popular adsorbent for this purpose.

REFRIGERATION

In the refrigeration process, cooling of unwanted substances, such as paraffin, causes them to solidify. In this condition they can easily be removed from the oil.

CONVERTING OIL FRACTIONS

After liquids have been separated by distilling, they can be changed into other products. Heavy fuel oils can be made into high octane gasoline. The conversion process is called **cracking**.

Cracking changes the heavy molecules of petroleum into small, light ones. There are two cracking methods:
- Thermal. The first to be developed, this method depends upon high temperatures and pressure.
- Catalytic. The more efficient of the two, this process also requires high temperatures. The pressure required is much lower than in thermal cracking. Catalysts, such as silica-alumina, speed up the chemical reactions. Catalytic processes include:
  a. Fluid catalytic cracking. The vaporized mixture of gas and oil and a powdered catalyst circulate together at high temperatures through a system of pipes and reactors.
  b. Hydrocracking. This process uses hydrogen and catalysts under high pressure. It yields more gasoline and less residue than ordinary methods.

ALKYLATION

Alkylation is designed to produce liquids closely related to gasoline from refinery gases. Its importance is not that it is another way to make gasoline. The product of alkylation is highly prized for its anti-knock properties. (Knocking is the "pinging" noise sometimes heard in automobile engines upon acceleration. It is caused by the fuel igniting in the cylinder too soon.) Adding the alkylated product to the gasoline causes detonation (burning) of the fuel to take place more slowly.

POLYMERIZATION

Like alkylation, **polymerization** is a way of making gasoline from refinery gases. However, only the olefinic gases react with the catalyst (material that causes the chemical reaction). Paraffinic gases are not changed by the catalyst which is usually phosphoric acid.

Polymerization was used during World War II to make isooctane for aviation gasoline. The concentrated polymers have antiknock qualities but they are not as good as those obtained from the alkylation process. They are used in the manufacture of lubricating oils and plastics.

ISOMERIZATION

Isomerization converts chemical compounds in refinery gases into its isomers. (An isomer is an ion or molecule of any substance.) The catalyst is usually aluminum chloride, antimony chloride, or platinum.

The product of isomerization is used as an additive in gasoline to increase its octane rating and reduce knocking.

BLENDING GASOLINE

Refineries add antiknock compounds to gasoline produced by fractionating. Usually about a teaspoonful to a gallon is enough. This would amount to about 1 part in 1300.

TRANSPORTING OF OIL

One of the advantages of petroleum and its various products is the ease of transport. The fastest and easiest method is through pipelines, Fig. 9-21. In the United States, pipelines are the largest single oil carrier. A pipeline network of more than 200,000 miles carries crude oil and refined petroleum products.

Fig. 9-21. Pipelines transport petroleum quickly and easily.

Pumping stations, spaced about 150 miles apart move the petroleum through the pipelines at speeds of two to three mph. Modern pumping stations have pushbuttons or computer controls.

Other modes of transporting oil include tankers, barges, railroad tank cars, and over-the-road tank trucks. See Fig. 9-22. Fig. 9-23 shows an oil unloading facility that has been constructed in the Gulf of Mexico for the use of super tankers. Called LOOP (for Louisiana Offshore Oil Port) it allows three tankers to unload their oil cargo at the same time. The oil is pumped ashore to refineries and storage. Because of a network of pipelines, oil unloaded in New Orleans can be piped as far north as Chicago and Buffalo.

OIL AS FUTURE ENERGY SOURCE

At present, slightly over 41 percent of our total energy consumption in the United States comes from petroleum. In 1990 about 16 million barrels per day were used. Reserves of petroleum (amount left in the ground) are getting scarce. Future energy supplies must come from other resources. Supplies are discussed in Chapter 4.

NATURAL GAS

As you have learned from a study of petroleum, natural gas is closely related to petroleum. You also know that all oil deposits contain natural gas. However, not all natural gas deposits contain oil.

EXPLORATION AND DRILLING

Gas prospecting is done in much the same way as the search for oil. Sometimes the gas is found seeping up from the underground deposits. Usually, geologists look for rock formations that suggest the presence of porous rock layers covered over by solid rock. This is the condition necessary for an oil or gas pool.

Fig. 9-22. Tanker delivers crude or fuel to port. (Amoco Corp.)

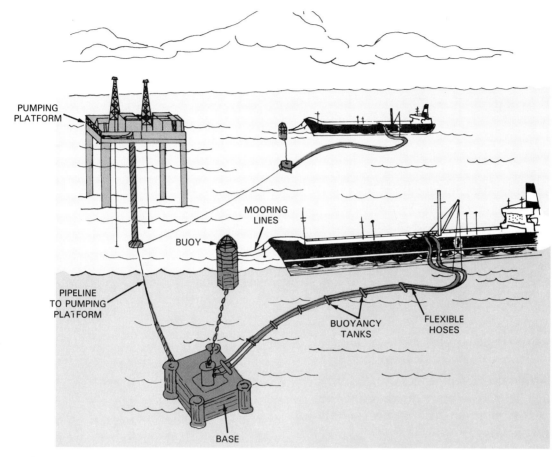

Fig. 9-23. Sketch of offshore oil unloading facility built in Gulf of Mexico. It unloads oil super tankers which cannot enter ports because of their huge size.

GAS DEPOSITS

Gas is often found trapped in natural underground reservoirs. There are three main types of reservoirs:
- **Sandstone.** In this type the reservoir is formed by sedimentary rock. This is rock formed of mineral deposits such as quartz.
- **Carbonate.** This type is made up of sedimentary rock that came primarily from the shells of sea life. Another name for it is limestone.
- Other rock formations. These are generally igneous (came from volcanoes) or metamorphic rocks. (The latter are formed from both igneous and sedimentary rock that have been changed by pressure, heat, water action, or contact with hot lava.) Also included in this category are formations of **shale.** This is a rock made up of mud, clay, or silt.

MAKEUP OF NATURAL GAS

Natural gas is a mixture of hydrocarbon compounds and small quantities of several nonhydrocarbons. It can be found as a pure gas or it may be mixed into a solution with oil in natural underground reservoirs. The hydrocarbon compounds include: methane, ethanes, propane, butanes, and pentanes. The nonhydrocarbons include: carbon dioxide, helium, hydrogen sulfide, and nitrogen. The fuel sold under the name "natural gas," is mostly methane. It may also contain small amounts of ethanes. Free natural gas that is found alone in its natural reservoir is called **nonassociated gas.** When it is found in the same reservoir as crude oil it is called **associated gas** whether it is separated from the oil or in solution with it.

GAS RESERVES

In 1974 the total natural gas reserves in the United States were thought to be about 237 quadrillion cu. ft. The largest reserves are found in Texas, Louisiana, and Alaska. These states account for about three-fourths of all the U.S. reserves.

Currently, the Energy Information Administration does not report reserves of natural gas. However, it does indicate underground storage of natural gas. Storage for 1990 was about 6.9 billion cu. ft.

Historically, Canada has been a major supplier of natural gas to the U.S. with Algeria supplying smaller amounts. There were about 265,000 wells in operation during 1990.

TRANSPORTATION OF NATURAL GAS

Natural gas is usually transported by one of three modes:
- Pipeline.
- Railroad tankers, Fig. 9-24.
- Truck tanker.
  The mode being used depends upon:
- The distance the natural gas must be moved.
- Whether the terrain is mountainous or suitable for truck or rail transport.
- How complicated the distribution system is.
- The kind of gas that is to be moved.
- The cost of building and operating the transportation system.

PIPELINE TRANSMISSION

Gas must be moved through pipelines by pressurizing it with pumps. These operate somewhat like the compressors used to pump up tires or provide air pressure for paint spraying.

GAS PRESSURE DURING TRANSPORT

Pressure in long distance pipelines can be anywhere from 500 to 1000 psi (3445 to 6890 kPa). Compressors are located about every 50 miles to restore the pressure lost by friction. The natural gas may travel at a speed of about 15 mph (24 km/h). Valves and regulators are also located along the way. These devices are needed to control pressure or to stop the flow entirely in case of a rupture (break) in the pipeline.

NATURAL GAS STORAGE

At certain times of the year, demand for natural gas as a fuel is greater than the ability to extract and transport it. To meet the demand, natural gas is pumped and stored for later use.

Usually, storage is underground. Natural reservoirs receive the gas from pipelines during the warm months when demand is lower.

TYPES OF STORAGE

Most gas storage is in old gas or oil fields. Gas remaining in these reservoirs is at such a low pressure that the wells are no longer in production. The gas is pumped into them through the wells originally drilled to extract the gas and oil. The gas is forced into the old wells with compressors.

Some gas is stored in **aquifers**, Fig. 9-25. An aquifer is a rock formation underground which holds large quantities of water that has filtered down from the surface. The rock is porous so that water soaks into the rock. The pores are full of water.

Gas pumped into the aquifer under pressure, displaces the water, forcing it to move deeper into the underlying rock.

The United States has 44 such aquifers. Most are found in the Midwest. Illinois has 19, Indiana 11, and Iowa 6. Several storage reservoirs have been established in abandoned coal mines. Others are in natural salt domes.

MANUFACTURED STORAGE

City gas companies use huge cylindrical tanks to store gas before it is distributed to users. The tanks are closed at the top and open at the bottom. The tanks float in pools of water and can rise or fall according to the supply of gas in them. The weight of the heavy tank keeps the gas under pressure at all times and forces it out into the gas mains for distribution to gas customers.

Fig. 9-24. Railroad tank cars are also designed to transport natural gas long distances.

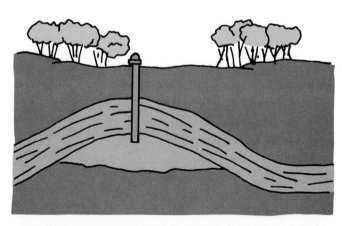

Fig. 9-25. Sometimes, natural gas is stored in natural underground aquifers such as this until it its used.

LIQUIFYING NATURAL GAS

Natural gas may be converted to a liquid for use in transport vehicles. Liquification reduces its volume and makes the gas more practical to move overseas as well as to store.

SUMMARY

In its natural state, petroleum is a liquid hydrocarbon. Its gaseous state is natural gas. Deposits of natural gas are found wherever oil is found. The first oil well was drilled in 1859. Need for gasoline to fuel the automobile led to development of the petroleum refining industry.

Since oil is buried beneath the earth, geologists must prospect to discover new deposits. Various instruments, gravimeters, magnetometers, and seismographs, are employed to find these deposits.

Wells must be drilled to reach oil deposits. Then a well casing is sunk in the drilled hole. Some wells are under great pressure and the petroleum is forced to the surface. In other instances, a pump is installed to draw up the petroleum.

There are three classes of crude petroleum: paraffinic hydrocarbons, aromatic hydrocarbons, and asphaltic base hydrocarbons. Major products of petroleum are: gasoline, kerosene, diesel fuel, fuel oil, asphalt, natural gas, lubricating oil, and coke.

The converting of oil fractions is called cracking. Cracking changes heavy petroleum molecules into light, small molecules. Two methods are employed. In thermal cracking, the petroleum is subjected to both heat and pressure. In catalytic cracking, catalysts are added to the process along with high temperature.

Petroleum and petroleum products are transported by a variety of methods. Among them are: pipeline, trucking, and rail. Similar transportation methods are used for natural gas. It is easy to store natural gas in old gas or oil wells and worked out salt domes.

DO YOU KNOW THESE TERMS?

aquifer
asphalt
associated gas
blowout preventers
bottom hole pressure
bubble caps
bubble tower
cable tool drilling
cable tool string

carbonate
cone bits
cracking
Christmas tree
crude oil
derrick
development wells
drill pipe
drive
electrodrilling
exploratory well
fishtail bit
flame drilling
fractionating tower
fractions
geologists
gravimeter
input well
isomerization
lift
magnetometer
mud
natural gas
nonassociated gas
pipe still
polymerization
refining
reservoir beds
roller bits
rotary drilling
sandstone
seismograph
shale
sonic drilling
source beds
test core
traps
tungsten carbide
turbodrilling
wet gas
wildcat wells

SUGGESTED ACTIVITIES

1. Using light wood pieces build a model of an oil well derrick and explain to the class how it works.
2. Secure literature from an oil company on the history of petroleum. Take part in a class discussion on why it became an important energy source.
3. Draw up a list of energy converters which depend on petroleum for their source of energy.
4. If your lab is equipped with different types of engines, discuss with your instructor methods of

testing the suitability of different fuels for each of the engines.

5. Referring to Fig. 8-1 in the text, construct a cutaway model of an oil-bearing rock formation.

TEST YOUR KNOWLEDGE

1. Geologists looking for oil consider the following geological formations as an indication that oil might be present. Check the correct answers:
   a. Dolomite.
   b. Limestone.
   c. Igneous rock.
   d. Sandstone.
   e. Shale.
   f. All of the above.
2. When drilling for oil a _____ _____ bit or one studded with tiny _____ is best when drilling through abrasive rock.
3. Oil raised to the surface by its own pressure is called artificial drive. True or False?
4. Which of the following are products derived from crude oil?
   a. Natural gas.
   b. Gasoline.
   c. Kerosene.
   d. Diesel oils.
   e. Fuel oils.
   f. Lubricating oils.
   g. Petroleum chemicals.
   h. All of the above.
   i. None of the above.
5. What is a pipe still?
6. Where is natural gas found?
7. List the three methods for transporting natural gas.

10

SOLAR ENERGY

The information given in this chapter will enable you to:
- State the current theory about the origin of the sun and explain the source of its energy.
- Define electromagnetic radiation and explain how it travels through space.
- List the types of radiation.
- Explain the terms: solar constant, insolation, entropy, Langley, and calorie.
- Give examples of how solar energy is stored.
- Cite examples of how solar energy is used as a direct source of energy.

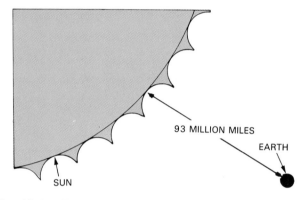

Fig. 10-1. The sun is 125 times the size of the earth. If the sun were 10 ft. in diameter, the earth would be 1 in. across.

As you learned from Chapter 1, most of our energy comes either directly or indirectly from the sun. "Sol" is a word for sun. Therefore, sun energy is usually called solar energy. However, very little of it is put to use directly. Explaining why is not easy. To put it simply, it is very hard to collect widely scattered sunshine. Further, it is expensive to build collector equipment to gather and put solar energy to use.

To understand the problems of collecting solar rays it is necessary to learn more about the nature of the sun. It is also helpful to understand the physical nature of **radiation of energy**. (This is the method by which solar energy travels.)

THE SUN

The **sun** is a luminous celestial body. The earth and other planets revolve around the sun and receive light and heat from it. The sun's diameter is 864,000 miles (1 390 000 km) while earth's is 6888 miles (11 085 km). Being made up of hydrogen, it has half the density of the earth and twice the mass (weight). See Fig. 10-1.

Light and heat radiating from the sun are caused by thermonuclear reaction taking place at its core. The temperature at the sun's center is 15 000 000°C or about 27,000,000°F. At the sun's surface, the temperature is 5 500°C (10,000°F).

From almost 93 million miles (150 million km) in space, the sun bathes the earth in huge amounts of energy. It is hard to imagine how much heat and light it provides. If it could be collected, two days of sunlight on earth could provide more power than all of the U.S. and Canadian fossil fuel reserves. Considering that the sun is expected to shine at least for the next 5 billion years, the supply is renewable.

Scientists estimate that the sun is over 5 billion years old. In that amount of time it has converted four million tons of solar matter every second into radiant energy.

Only one part of every 2 billion parts of the sun's rays reaches the earth. This is easier to understand if you consider how much smaller earth is than the sun. See Fig. 10-2. Radiation moves outward from the sun in all directions just like light from a light bulb when it has no reflector or globe. The earth can intercept (catch) only a small amount of the energy.

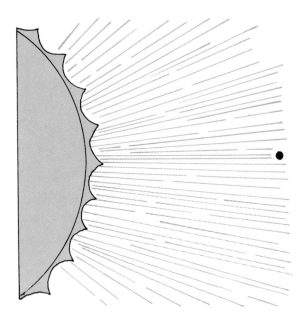

Fig. 10-2. So immense in size is the sun that only 1/2,000,000,000 (one-two-billionth) of its rays even strike the earth.

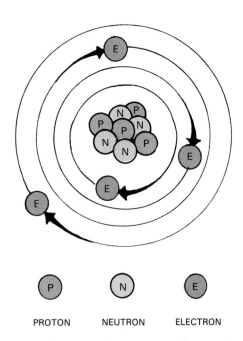

Fig. 10-3. A single atom has a nucleus (core) of particles called neutrons and protons. Around this core are several orbiting electrons that are always in motion.

NUCLEAR FUSION

A joining of the nuclei (centers) of atoms creates the heat and light given off by the sun. The process is called **nuclear fusion.** It starts up when gravity creates great pressure on the hydrogen atoms. They begin to heat up. The molecules of hydrogen move faster and faster due to the heat buildup. They collide with each other with such force that their nuclei fuse. As this occurs, a small fraction of their mass is released as energy. This energy pushes to the surface of the sun and radiates outward.

HOW SOLAR ENERGY TRAVELS

How is it that solar energy can move 93 million miles through space and still have enough "punch" left to produce heat and light? The answer lies in the way heat and light waves are thought to travel through empty space. This theory is known as **electromagnetic radiation.** You read about it in Chapter 1.

HOW ELECTROMAGNETIC RADIATION IS CREATED

As you know from Chapter 1, an atom is made up of electrically charged particles. A cluster of positively charged particles make up the **nucleus** (center). The protons attract negatively charged particles called **electrons** which travel around the nuclei in elliptical or circular paths as pictured in Fig. 10-3. In some instances,

electrons may have orbits other than circular or elliptical; however, this is not important for an understanding of how electrons act in our physical world.

Not every electron travels the same distance away from the nucleus. The paths are at different levels, Fig. 10-4.

A burst of energy occurs whenever one of the orbiting electrons moves to another level. This starts a chain reaction as other electrons move to different levels and even into different atoms, Fig. 10-5.

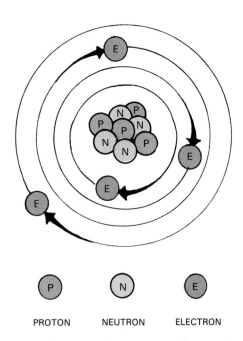

Fig. 10-4. The orbits of electrons around the nucleus of the atom are not along the same paths or at the same distance from the nucleus.

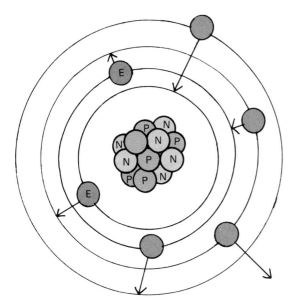

Fig. 10-5. When orbiting atoms leap from one level to another, energy is released. This is the nature of solar energy. Changing of levels creates a chain reaction. Some electrons may even leap out of one atom to join another atom.

HOW RADIANT ENERGY TRAVELS

The electromagnetic energy just described produces both electricity and magnetism. The electrical and magnetic energy moves outward in waves. The waves are not all the same length. Short waves have a higher **frequency** than long ones. Frequency means the time lapse between waves reaching a receiver. You can think of an electromagnetic wave as part of a wavy line, a series of small hills and valleys. What makes the waves different from each other is the **wavelength**. This is the distance from one crest to the next. Fig. 10-6 shows a short and long wavelength.

Most electromagnetic waves are invisible. Wavelength determines whether or not you can see them. The different waves from shortest to longest wavelength are:

- Gamma rays. These are only about one-250 trillionths of an inch long.
- X rays.
- Ultraviolet rays.
- Light rays.
- Infrared rays.
- Microwaves.
- Radio waves.
- Electric waves.

As Fig. 10-7 indicates, some wavelengths are very long. Others are very short. You can also think of the travel of electromagnetic waves being like the small waves created when a pebble is dropped into water. See Fig. 10-8. Ripples or waves of water move out from

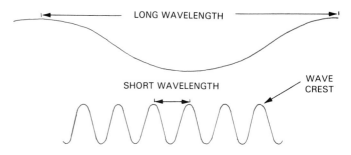

Fig. 10-6. A wavelength is the distance from the top of one wave to the top of the next.

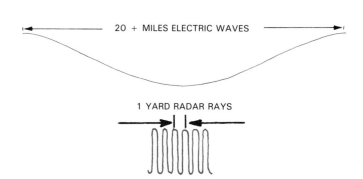

Fig. 10-7. The longest wavelength belongs to electric waves which are 20 or more miles long. Radar wavelengths are only 3 ft. long.

Fig. 10-8. Electromagnetic waves of energy travel outward from the source somewhat like the ripples travel outward when a rock is dropped into water.

the point where the pebble was dropped in ever-widening circles. Eventually the waves reach the shore.

Energy of rays is also related to wavelength. The shorter the wavelength, the more energetic its radiation. Most solar energy is made up of short wavelengths. Forty-six percent are visible light. This is the wavelength to which the human eye is sensitive. Forty-nine percent are infrared radiation. We experience it as heat. The rest of the sun's radiation is in the ultraviolet wavelengths. These waves are shorter than the visible violet wavelength. All wavelengths travel through space at the same speed.

METRIC UNITS OF MEASURE

Three metric units are used for measuring solar radiation.
- **Langley.** The langley equals 1 calorie per square centimeter (cm²). It is also equal to 0.001 163 Whr/cm² or 4.8 J/cm².
- **Calorie.** A unit of heat measure. One calorie per cm² is equal to 0.0697 W/cm².
- **Watts.** A unit of power based on energy/area/time. A watt per square meter is equal to 3.152 Btu/sq. ft./hr.

THE SOLAR RESOURCES

As was stated earlier, only one-two-billionths of the radiation released by the sun reaches the earth. Even so, this is a vast amount of energy, as stated earlier.

We are able to measure the amount of solar radiation we receive every year. These figures are important in a study of solar energy.

SOLAR CONSTANT

The rate at which solar radiation falls on the outer limits of our atmosphere (the power input) has been set at 1.36 kilowatts per square meter (kW/m²). This is about 130 watts per sq. ft. Actually, the solar radiation varies a little because of the elliptical orbit of the sun. At times, it is as high as 1.4 kW/m²; at other times as low as 1.3 kW/m². See Fig. 10-9.

INSOLATION

Insolation is a shortened form of incident solar radiation. It is a term for the amount of radiation falling on the earth's surface at any given time. Insolation is usually expressed in Btu per square foot per hour. It is written as: Btu/sq. ft./hr.

There is no constant for insolation as there is for the amount of energy reaching the outer atmosphere. A number of things happen to sunlight as it enters earth's atmosphere. While the radiation is able to pass through airless space quite easily, it is dispersed (split up) upon entering the atmosphere.

Some of the rays are scattered by atmospheric conditions and are called **diffused radiation**. Dry air molecules, dust, water vapor, and thin layers of clouds absorb some of it. Whenever there is heavy cloud cover,

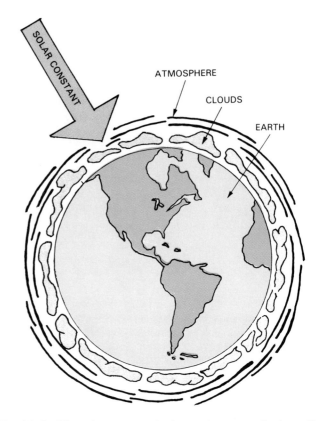

Fig. 10-9. The solar constant is the total amount of solar radiation reaching the earth's outer atmosphere. How much of it reaches the ground depends on the clouds, diffusion, latitude, and other conditions.

all but the scattered radiation is absorbed or reflected by the clouds. See Fig. 10-10.

About 15 percent of the sunlight which strikes the earth is reflected back into the atmosphere. Another 5.3 percent is absorbed by the ground. About 2 percent is absorbed by marine vegetation. Land vegetation absorbs .2 percent. Huge quantities are used to evaporate water and lift it into the atmosphere.

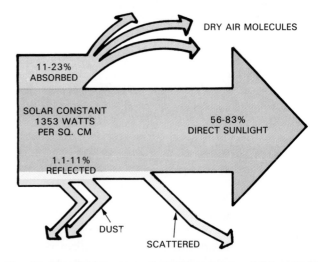

Fig. 10-10. Earth's atmosphere breaks up sunlight so that 17 to 44 percent of it never reaches ground level at all.

The amount of solar energy reaching the ground depends not only on weather conditions but on latitude and the season. However, in the U.S., the average over 24 hours is 177 watts per square meter (W/m²). A collector 1 meter square would provide enough energy for a 200 W light bulb. This does not seem like much energy but it adds up. Over 24 hours this amounts to a little more than 3770 calories per square meter. There are 65 calories in a British thermal unit, thus:

$$Btu = \frac{number\ of\ calories}{65}$$
$$Btu = \frac{3770}{65} = 58$$

The solar energy falling on each acre of the lower 48 United States every day is equal to the energy of 10 barrels of oil. Multiplying this by the 2310 million acres in the contiguous (touching) 48 states, yields the equivalent of a little more than 23 billion barrels of oil. This is nearly four times the present consumption of oil. Fig. 10-11 shows the insolation for the United States in the months of January and July.

ENTROPY AND SOLAR ENERGY

You may recall that Chapter 1 discussed the losses of energy due to entropy. Whenever energy is converted from one form to another it loses some of its ability to do work. This reduces the efficiency of the power system. Direct solar energy has an advantage. It goes through but one conversion from radiant energy to useful heat. Fossil fuels must go through several different conversions before they become a usable fuel. Several additional conversions are necessary before they can produce useful energy.

Consider the light from an electric light bulb. The energy to produce the light was originally solar energy. The solar energy was transformed into plant life. When the plants died they decayed. The decayed vegetation became coal. The coal gives up its energy as heat when burned. The heat creates steam for running a turbine. The mechanical energy of the turbine drives a generator that creates electrical power. The electric power is sent to the customer where it will be converted to light, heat, or mechanical energy. In all, six conversions took place from sunlight to fossil fuel to electric light, Fig. 10-12. The efficiency of the system, having gone through so many conversions, is very low.

NATURE'S SOLAR STORAGE

Large quantities of solar energy are absorbed by the earth through natural systems. Some is stored for only short periods of time. Examples of this are the radiation stored in the form of heat in earth mass such as soil and rock. Large bodies of water also act as storage. The water traps and holds the energy as heat.

Another type of solar storage takes place when plant life interacts with sunlight. Solar rays are stored in plant fiber. Some plant life provides food and energy for animal and human consumption. Some plants provide wood as a fuel. As you learned in your study of fossil fuels, a great deal of energy has been stored underground in the form of solid and liquid carbon.

NATURE OF SOLAR RADIATION

Radiant energy passes through millions of miles of space with little effect on the temperature of the space. When it strikes matter three things can happen:
- The material may intercept (absorb) the energy and convert it to heat energy, Fig. 10-13.
- The surface of the material may reflect the radiation back into the air, Fig. 10-14.
- The material may transmit the radiant energy to surrounding space or matter, Fig. 10-15.

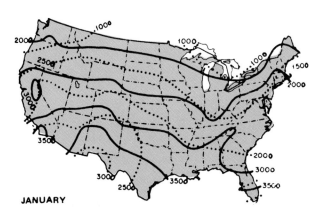
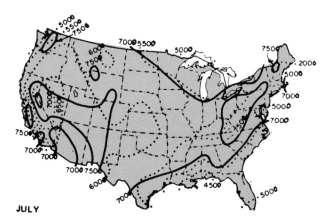

Fig. 10-11. Insolation in the United States for January and July. Areas of equal daily solar energy (on cloudless days) are shown as solid lines. Dotted lines represent equal energy totals for cloudy days. Totals are in kilocalories.

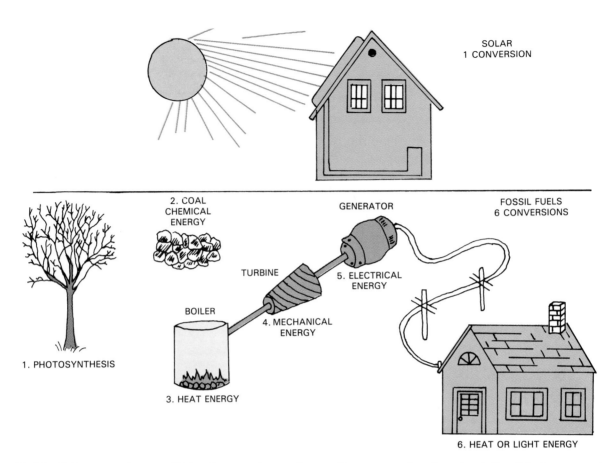

SOLAR
1 CONVERSION

2. COAL CHEMICAL ENERGY

GENERATOR

FOSSIL FUELS
6 CONVERSIONS

TURBINE

5. ELECTRICAL ENERGY

BOILER

4. MECHANICAL ENERGY

1. PHOTOSYNTHESIS

3. HEAT ENERGY

6. HEAT OR LIGHT ENERGY

Fig. 10-12. Solar energy is very efficient because it goes from radiant energy directly to heat. Fossil fuels may go through as many as six conversions to produce heat or light.

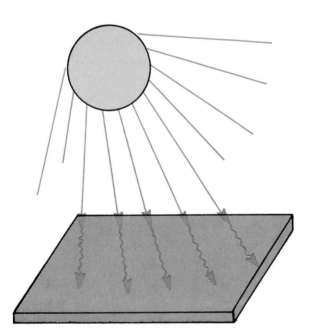

Fig. 10-13. Dark surfaces that are rough will absorb more solar radiation.

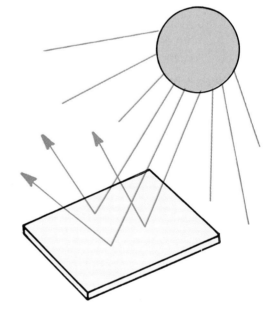

Fig. 10-14. Smooth, shiny surfaces will reflect more sunlight than dull, dark, or rough ones.

When a substance captures the solar radiation striking it, the light energy becomes heat energy. The molecules of the substance begin to move faster and the temperature of the substance rises.

HEAT TRANSFER

As the temperature of a substance increases, it tries to pass along some of its heat to its surroundings. It

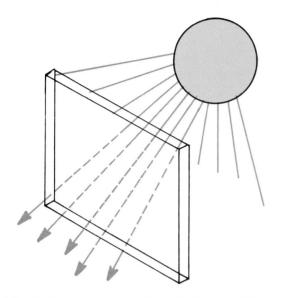

Fig. 10-15. Transparent material will allow sunlight to pass through it.

is one example. Materials which transmit light readily do not necessarily transmit heat waves. Glass will absorb most thermal radiation striking it. Yet, when sunlight travels through glass and is absorbed by materials in enclosed space, the heat given off will not readily pass out through the glass, Fig. 10-16. This heat trapping process is known as the **greenhouse effect**. You have probably experienced the results of the greenhouse effect when an automobile is left standing in the sun for a few hours. The interior becomes much warmer than the air outside, Fig. 10-17.

Polished or shiny surfaces reflect solar radiation to a great extent. Light-colored surfaces also tend to reflect some radiation. A light-colored roof on a house, for example, will not absorb as much heat as a black one. It tends to reflect solar radiation back into the atmosphere.

is important to remember what you learned earlier. Heat will not normally move out of cooler bodies into warmer ones. However, heat always moves from a warm body to cooler surroundings. This movement will continue until the temperature of the cooler substance is the same as its surroundings. At this point the substance is said to have reached equilibrium (sameness) with its surroundings.

This transfer of heat is possible because of the three natural processes discussed in Chapter 1: conduction, convection, and radiation. These processes are important in the operation of systems which use the energy of sunlight.

GREENHOUSE EFFECT

Substances which pass along all, or nearly all, of the visible rays they receive are said to be transparent. Glass

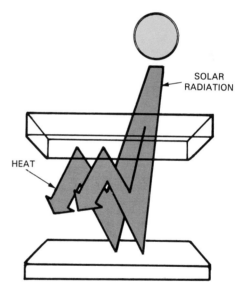

Fig. 10-16. Solar radiation will pass readily through glass. However, heat energy has a different wavelength and will not pass back out through glass as readily.

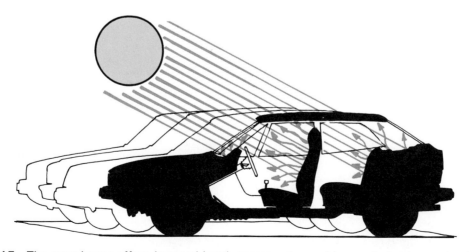

Fig. 10-17. The greenhouse effect (trapped heat) causes autos and homes to heat up naturally in sunlight.

HISTORY OF SOLAR POWER SYSTEMS

Attempts to use sunshine directly to heat space and power machines are not new. Nearly 2200 years ago, the ancient Greek mathematician, Archimedes, supposedly burned the ships of attacking Romans at Syracuse. A version of the story suggests that he did it with a large hexagonal (six-sided) mirror, concentrating the sun's rays on their sails. In another version, he lined up soldiers carrying polished shields. Solar experts doubt that it could have been done with shields which are convex. There would have been no way of concentrating the solar rays so they could create the heat needed.

One expert believes he could have done it with a number of handheld mirrors, highly polished, all trained on a single target, Fig. 10-18.

SOLAR ENERGY FOR STEAM

Other early inventors hoped to get more than warmth directly from the sun. They tried to use solar radiation for running steam engines. Among the first experimenters with solar steam was John Ericsson, the designer of the warship Monitor. In 1868 he began with solar-heated water, then switched to heated air because hot air was easier to contain and less dangerous. He built the largest solar engine known up to that time. It had 200 sq. ft. of collectors and developed nearly 2 horsepower.

Meanwhile, similar solar engines were built in India and France. In 1901, A.G. Eneas designed a parabolic collector, Fig. 10-19, which directed reflected sun onto a boiler. According to a newspaper report at the time it resembled ". . . a huge umbrella, opened and inverted at such an angle as to receive the full effect of the sun's rays on little mirrors lining its inside surface. The boiler (of the engine) which is thirteen feet and six inches long, is just where the handle of the umbrella ought to be . . . From the boiler, a flexible metal pipe runs to the engine house near at hand."

Fig. 10-18. Ancient Greeks were said to have used the sun to set fire to the ships of Romans invading Syracuse nearly 2200 years ago.

Fig. 10-19. Eneas' solar engine had a collector shaped like an upside-down umbrella. A boiler sat just where the handle should be. A flexible pipe carried the steam to the engine.

During the early 20th century there was another burst of solar steam activity. H.E. Wilsie and John Boyle, Jr. experimented with several designs. Ammonia, ether, and sulfur dioxide were used to raise steam at a lower boiling point. The two inventors designed and built engines of 15 to 20 horsepower and worked out methods of storing heat from the collector to be used as needed. Their work came to nothing. The solar engines, at $164 per horsepower, were not as cheap as conventional steam engines.

In 1911, Frank Shuman tested a 100 hp (74.6 kW) engine on low-pressure steam heated in 10,000 sq. ft. (92,000 dm²) of troughs lined with reflectors. Later, with the help of Professor E.C. Boys, he moved the engine to Egypt intending to use it for irrigation. He constructed seven more troughs each with more than 13,000 sq. ft. of collector surface. The system worked, developing 50 to 60 horsepower (37 to 48 kW). It was abandoned during World War I, however, because of cheaper human labor.

For the next 60 years, little effort was made to find ways of using solar energy. Gas and oil were so cheap that few bothered. There were some notable efforts, however. During World War II the U.S. military used a small solar still. It was developed to desalinate (take the salt out of) small quantities of sea water and to purify contaminated water.

Simple solar water heaters were common in the southern U.S. until around 1940. Again, cheap and plentiful fossil fuels caused a decline in their use.

STORAGE OF SUNLIGHT

Sunlight can be stored in different ways for later use. All solar energy systems depend on trapping and holding the energy in matter for some period of time.

One way of doing this is to place a bulky material in sunlight where it can be heated. Another is to carry heated air to the material. A third way is to convert the sunlight into electricity. All of these systems will be discussed later.

Some materials can hold more heat than others. Such materials have high specific heat. They are, therefore, more desirable for storing heat.

SPECIFIC HEAT

The capacity of a material to hold heat is called its **specific heat.** The term is used as a measure of ability to store heat. Specific heat is the amount of heat, in Btus, a substance can hold when its temperature is raised one degree Fahrenheit.

In the building trades, however, the amount of a substance is usually given in cubic feet, not pounds.

Heat capacity is then given in volume. The heat capacity is called the **volumetric heat capacity.**

The volumetric heat capacity of one cubic foot of a substance is found by multiplying its specific heat by its density (number of pounds per cubic foot). Fig. 10-20 compares the specific heat, density, and heat capacity of water with some common building materials.

Material	Specific Heat (Btu/lb/°F)	Density (lb/ft³)	Heat Capacity (Btu/ft³/°F)
Water	1.00	62.5	62.5
Concrete	0.27	140	38
Brick	0.20	140	28
Gypsum	0.26	78	20.3
Marble	0.21	180	38
Asphalt	0.22	132	29.0
Sand	0.191	94.6	18.1
White Pine	0.67	27	18.1
Air (75 °F)	0.24	0.075	0.018

Fig. 10-20. Specific heat and heat capacity of some building materials are compared to water. Water is one of the best materials in which to store heat. (HUD)

Problem: A certain building material has a specific heat of 0.27 Btu/lb/°F, a density of 129 lb/ft³. What is its volumetric heat capacity?

$$0.27 \times 129 = 34.8 \text{ Btu/ft}^3/°F$$

How sunlight is stored for use in heating buildings depends on what type of system is used for collecting the solar energy. Materials commonly used are concrete, concrete block, water, ceramic tiles, wood, rocks, and air. Another material showing promise are salts which can be made to change from a solid state to a liquid state. They are called **change of state** or **phase-change materials.**

When salts such as sodium sulfate decahydrate are heated, they change to a liquid at temperatures around 90°F (32°C). When the heat is drawn off, it becomes a solid once more. In the process of liquefying and solidifying, the salts can collect and then release huge quantities of heat. Thus, a very small amount of the salts can store great quantities of heat energy.

SOLAR HEATING OF BUILDINGS

On a clear day, sunshine delivers about 250 Btu (264 kJ) of radiant energy per square foot to the surface of the earth. This radiation produces enough energy to heat buildings.

Two basic solar heating systems are used:
• Passive solar systems.
• Active solar systems.

PASSIVE SOLAR HEATING

Passive systems have no moving parts. They collect and distribute heat inside a building without the aid of any mechanical means such as fans, motors, or pumps. The heat flows by natural radiation, conduction, and convection.

Two things are necessary for a passive solar system.

- Glass through which the solar radiation passes. The glass traps the heat by preventing it from being reflected back to the atmosphere.
- A thermal mass to collect the heat, store it, and then pass it on via radiation to the air space in the building. Fig. 10-21 shows the basic elements of the passive system.

There are several types of passive systems. Most of them have one thing in common. The elements of the system are part of the structure of the building.

The simplest and least expensive is the **direct gain solar heating system**. The space to be heated receives direct sunlight, Fig. 10-22. Sunlight, streaming through large south-facing windows falls on floors and/or walls. These parts of the house must be constructed of materials which can do a good job of storing the heat.

One of the best building materials for heat storage is concrete masonry and units such as brick, block, stone, and adobe. Solar radiation strikes the dark surface of the thick masonry. The radiation is absorbed as heat. The heat will remain in the masonry until the room cools. You will recall that it is the nature of warm

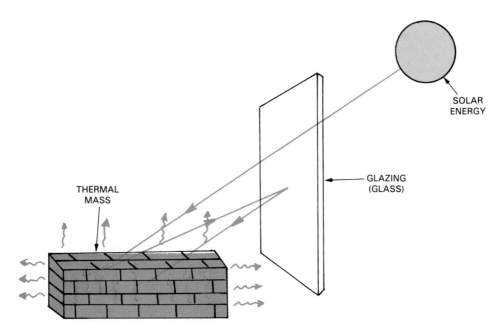

Fig. 10-21. Passive solar energy operates wherever the sun is shining through glazing and the sunlight is stored in thermal mass such as a masonry floor.

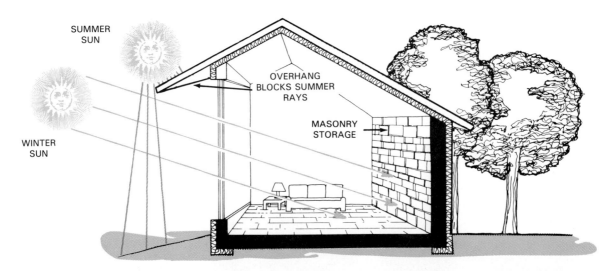

Fig. 10-22. In a direct gain passive solar system, solar heat is collected and stored right in the living space.

bodies to pass off heat to cooler surroundings. This principle goes into action as the heated masonry materials pass their heat to the cooler room air. It is normal for such storage walls and floors to be from 8 to 12 in. (203 to 304 mm) thick. The masonry surface must be exposed so it can better receive and radiate heat.

INDIRECT GAIN

Indirect gain solar heating is used to describe a second type of passive solar heating. Sunlight strikes some kind of thermal mass first. The heat is not stored in the living space. It is first absorbed by the mass and then transferred to the living space.

One way to get thermal mass is to put it into a wall. Some indirect gain systems use water for storage. Some use masonry. Another way is to store water on the roof of the building. This is called a **pond**. Sunlight heats the water during the day. At night the heat of the water will be transferred to the space beneath it.

Fig. 10-23 shows a masonry storage wall for an indirect gain system. It is called a Trombe wall after its designer, a French scientist, Felix Trombe.

Trombe walls are usually constructed of brick, concrete block, or solid concrete. Wood may be used in some cases. The side of the wall facing the sun is painted black or a dark color so it will absorb solar radiation better.

A second necessary element of the Trombe wall is an area of glazing (clear glass or plastic) in front of it. The glazing allows the sunlight to pass through but prevents most of the heat from escaping back to the outdoors. The glazing is usually about 4 in. (102 mm) from the wall.

As the sun streams through the glazing, it heats both the wall and the air between the glazing and the wall. Some Trombe walls take advantage of the heated air to give the living room immediate heat. Openings called vents are cut in the wall at ceiling level and at floor level. Air, heated in the space between the glazing and the wall expands and becomes lighter than cold air. It rises, passes through the vents at the ceiling, and provides heat to the living space. Meanwhile, cooler air is pulled into the air space between the glazing and the wall through the vents near the floor. This air movement is called **thermosiphoning**. Fig. 10-24 shows this flow of air.

The Trombe wall heats living space in another way as well. Solar radiation heats up the thick wall during daylight hours. First the side facing the sun heats up. Then the heat conducts through the wall until the entire wall is the same temperature. When the side facing the living space is heated up, it will pass off heat to the cooler living space.

If the wall is thick enough, this will happen after the sun has gone down. The wall will continue to give off its stored heat as long as it is warmer than its surroundings. The thicker the wall, the longer it will take to heat it up and the longer it will provide heat to its surroundings.

You can see, then, that having the proper wall thickness is important. If too thin, the wall will start to heat the living space too soon. The room may become uncomfortable while the sun is still shining. It may also pass off its heat too soon and the living space will become cold before the night is over. A wall that is too thick will not be able to pass along its stored heat until well into the night. It will still be passing along heat during the day when sun has begun to heat the room air through the circulation system. Thickness of a Trombe wall may vary from 8 to 16 in. (203 to 406 mm).

As we know from the chart in Fig. 10-20, water is a better storage medium than masonry. It can be stored in barrels, Fig. 10-25, jugs, tanks, or free-standing vertical tubes.

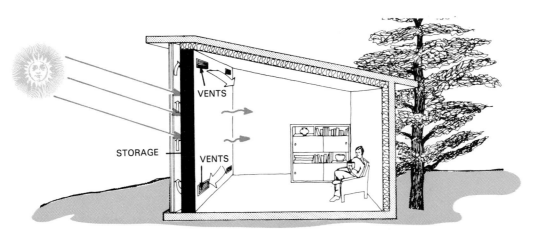

Fig. 10-23. In an indirect gain solar system, heat is stored in some type of thermal mass before it enters the living space. Here a Trombe (masonry) wall has been placed directly behind glass.

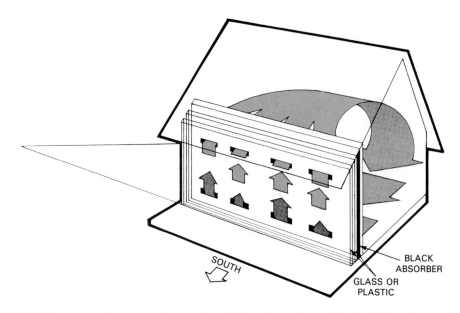

Fig. 10-24. Vents, small rectangular openings in Trombe wall, provide circulation of sun-warmed air into living space. The vents can be closed at night or when heated air is not wanted.

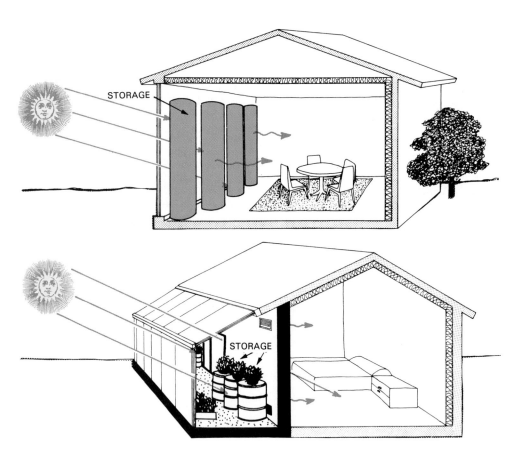

Fig. 10-25. Water wall stores heat and passes it along to living space by convection. Water has twice the heat storage capacity of masonry. Top. Free-standing water wall made up of tubes. Bottom. Water storage with an isolated gain system. Solar heat is collected and stored in a wall, a floor, and water drums in an area set apart from the living space.

While water will store more energy per pound than masonry, it is hard to contain. Leaking containers not only cause loss of water; they might damage furniture, furnishings, or the structural members of the house.

We mentioned before that water can also be stored in roof "ponds." Here it can be contained in tough plastic, Fig. 10-26. During winter nights the pond is covered by movable sheets of insulation. During the

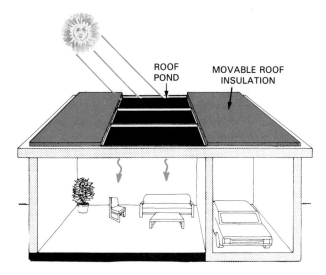

Fig. 10-26. Roof-level heat storage in a "pond." Movable insulation covers the roof during a winter night.

day the insulation is removed and the pond is open to the sunlight so the water can absorb solar heat. The heat is conducted to the living area later.

ISOLATED GAIN

A third method of passive solar heating separates the collecting and storing of heat from the living space. This type is called isolated gain. Heat can be contained in the system and is drawn from storage only when it is needed. A system of this type is shown in Fig. 10-25, bottom view. The greenhouse attached to the living space collects the solar heat. Storage mass includes the wall separating the greenhouse from the main dwelling and containers of water. A rock bin is another storage area sometimes used. See Fig. 10-27.

Other common methods of getting solar heat by isolated gain are:
• The convective loop.
• The window solar collector.

Both of these depend on thermosiphoning to make them work. (Thermosiphoning is the action of rising warm air or water pulling in cooler air or water after them.)

The convective loop, Fig. 10-28, has a flat plate collector connected to a storage tank or rock storage. As water or air in the collector heats up, it rises and enters the storage area. At the same time, cooler water or air from the bottom of the storage container is pulled into the collector. This natural convection (movement) takes place as long as the sun is shining. The flat plate collector is described later in this chapter.

The window collector works on the same principle. It is similar in operation to the convective loop except that:
• It does not use ductwork. Instead radiation pouring through a pane of glass falls on black absorbent material and heats the air. It rises and pulls cool air in behind it.
• The warm air is moved directly into the living space rather than into storage. Fig. 10-29 illustrates a window collector.

CONTROLLING PASSIVE HEATING

Buildings with passive solar heating systems need some methods of control to block out sunlight when it is not wanted. Also, at night some protection is needed to prevent heat losses to the cool night air. Fig. 10-30 shows how the path of the sun varies in summer and in winter. A large roof overhang will cut off direct rays of the sun in the summertime but a low winter sun can easily shine in. See Fig. 10-31.

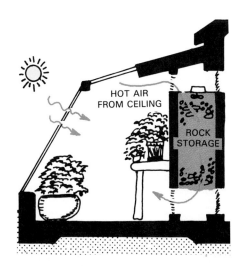

Fig. 10-27. Isolated gain solar heating may store heat in a rock bin.

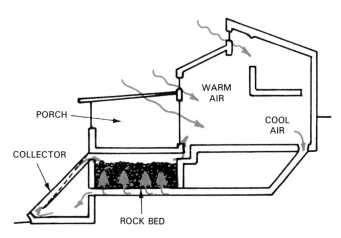

Fig. 10-28. Convective loop system depends on solar heat to drive warmer air or water to storage or living area. (U.S. Dept. of Energy)

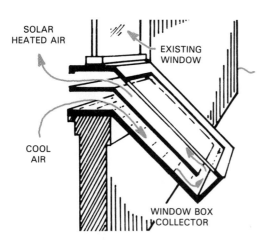

Fig. 10-29. Window flat plate collector uses thermosiphon principle. Heating of air in upper section of panel causes the air to rise, drawing in cool air after it through the lower panel.

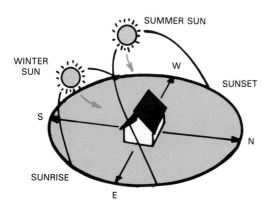

Fig. 10-30. Passive solar construction must take into account the seasonal path of the sun so that the dwelling takes advantage of the low winter sun but shades living space against high summer sun. (Illinois Institute of Natural Resources)

Greenhouses can be vented to release unwanted heat to the atmosphere. Movable vents can be opened or closed as necessary. At the same time, vents in greenhouse walls leading to living space, can be closed in warm months.

INSULATION AND SHUTTERS

Both direct gain and isolated gain designs require movable insulation and/or shutters. Spaces with large expanses of glass will lose heat to the cooler night air, Fig. 10-32. Insulating blankets and closures will help control this loss. See Figs. 10-33 and 10-34.

ACTIVE SOLAR HEATING SYSTEMS

Active solar heating systems are very different from passive systems. Passive systems depend upon natural

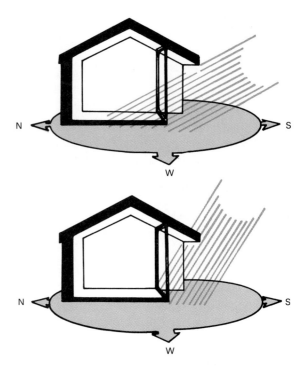

Fig. 10-31. Properly designed overhang on roof blocks summer sun and allows winter sun to shine in, heating living space. (Solar Energy Research Institute)

Fig. 10-32. Glazing, which readily takes in solar heat during the day, also readily loses heat through conduction to the cool night air. (Illinois Institute of Natural Resources)

radiation and convection to distribute the solar heat. Active systems depend on several mechanisms to distribute the heat. Moreover, whereas passive structures are usually part of the architecture of the building, active systems are separate from the supporting structure of the building.

PARTS OF AN ACTIVE SOLAR SYSTEM

The active solar energy heating system, Fig. 10-35, is designed to:
• Collect and trap solar rays.

SWINGING SHUTTER
WITH PULL DOWN CURTAIN

Fig. 10-33. Shutters and pull down curtains will help keep out unwanted solar heat.

RIGID INSULATION

Fig. 10-34. Insulated shades can be used to keep out unwanted heat during the day or keep indoor heat from escaping. (Illinois Institute of Natural Resources)

Fig. 10-35. South facing panels of an active solar heating system. Flat panel collectors are used to heat air which is moved away from the panels with blowers.

- Pick up heat from the collector and move it to where it can be used.
- Store heat for later use when the sun is not shining.

The various mechanisms which provide all these functions include solar collectors, piping or ductwork, fans, pumps, motors, heat exchangers, and storage bins or storage tanks. See Fig. 10-36.

SOLAR COLLECTORS

The portions of active solar heating systems which absorb the sunlight are called **solar collectors**. There are three basic types:

- **Flat plate collectors.** These are best suited to residential use.
- **Evacuated tube collectors.** These are more expensive and produce higher temperature heat.
- **Parabolic concentrators.** This type is capable of pro-

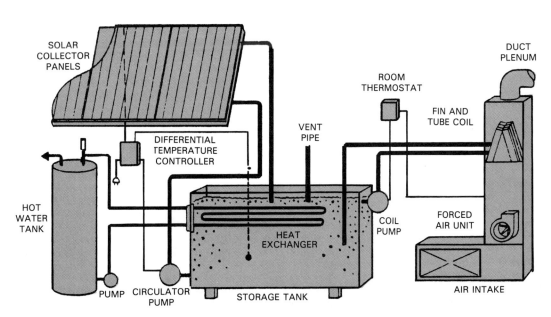

Fig. 10-36. Schematic of an active space and water heating system.   (Department of Energy)

ducing very high temperatures.

Basically, a flat plate collector is a closed, insulated box with a clear covering. The sun can shine into the box heating up the black interior, Fig. 10-37.

The surface of the solar collector which absorbs the heat of the solar radiation is called the **absorber**. It should be painted black or a very dark color. The texture of the absorber surface should be slightly rougher than an eggshell.

In systems using fluids to absorb the collected heat, the collector will have several fluid tubes running lengthwise through it as shown in Fig. 10-38. The sun heats either a plate that conducts heat to the tubes or the tubes are heated. Liquid flows through the tubes picking up the heat.

Coatings of lamp black or fine carbon powder produce a dark collector surface that absorbs infrared, ultraviolet, and visible rays of solar energy. Dark surfaces placed in the sun will always get much hotter than white or shiny surfaces. Dark colored objects will heat up to about 253°F (123°C).

A transparent cover serves as a trap for heat. Being clear, it allows the solar radiation to pass through but traps heat.

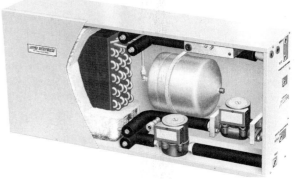

Fig. 10-38. Two components of a domestic hot water system. Top. Solar collector. Bottom. Transfer module. It contains the heat exchanger and the pumps for moving the collected heat to a hot water tank which is the third element of the system. (Lennox Industries, Inc.)

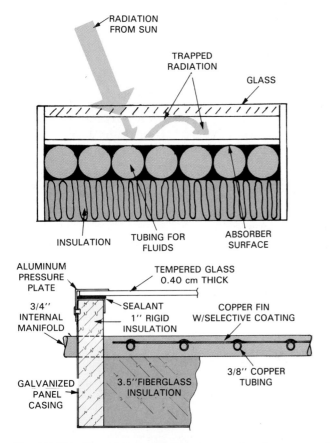

Fig. 10-37. Two cross section views of collector panels designed for a liquid system. Top. How heat is trapped and absorbed. Bottom. Construction details. (W.L. Jackson Mfg. Co., Inc.)

The cover material is usually glass. Glass is the preferred material because it permits more of the radiation to pass through it than other materials. Further, glass remains clear indefinitely. This is important since a clear material will transmit more radiation. If collectors are used in cold climates, double glazing may be used. The flat plate is the most commonly used collector. Fig. 10-39 shows a collector installed at an expressway rest area to heat water. Note also the photovoltaic cells at the top of the center collector panel.

EVACUATED TUBE COLLECTORS

In the evacuated tube collector, the absorber is a tube within a tube. The two are separated by a vacuum. There is almost no air to carry heat from the inner glazing to the outer glazing. Sunlight can enter and pass through the vacuum easily but heat does not move easily through a vacuum. This makes the evacuated tube collector very efficient. It is especially good in cold

Fig. 10-39. These flat plate solar collectors heat water at an expressway rest area.

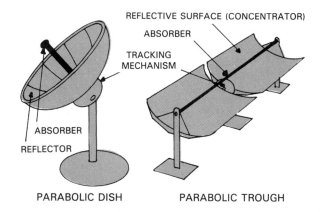

PARABOLIC DISH   PARABOLIC TROUGH

climates where the temperature difference between the hot absorber and the outside air is great.

The absorber tube is coated with a dark heat-absorbing coat of paint. Inside is an absorber plate of copper and fluid tubes to transfer the absorbed heat. Fig. 10-40 shows a simple drawing of the evacuated tube collector. It works best where temperatures of 200°F (93°C) or more are needed.

CONCENTRATING COLLECTORS

The concentrating collector gathers in solar rays and reflects them onto a very small absorber area. The absorber heats up faster and becomes much hotter. There is much less heat loss because the absorber has less surface area.

Concentrating collectors are easy to recognize. One type is shaped like a dish, Fig. 10-41. Fig. 10-42 shows two types of parabolic troughs. Both dish and trough

Fig. 10-41. This type of concentrating collector is called a parabolic dish. Top. Basic design of collector and name of its parts. (HUD) Bottom. View of a parabolic dish unit shows construction of the reflector and absorber.

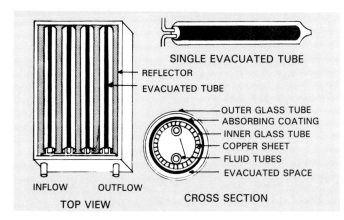

Fig. 10-40. Evacuated tube collector. Tube-within-a-tube surrounded by vacuum makes this design very efficient. Solar radiation can travel through the vacuum to the inner tube, but heat cannot escape to the atmosphere through the vacuum. (HUD)

are lined with a highly reflective metal coating. In the deep trough the light is refracted (bent) through a lens that works roughly like a magnifying glass. The amount of energy directed onto the absorber can be as high as 50 times that of a flat plate collector. Fig. 10-43 shows a trough collector used for domestic water heating.

There are several drawbacks to concentrating collectors:
• They will work only in direct sunlight.
• The focusing must be very exact. This requires an expensive tracking system that needs maintenance.
• Concentrators need heavy frames to control vibration. The weight of the frame and the dish or trough make it unsuited for many applications.

Both concentrator and the evacuated tube are more expensive than the flat plate collector. Few of them are being manufactured. The chart in Fig. 10-44 gives suitable applications for each type of collector.

APPLICATION	COLLECTOR
Agriculture and Pool Heating	Unglazed or Single-glazed* Flat-Plate (often plastic)
Space and Hot Water Heating	Flat-Plate (metallic)*, Evacuated-Tube, Non-Tracking Concentrator
Solar Air Conditioning	High Performance Flat-Plate*, Evacuated-Tube, Concentrator (tracking or non-tracking)
Industrial Process Heat (200-400 °F)	Evacuated-Tube*, Concentrator (tracking or non-tracking)
Industrial Process Heat (over 400 °F)	Tracking Concentrator
Solar Thermal Electric Power	Heliostats*, tracking Concentrator

*Most common choice
Note: All diagrams are descriptive illustrations, not working drawings.

Fig. 10-44. Suggested applications for different designs of collectors. (HUD)

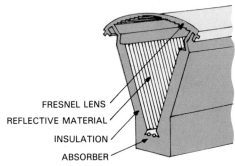

FRESNEL LENS
REFLECTIVE MATERIAL
INSULATION
ABSORBER

Fig. 10-42. Two types of parabolic trough collector. Top. Shallow trough. Bottom. Deep trough collector. Note absorber is in the floor of the trough. (HUD)

located on the roof, on a wall, or on the ground. Ducts carry the warmed air away from the collector and bring in cool air to replace it. A basic air system is shown in Fig. 10-45.

Heated air that is not needed immediately is stored. The storage medium is usually solid material such as a bin of rocks. The air is blown into the bin where it flows over and around the rocks. The hot air gives up its heat to the rock. At the top and bottom of the bin are empty sections for air movement. These are called **plenums**. Warmed air enters the bin through the top. Cooler air is drawn off the bottom and returns to the collector. During cloudy days and at night the system draws heat from the rock bin storage.

Fig. 10-43. Parabolic trough collectors on a house track the sun. Attached to the south exposure of a roof, they supply domestic hot water for the home.

TRANSFER SYSTEMS

Active solar collectors need devices to carry away the collected heat. Two systems are used:
• Air system.
• Liquid or hydronic system.

AIR SYSTEMS

Air systems use blowers and ducts to circulate air heated by the flat plate collector. The collector can be

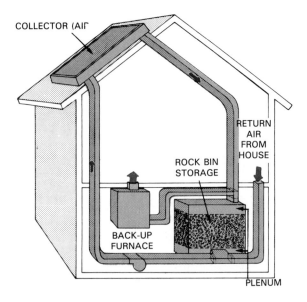

COLLECTOR (AIR

RETURN AIR FROM HOUSE

ROCK BIN STORAGE

BACK-UP FURNACE

PLENUM

Fig. 10-45. Diagram of an active solar collector. Air is used as the medium for transferring heat. (HUD)

Bins are usually constructed indoors at ground level or in a basement. They may also be buried in the ground outside the building. This is not recommended, however. Water may seep into the bin and ruin the insulation. Then too, it would introduce moisture-laden air into the building.

Rocks used for heat storage should be from 3/4 to 1 1/2 in. (19 to 38 mm) in diameter. River gravel is the usual choice. Other storage material such as cast iron, bricks, or ceramic blocks are sometimes used.

Rock bins can be constructed of either masonry or wood. They must be airtight to prevent heat leakage. Wood storage bins must be insulated to a value of between R-19 and R-44.

LIQUID SYSTEMS

A typical liquid system is shown in Fig. 10-46. The heat energy absorbed in the collector is picked up and carried away by water which circulates through tubing attached to the absorber element of the collector. Piping carries the heated water away and brings cool water into the collector to be heated.

The piping is connected with the heating system and with a water storage tank. From 1 to 2 1/2 gallons (4 to 9.5 L) of water are needed in the system for every square foot of collector area.

The tank used to store the water may be made of concrete, steel, or fiberglass. Steel tanks should be glass or epoxy lined. Because of its porosity, concrete should be rubber lined.

In cold climates, antifreeze solutions are used instead of plain water. Then a heat exchanger must be used.

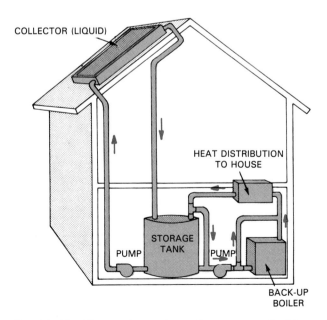

COLLECTOR (LIQUID)

HEAT DISTRIBUTION TO HOUSE

STORAGE TANK

PUMP

PUMP

BACK-UP BOILER

Fig. 10-46. This active solar collector uses liquid as the medium for heat transfer.

HEAT EXCHANGERS

A **heat exchanger** is a device used in a heating or cooling system. It moves or transfers heat energy from one medium to another.

The radiator of an automobile is one type of heat exchanger that is familiar to almost everyone. Heat generated by the operation of the engine transfers to water circulating through the engine's water jacket. As the water flows through the radiator, its heat transfers to the ambient air moving through the radiator.

Heat exchangers in a liquid solar system work the same way. There are several types. The most common are:
- Air to liquid.
- Liquid to air.
- Liquid to liquid.

Fig. 10-47 shows a simple sketch of a liquid to air and air to liquid exchanger. Fig. 10-48 illustrates a liquid to liquid exchanger.

PHASE-CHANGE STORAGE MEDIUMS

Experimental storage units using phase-change materials are being tested. You may recall from earlier discussion that these are materials which change from one state to another as they are heated or cooled. For example, butter is hard when it is cold, but becomes liquid when it is heated.

When used in a heat storage system, phase-change materials are generally contained in tubes or trays. Containers must be tightly sealed so the material does not leak out, transfer moisture, or dry out. Large numbers of the storage units are needed to provide a large surface area for heat exchange. This makes phase-change materials very expensive. Fig. 10-49 compares the heat storage capabilities of water, rock, and phase-change materials.

SOLAR HOT WATER HEATERS

The most popular and simplest solar energy systems are those used for heating water. The basics of the systems are the same as described earlier in this chapter for space heating. The solar hot water heater, however, is much simpler. There are three systems:
- The direct open loop system. This is a passive system and is described under the passive solar section of this chapter.
- The direct pumped system.
- The indirect (closed loop) system.

In the direct pumped system the solar heated water is emptied directly into the supply tank of a conventional hot water heater. Cold water from the house

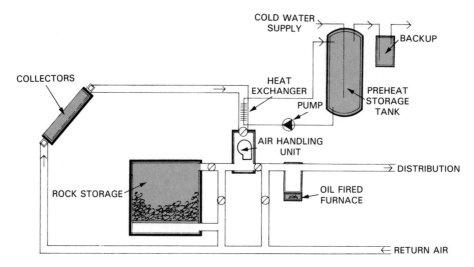

Fig. 10-47. Some solar collector systems use heat exchangers so the transfer material and the medium being heated remain separated. This is a diagram of an air-to-liquid heat exchanger. (HUD)

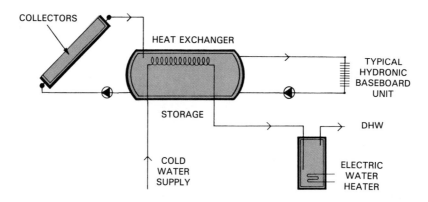

Fig. 10-48. Schematic of a liquid to liquid heat exchanger.

STORAGE MEDIUM	ADVANTAGES	DISADVANTAGES
Water	—least expensive for retrofitting because of compactness	—possible corrosion —some loss of efficiency because of the need for heat exchangers when nonfreezing liquids or corrosion inhibitors are used —leakage can be destructive
Rock	—no heat exchanger needed between collectors and storage —leakage, though not good for efficiency, is not destructive	—location of any air leakage difficult to detect —retrofitting expensive due to large size storage bin and of ducts
Phase-change materials	—very compact storage and good for retrofitting	—expensive container/heat exchanger —long term reliability not proven

Fig. 10-49. Comparison of three materials used for heat storage in a solar heating system.

supply replenishes the water supply in the solar collector, Fig. 10-50.

An indirect system gets its name from the fact that the solar collector does not directly heat the water you use. The fluid heated by the collector never comes in contact with the hot water. The solution is in a closed loop. The loop runs through both the collector and the supply tank or the hot water heater. The heat is

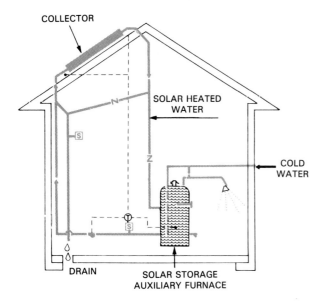

Fig. 10-50. In a direct pumped system, a pump moves water through the collector on the roof.

transferred through a heat exchanger. The exchanger is usually a coil of tubing inside the water supply tank. As the heated fluid passes through the coil it gives off its heat to the water in the tank. Fig. 10-51 shows a typical closed loop, indirect system.

HEAT PUMPS AND SOLAR ENERGY

The heat pump is a machine that is designed to draw heat from one source and release it at another place at a higher temperature. It can be connected to solar heating systems to extract more heat from a solar heat system.

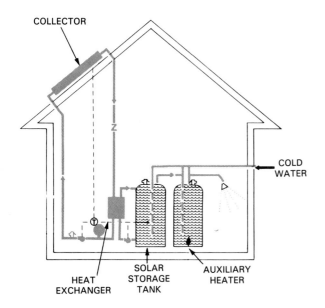

Fig. 10-51. The heated fluid coming from the roof collector on this system passes off its heat to the hot water supply through a heat exchanger.

The pump draws heat from the tank through the transfer fluid and can release it in a room at a much higher temperature. By turning a valve, the process can be reversed so the heat will be pumped from the room to cool it. The heat is then passed off to the water in the tank where it can be used as hot water supply. Fig. 10-52 is a diagram of a solar-assisted heat pump system.

HIGH TEMPERATURE SOLAR ENERGY

The heating of buildings is done with solar-generated heat at a low temperature. High temperatures are hard, but not impossible, to get through solar radiation. It requires the use of a solar furnace. The furnace gathers the solar energy from a large area and concentrates the rays in a very small area. This is done with an arrangement of many mirrors or heliostats to catch the sunlight and reflect it into a special collector.

The principle of high-temperature solar collection has been known for a long time. However, no attempt was made to develop solar furnaces until a few years ago. The first large furnace was designed and constructed at Mountlouis, France in 1952. Later, the French built a second one at Odeillo. A schematic of this installation is shown in Fig. 10-53.

Sixty-three mirrors are arranged on a terrace opposite the target area. The mirrors are movable so they can track the sun as it moves through the sky. The sunlight strikes the mirrors and is reflected into a 130 ft. high parabolic collector. It, too, is made up of small mirrors. Because of the shape of the collector, the reflected sunlight is pinpointed on a small part of the target where the boiler is located, Fig. 10-54.

The 9500 mirrors making up the concentrator are flat. Each must be focused by hand so that the area in which the reflected sunlight is gathered measures only about 17 cm (about 7 in.) in diameter. About 270 kW of energy are concentrated in this area.

No commercial use has yet been made of solar furnaces. They are still in the development stage. The Mountlouis furnace became the model for others. These were built at:
- Natick, Massachusetts (later moved to the Sandia Laboratories at White Sands, New Mexico).
- Sendai, Japan.
- Odeillo, France.

If the solar furnace proves successful, the heat produced could be used to create steam for generating electric power. The steam might also be used as process heat in manufacturing.

One of the chief disadvantages of the solar furnace is the need to collect the solar energy from a large area. This would require hundreds of heliostats covering many acres of ground.

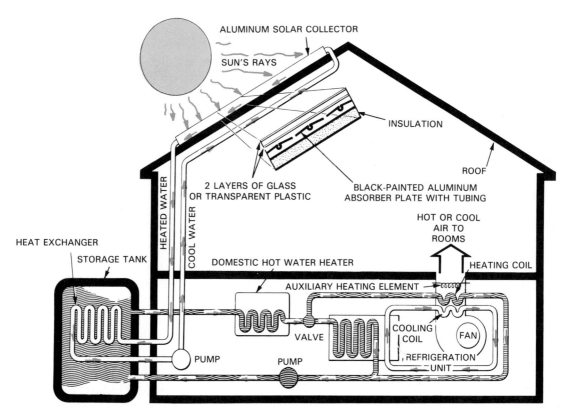

ALUMINUM SOLAR COLLECTOR

SUN'S RAYS

INSULATION

ROOF

2 LAYERS OF GLASS
OR TRANSPARENT PLASTIC

BLACK-PAINTED ALUMINUM
ABSORBER PLATE WITH TUBING

HEAT EXCHANGER

STORAGE TANK

HEATED WATER

COOL WATER

HOT OR COOL
AIR TO
ROOMS

HEATING COIL

DOMESTIC HOT WATER HEATER

AUXILIARY HEATING ELEMENT

VALVE

COOLING
COIL

FAN

REFRIGERATION
UNIT

PUMP

PUMP

Fig. 10-52. Because heat pumps work better with warmer temperatures, they are often teamed up with solar collectors. In such systems, the heat pump collects its energy from a hot water storage tank.

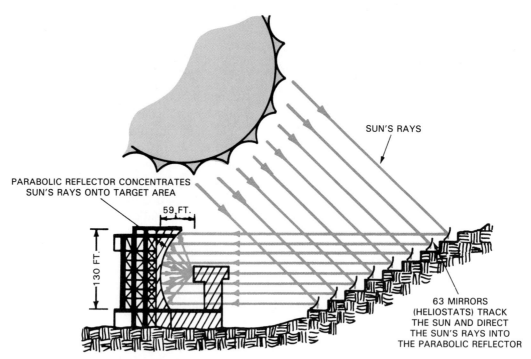

SUN'S RAYS

PARABOLIC REFLECTOR CONCENTRATES
SUN'S RAYS ONTO TARGET AREA

59 FT.

130 FT.

63 MIRRORS
(HELIOSTATS) TRACK
THE SUN AND DIRECT
THE SUN'S RAYS INTO
THE PARABOLIC REFLECTOR

Fig. 10-53. Simplified sketch of a solar furnace. The mirrors reflect sunlight back onto a parabolic dish concentrator. It pinpoints all the light to a target area that holds a boiler.

BLACK BODY RADIATION

Black body radiation is reverse radiation that occurs at night. The earth and other substances which have absorbed heat during the day, give up heat to the atmosphere. That is why in winter, for example, clear nights are cold. There are no clouds to prevent the escape of solar heat.

Fig. 10-54. This is a model of a parabolic dish collector. This type can be used for a solar furnace. The curved surface of the dish is made up of many individual mirrors.

a white aluminum plate inside a tub which rested on the roof of their laboratory. A clear plastic sheet covered the tub. When the temperature at the plate was on the freezing mark, 32°F (O°C), on a clear night, the temperature at the film was -22°F (-40°C). Scientists are now working to see if black body radiation can be used to refrigerate food and medicine in areas where there is no electric power.

The process has already been successfully used in South America to purify water. Chilean scientists experimenting with black body radiation froze water at night to separate it from undesirable minerals such as boron and arsenic. They were able to produce water suitable for drinking and for irrigation.

SOLAR CELLS

Generation of electricity directly from solar energy was possible as early as the 19th century. Little use was made of the technology, however, because the devices produced only small amounts of electricity. Chapter 1 introduced the solar cell, also known (more accurately) as a **photovoltaic cell** or a PV cell. See Fig. 10-55. The solar cell is a small crystal of silicon that has been "doped" with other elements such as boron and phosphorous. Such cells can convert light energy into electrical energy. Electrical energy is the motion of electrons through conducting materials.

A solar cell is basically a diode. As was discussed in Chapter 7, a diode allows free electrons to move easily in one direction while making movement difficult in the opposite direction. Solar radiation striking the cell heats the material by jostling the atoms of the semiconductor material. With addition of enough heat,

This principle can be used to provide cooling where electric power is not available. It has been used to keep buildings cool. The palace of the sultan of Zinder in Nigeria, Africa was said to be constructed for this purpose. It was built around a central patio and during the nights the roof radiated heat to the atmosphere while cold air was pulled into the patio and the rooms under the roof to keep the house cooler in the heat of the next day.

French scientists have experimented with this principle for many years. In one experiment they placed

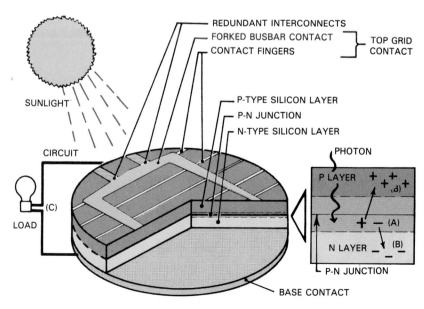

Fig. 10-55. This simplified sketch shows how a solar cell is constructed and how it works.   (Solar Power Corp.)

electrons leap free and move into the cells conduction band. If the cell or a group of cells are connected into a circuit, the electrons will flow through the circuit to operate electrical or electronic devices.

MANUFACTURING THE SOLAR CELL

Solar cell technology has developed slowly, but significant advances have been made in recent years. Some of the first successful cells cost about $7000 a square meter. Their efficiency was about 11 percent.

Several methods of manufacture are used: single crystal silicon, amorphous silicon, polycrystal silicon, and ribbon silicon. The two main types of cell are the crystalline silicon, and the amorphous silicon. Both work well out of doors although the amorphous cells have been widely used on calculators and watches because they respond to fluorescent light better than to natural light. Crystalline cell production was the first to be developed.

This method is slow and expensive. The basic material for the process is crystalline silicon. The raw material is first melted. Then the molten material is slowly drawn from the crucible (pot), forming a single crystal in the shape of circular rods or ingots. The crystal, which is very hard, has to be sliced into thin discs 1/25 in. thick. Since silicon is very hard the cutting has had to be done with a diamond saw. Most of the precious, costly material is wasted since the saw cut is thicker than the resulting discs.

After being cut the discs must be polished. Then,

since silicon is a poor conductor of electricity, impurities (such as boron and phosphorus) are added while the cells are reheated. The impurities give one side of the wafer a positive charge while the other side gets a negative charge. A thin area in the middle of the wafer is not touched by the impurities. (What has just been described in this paragraph is a basic diode.) The positive side of the cell is called the P layer. The negative side is called the N layer. The neutral area is known as the **PN junction**. The junction does not conduct electricity; it insulates the P and N layers from each other.

ETCHING

Solar cells also require an etching process. This process applies a metal grid or coating. This metal coating allows the electrical charges to travel out of the cell when conducting wires are attached to the cell.

Individual cells are connected to other cells in series (positive terminal to negative terminal) as shown in Fig. 10-56. A group of interconnected cells is called a **module**. Any number can be connected to get the desired voltage.

Modules are housed in a protective base and covering. A rigid frame provides additional protection as well as a means of mounting the module, Fig. 10-57.

COST AND EFFICIENCY IMPROVEMENTS

The manufacture of solar cells has been expensive. In 1974 the price of a solar cell module was $50 per

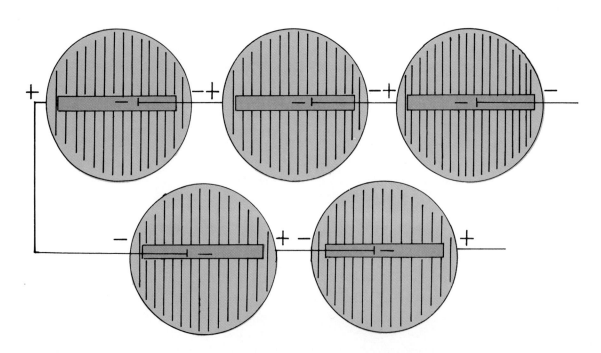

Fig. 10-56. Diagram of a solar cell array connected in series. Total voltage will be the sum of the individual cell voltage.

Fig. 10-57. Solar cells are combined into modules called arrays. The cells must be enclosed to protect them from impact and corrosion. Strings of connected cells are laminated or sandwiched between sheets of tempered glass or plastic and a white-coated metal backing. (PHOTOCOM, INC.)

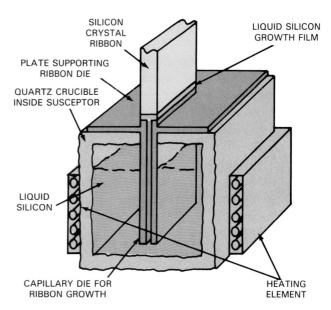

Fig. 10-58. This is a simplified drawing showing the edge-defined film-fed method of "growing" silicon solar cells.

watt of peak output. **(Peak output**, sometimes called peak power, is the maximum output of a photovoltaic cell at noon.) This output is used to indicate the size of a photovoltaic power plant. The same module today costs about $5 per watt of peak output. In 1988 the Sandia National Laboratories in Albuquerque, New Mexico developed a two-layer solar cell that gave an efficiency of 31 percent. An upper layer of gallium-arsenide converted 27.2 percent of the light striking it into electricity; unabsorbed light passed through to the lower layer of silicon crystal which accounted for an additional 3.7 percent of efficiency.

It is predicted that in time the cost per kilowatt will drop below 50 cents a kilowatt. This is still much more expensive than the cost of electricity produced by nuclear or fossil-fueled systems. Still, experts believe the year 2000 will see the cost of solar cells further reduced to the range of 10 to 30 cents per kilowatt.

Experiments and research continue looking for more efficient and cheaper solar cells. Among the alternative manufacturing processes are:

• Amorphous silicon cell development that produces the wafers (discs) in sheets that do not need to be sawed. The process is faster and produces only minimal waste, Fig. 10-58. If solar cells can be manufactured this way, they are certain to be more competitive in the energy-producing market. While the efficiency of the cells made by this method is low, their cost is much less. Another advantage of amorphous silicon manufacturing technology is the ability of the process to produce larger cells. Several companies are now making commercial modules that are from a foot to 4 feet square.

• Thin-film solar cells. In this process, silicon in the form of a vapor or liquid is deposited on a base such as glass plates or foil. Though less efficient at 4 to 6 percent than crystal silicon cells, thin-film cells are cheap. Thinner by up to 10 times than crystalline cells, they can be stacked one on top of the other to improve efficiency.

• Concentrators can be used to increase efficiency. **Concentrators** are focusing lenses or mirrors that intensify the sunlight on the solar cells. While this causes heat buildup that reduces efficiency, the heat can be siphoned off and used for other purposes. Concentrators are expensive since they require tracking systems to keep them aimed directly at the sun.

• Ribbon cells. Silicon is drawn out into thin ribbons and cut with a laser to reduce waste. The rest of the process is similar to that used for producing crystalline cells.

HOW SOLAR CELLS WORK

Sunlight falling on the solar cell delivers bundles of energy called **photons**. As they strike the solar cells they dislodge electrons. The electrons flow through the grid and can move out of the module through the circuit conductors. It is thought that one photon of solar energy frees one electron.

Only a very small amount of current can be gotten from a single solar cell, but, when many cells are connected, considerable amounts of power can be obtained.

An array of solar cells is sometimes called a solar battery. It does function much like a battery, but there is one big difference. The battery will eventually wear out while the solar cell will produce current indefinitely as long as the sun shines.

APPLICATIONS OF PHOTOVOLTAIC CELLS

As their price has dropped, solar cells have become practical enough for more general use. Their first use, to provide electric power in remote areas where other sources are not available continues to be important. Besides the space program, Fig. 10-59, where solar cells supply electricity 24 hours a day for satellite operation, this exciting power source is currently used for:

• Portable systems for water purification.
• Operating small battery chargers.
• Telephone communications.

The other popular, universal application of the solar cell is in photography. Small units are found in many light meters and cameras to indicate or control the proper exposure of film.

Newer applications have followed the drop in price of the cells. These include powering of:

• Water pumps.
• Signals for automobile and railway traffic.
• Lightweight generators that provide solar-powered electricity for boats and land vehicles such as golf carts. See Fig. 10-60.

An interesting and exciting experimental application of the solar cell was its successful use to power flight. In December, 1980, Solar Challenger, a specially built aircraft weighing but 185 lb. (84 kg), carried a pilot safely over a long distance using only sunshine for

Fig. 10-59. Solar arrays provide electrical power to operate electronic equipment in space satellites. (NASA)

power. The plane made about 45 flights. The longest was 18 miles, Fig. 10-61.

The Northeast Sustainable Energy Association, located in Greenfield, Massachusetts, promotes interest in solar energy with its organization of an annual solar-

A

B

C

D

Fig. 10-60. Solar cells can supply electric power for a number of purposes where large amounts of electricity are not required. A—Billboard lighting. B—Flashing light for a school speed zone. (Bill Roush, Solar Electric Systems of Kansas City, Inc.) C—Cabin lights on a river boat. (Hank Meels) D—Lighting for an RV.

Fig. 10-61. First solar flight was made by Solar Challenger in 1980. Weighing but 185 lb., the craft had 15,000 photovoltaic cells in its wings and tail. It was developed by a design team headed by Dr. Paul McCready, a California scientist.

powered car race. The American Tour de Sol, an annual race from Albany, New York to Plymouth, Massachusetts, is designed to attract young people into solar engineering. See Fig. 10-62.

An experimental PV cell power station has been operating in California since 1983. The PV cells are mounted on motor-driven trackers. Fully automatic, it turns itself on in the morning and off at night and will provide electricity for about 10,000 people when all units are operating.

SATELLITE PHOTOVOLTAIC SYSTEMS

As early as 1968 scientists discussed the possibility of orbiting satellite solar power stations. They would consist of huge arrays of photovoltaic cells to generate

A

B

C

D

Fig. 10-62. Considerable attention has been directed to use of solar energy in transportation. A—Solar powered automobile exhibited at the 1982 World's Fair. B—One of the entries in the 1992 Tour de Sol competition from Albany, New York to Greenfield, Massachusetts. C—Winner of the American Tour de Sol in 1990 collecting energy at one of the stopover points. (Mark Morelli) D—The GM Sunraycer was the winner of 1950-mile World Solar Challenge race from Darwin to Adelaide, Australia, completing the run in an elapsed time of 44 hours and 54 minutes. (GM Hughes Electronics) electricity and microwave converters with antennae to beam the electric power to earth stations. The satellites would be placed in geosynchronous orbit 22,245 miles (35 800 km) above the earth. Being geosynchronous, it would be traveling fast enough to stay in constant touch with its earth collector station.

From their positions high above the earth the satellites would have sunlight 24 hours a day. Occasionally the sunlight would be blocked because of eclipses but only for short periods before and after the equinoxes (change of seasons).

The solar panels would be arranged to face the sun continuously. The antennae would be designed to always point toward earth.

SUMMARY

While indirect solar energy provides a large part of our usable energy in the world, comparatively little use is being made directly of solar energy. While it is not certain how the sun came into being, there is a theory that a huge ball of energy became matter. Hydrogen and helium atoms formed creating the sun.

Solar energy travels through space theoretically through magnetic radiation. These waves of radiation are somewhat like the waves created when a pebble is tossed into still water. Even though only one-two-billionth of the sun's total energy strikes the earth, annual solar radiation reaching us is huge.

Solar energy is presently in limited use to heat buildings and produce electricity. It has been demonstrated that it can be used to drive vehicles such as cars and planes. While the electrical output of solar cells is low and costly compared to nuclear and conventional electric power systems, new technology is reducing costs to the point where, by the year 2000, solar power stations will be probable.

While solar energy is virtually free, the technology to capture and utilize it is expensive. Technology is developing, however, to the point where solar energy can compete with fossil-fuel systems.

DO YOU KNOW THESE TERMS?

absorber
calorie
change of state materials
concentrator
diffused radiation
direct gain solar heating systems
electromagnetic radiation
electrons
evacuated tube collector
flat plate collector

frequency
greenhouse effect
heat exchanger
insolation
langley
module
nuclear fusion
nucleus
parabolic concentrators
peak output
phase-change materials
photons
photovoltaic cells
plenum
PN junction
pond
radiation of energy
solar collectors
specific heat
sun
thermosiphoning
volumetric heat capacity
watts
wavelength

SUGGESTED ACTIVITIES

1. Construct a device or toy capable of converting solar radiation to heat.
2. Construct a device which can concentrate the sun's rays on a small area.
3. Build a device which can convert solar radiation to motion.
4. Build, assemble, or demonstrate a device which can convert solar radiation to electricity.
5. Build a device that uses the sun to purify water.
6. Survey your community and report on the number and types of devices which use the sun to save energy. Select one of the types and write a report on how it works. Discuss with the owner the savings experienced in energy.

TEST YOUR KNOWLEDGE

1. Light and heat coming from the sun is caused by _____ taking place at its core.
2. If the sun is 93,000,000 miles away and light travels at 186,000 miles per second, how long would it take the light to reach the earth?
3. Describe how electromagnetic waves travel.
4. List the types of electromagnetic waves. Which are visible?
5. Solar constant is all of the radiation falling on an object inside the earth's atmosphere. True or False?

6. The average insolation of (amount of energy reaching) the earth is _____ kWh per sq. meter per year.

7. The greenhouse effect is caused by: (select correct answer)
   a. Glass protects against the effects of cold winds; therefore it is warmer behind glass.
   b. Solar radiation passes readily through glass but heat energy does not readily pass through.
   c. Glass magnifies the intensity of solar radiation.

8. _____ _____ is the amount of heat, in Btus, a substance can hold when its temperature is raised one degree F.

9. What are phase-change materials? Of what use are they to solar energy?

10. On a clear day, the sun delivers about _____ Btu per sq. ft. of radiant energy to the earth's surface (25, 250, 2500).

11. In an isolated gain type of solar heating system, heat is stored and is drawn off as it is needed. True or False?

12. List the three basic elements of an active solar heating system.

13. List three areas where solar cells are used today.

14. At present, the range of efficiency of solar cells is from _____ to _____ percent (5 to 10, 16 to 25, 18 to 42).

A

B

Fig. 11-1. Mechanical or kinetic energy can be produced by converting other forms. A—An internal combustion engine converts fuel (chemical energy) to produce mechanical energy. B—The engine's mechanical energy powers a land vehicle. (Ford Motor Co.)

11 CHAPTER

KINETIC ENERGY

The information given in this chapter will enable you to:
* List the sources of kinetic energy.
* List five sources of kinetic energy in nature which can be harnessed to produce power.
* Describe how wind is created and where prevailing winds are located.
* Explain the nature of wave energy.
* Discuss the origin of tides and list likely sites for tidal power stations.
* Discuss how the various forms of natural kinetic energy can be converted to electricity or stored forms of energy.

As you have already learned in Chapter 1, kinetic energy is the energy of matter in motion. It is also called **mechanical energy**. Kinetic energy is made available as a result of:
* Conversion of other sources of energy.
* Harnessing of a natural source.

Gasoline is a good example of an energy source converted to produce motion. The gas engine burns the fuel to move a piston. The motion of the piston may move wheels as on an automobile, Fig. 11-1. A similar but smaller engine could turn a cutting bar on a lawn mower, Fig. 11-2.

However, this chapter is really more concerned with the harnessing of natural kinetic forces. These forces are renewable.

KINETIC ENERGY IN NATURE

Most forms of kinetic energy found in nature are the effect of solar radiation. Therefore, we can call them indirect solar energy. Indirect solar sources include the motion of:

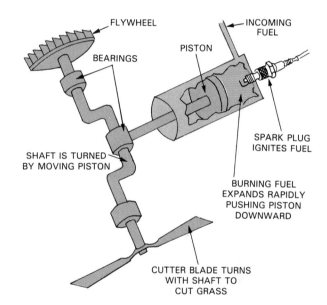

Fig. 11-2. It is easier to see how a lawn mower converts fuel to drive a blade.

* Wind, Fig. 11-3. Heating action of the sun creates air motion which can sometimes reach great force.
* Waves, Fig. 11-4. Winds drive water with great energy.
* Tides. Gravitational attraction of the moon and the sun cause the levels of seas to rise and fall on a regular basis.
* Falling and flowing water, Fig. 11-5. Rain falls on higher ground and flows to lower levels as the result of the gravitational attraction of the earth.

WIND ENERGY

Wind is air in motion. All winds, from the gentlest breeze to the awesome force of a hurricane, are caused by three natural events:
* Differences in temperature of the atmosphere.

Fig. 11-3. Wind is kinetic energy resulting from air in motion. It cannot be seen but its effects are visible.

Fig. 11-4. Waves have considerable energy that is dissipated as they pound the shore.

- Rotation of the earth.
- Unequal heating of the oceans and land.

ATMOSPHERIC TEMPERATURE AND AIR PRESSURE

Air is extremely light. However, it has weight and can expand or compress like other gases. One cu. ft. of air (0.28 m³), on a warm spring day, weighs about 1.22 oz. (35 g).

When air is heated by the sun it starts to move. First it expands and becomes lighter. Then it rises. Cooler air, being heavier sinks and moves in under it, Fig. 11-6.

One of the reasons this happens is that when warm air rises it causes a lower pressure (partial vacuum) underneath. The colder air around it is at a higher pressure and so it moves into the space the warm air has left. Thus, when atmosphere at a certain part of the earth is made warmer than atmosphere at other parts, the colder air moves into the warmed region. This movement creates wind.

World-wide, there are great differences in temperature. For example, the tropics can have

Fig. 11-5. Rivers collect water from higher ground and deliver it downstream. An abrupt change in water level or ground level results in a waterfall that can be used to provide usable energy.

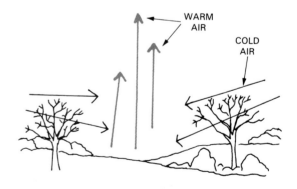

Fig. 11-6. Heated air expands, grows lighter, and rises. Low pressure below the rising air causes cold air to move in from any direction, creating wind.

temperatures above 100°F (38°C) while in the polar regions the temperature is well below freezing.

The differences in temperature between these regions are mainly the result of the angle at which the sun's rays strike the equator and how they strike the polar regions. At the equator the sun's rays fall on the earth from almost directly above. At the poles, however, the sunlight strikes the earth at such an angle that the heat is not as intense nor absorbed nearly as well. As Fig. 11-7 shows, sunshine striking the earth at right angles is better absorbed by the earth's surface. When the sunlight strikes earth at the poles, most of the heat is reflected back into space.

TEMPERATURE'S EFFECT ON WIND

The great differences in temperature between the equatorial zone and the polar regions give rise to massive movements of air. Hot, light air from the tropics rises and flows toward the poles. At the same time, colder, heavier air from the poles moves under the warm air toward the equator. Fig. 11-8 shows this. This mass movement of air is called a **planetary wind**.

WIND FROM EARTH'S ROTATION

Air movement, caused by uneven heating of the earth, is strongly affected by earth's rotation. The speed of rotation is about 1000 mph (1600 kmh) at the equator and 0 mph at the poles.

Strictly speaking, rotation of the earth does not create wind, but it does deflect wind caused by other factors. If the earth did not rotate, the colder winds from the north and south poles would blow directly

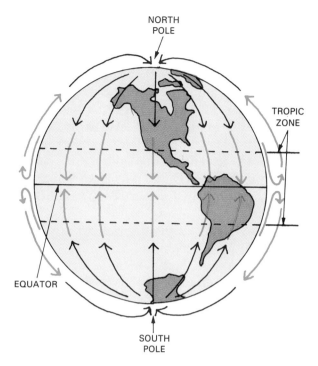

Fig. 11-8. Uneven heating of the earth by the sun creates world-wide wind patterns. Low-level winds flow toward the equator from north and south.

toward the equator. Warmer air would rise at the equator and flow both north and south toward the poles.

Because of the rotation, air moving toward the equator is skewed to the west. Air moving toward the poles is turned to the east. See Fig. 11-9.

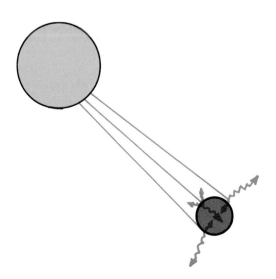

Fig. 11-7. Why some areas of earth warm up more than others. The earth's surface absorbs direct rays better than those rays striking it at a glancing angle. Much of the heat is reflected into space.

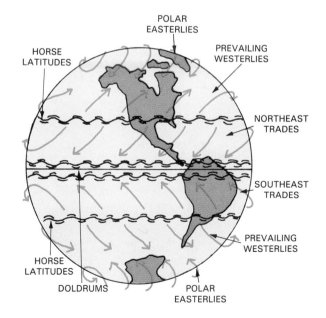

Fig. 11-9. The earth's rotation deflects planetary winds from the poles. This pattern results in bands of wind and bands of calm. Horse latitudes and doldrums are areas of little or no wind.

WIND PATTERNS

Look at Fig. 11-10. You will notice that there are three distinct areas of wind patterns both north and south of the equator. These areas are also called **cells**. At the equator is the tropical cell. It extends about 30 degrees both north and south. Air at the earth's surface in this cell is rising and is always hot and sultry. There is very little wind, but light breezes may come from any direction. The sky is overcast while showers and thunderstorms are frequent. Air pressure is always low.

North and south of the tropical cell are the **horse latitudes**. These are high pressure belts where the air is sinking to the surface. The air is dry and warm. Winds are light and changeable or there may be no wind at all for several days. Most of the world's deserts are in the horse latitudes. (The name "horse latitudes" came from a practice of crews of sailing ships. During the time when colonies were being established in America and other parts of the western hemisphere, ships were often becalmed by the lack of wind. Horses were thrown overboard to conserve the supply of drinking water.)

TRADE WINDS

Trade winds are planetary winds. They move toward the equator. Their name comes from the old meaning for "trade" which is a "steady track." The trades are the most constant of winds for they often blow for days or weeks with but slight change in speed or direction.

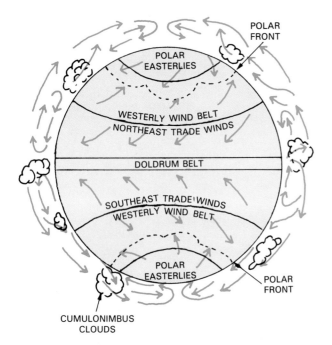

Fig. 11-10. There is a predictable general pattern of winds for the whole world.

EFFECT OF LAND ON WIND

Because land and water heat and cool at different rates, the size and shape of land masses have a great effect on the direction and force of winds. Land masses gain and lose heat faster than large bodies of water. Ocean waters are always being stirred up so that heat losses and heat gains at the surface are spread out over a large volume of water, Fig. 11-11. Therefore, the actual changes in temperature of the water surface are not as great. The temperature differences between day and night and winter and summer are greatly reduced.

On land, however, not much heat penetrates into the soil. The heat that does enter the earth's surface does not become distributed through the soil to any great depth. As a result, what little heat the earth's surface absorbs during the day is quickly lost during night and during the winter months. The daily and annual difference in temperature of land masses is much greater than that of bodies of water.

LOCAL WINDS

Other winds which we encounter from day to day are caused by local temperature differences and topography (lay of the land). These are called **local winds**. We all know about the sea and land breezes that blow around seacoasts and lake shores. During the day the air above the land warms up much more rapidly than the air above the water. The heated air over the land rises while cooler air from above the water moves inland, Fig. 11-12. At night, the land cools more rapidly cooling the air above it. Meanwhile, the air above the water is warmed by the stored heat in the water. The air movement reverses. Cooler land air moves in under the rising warm air above the water, Fig. 11-13.

In hilly and mountainous regions, winds are created by unequal heating of the sides and bottoms of the valleys. Down-slope winds occur on clear nights because the air on the upper slopes cools and becomes heavier than the air below. It therefore slides into the valley. See Fig. 11-14.

Some local winds occur so often and blow so regularly that they are given names. For example, the name **mistral** is given the strong cold wind that blows down the Rhone Valley in France.

When wind blows across a range of mountains the air on the leeward (downwind side) is compressed and heats up. It then blows across the plain as a dry hot wind. An example of this kind is the Santa Ana wind of southern California. It starts in the high desert and spreads across the coastal plain as a dry, dusty wind. Such a wind is also known as a **foehn**.

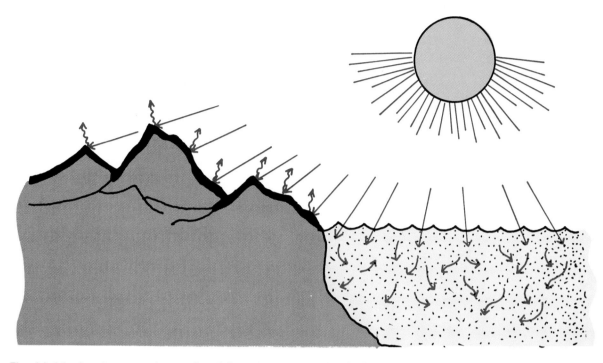

Fig. 11-11. Land masses do not absorb heat to any great depth. Through thermal currents, the heat absorbed by water is distributed to a greater depth. Thus, water can store more heat and releases it more slowly.

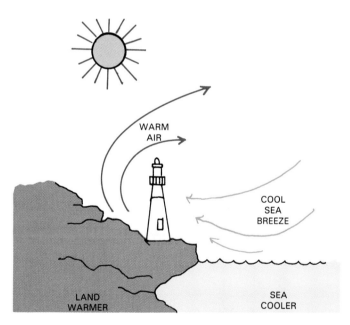

Fig. 11-12. Because land masses heat faster than water, heated air rises over land mass during the day. Local winds are created as cooler sea or lake air moves over the shore to replace rising air.

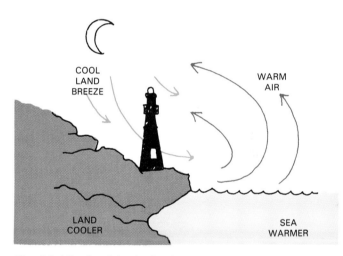

Fig. 11-13. At night the land mass is cooler than the water. Cooled air moves out over the water replacing the rising warmer air.

NORTH AMERICAN WIND PATTERNS

The Rocky Mountains, which run north and south along the west coast of the United States, Canada, and Mexico, stop most of the strong westerly winds from the Pacific. With no sea winds, inland areas of the United States get cold air moving in from the north pole and warm, moist air from the south. In the Midwest, the prominent winds come from the northwest during winter and from the southwest during summer. See Fig. 11-15.

AMOUNT OF WIND ENERGY IN U.S.

About 2 percent of the solar radiation falling on the earth produces all the earth's wind energy. The total amount of wind energy available in the lower 48 states of the U.S. is about 14 times the energy used by the United States in 1973. About 30 percent of the wind

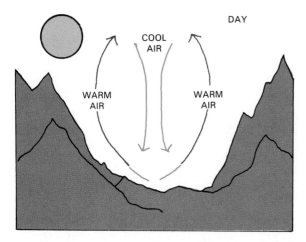

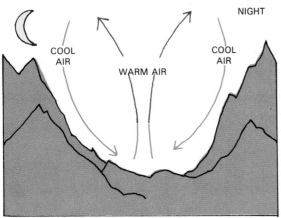

Fig. 11-14. Mountain and valley breezes are mainly updrafts and downdrafts. These are caused by heating and cooling of the land surface.

is high above in the atmosphere where it cannot be reached and used. However, there is still a large amount of wind at and below the 100-ft. level.

TYPES OF WIND

According to weather statistics, there are only a few days a month in most parts of the world when there is no wind. Some, winds, however, are not useful for producing energy. You can classify winds by their ability to produce energy.

• Energy winds.
• Prevalent (prevailing) winds.

Energy winds are those which blow at a speed of 10 to 25 mph (16 to 40 kmh). This type may be available on an average of two days a week. Prevalent winds (from the word prevail) are weak. They travel at about 5 to 15 mph (8 to 24 kmh).

Strong energy winds are found mostly in the temperate and polar regions. Weak or prevalent winds are found in the tropics.

WAVE ENERGY

When the wind blows over large bodies of water such as lakes or oceans, it drives the water before it in waves. The water moves in huge ripples toward land with great energy. The energy is given up as the water dashes against a shore, Fig. 11-16.

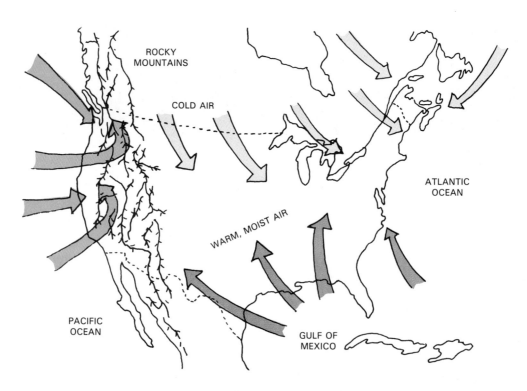

Fig. 11-15. Prevailing wind patterns in the United States. Mountains turn away strong west winds from the Pacific. Cold air sweeps down across the central plains from the Arctic. Moist warm air moves northward from the Gulf of Mexico.

Fig. 11-16. Waves are the result of wind blowing across a large expanse of water.

PROPERTIES OF WAVES

All waves, whether traveling in air or water, have similar properties. One of these is amplitude. This is the maximum height of the wave. In water, the amplitude is the maximum height of the crest of the wave above the still water level.

Another property is called wavelength. This is the distance from one crest to the next one. Still another is velocity or the speed at which the waves are traveling. There are other properties but these are the ones which are important to the harnessing of the energy of ocean waves.

TIDES

Tides are a periodic change in the water levels of seas, oceans, and large lakes. When the water level is high it is known as high tide. When the water is at its lowest level it is called low tide. There is some tidal activity in all large bodies of water but some is so slight that it is hardly noticed.

CAUSES OF TIDES

Rise and fall of tides happens daily on a predictable schedule. Time of a high tide at any place can be established by adding 50 1/2 minutes to the time of high tide for the previous day, Fig. 11-17. This delay coincides (happens at same time) with the daily change in moonrise. It is easy to see, then, that the moon must be the cause. More specifically, it is the gravitational attraction of the moon for the earth.

This gravitational attraction "pulls" at the earth and actually causes a slight bulge in the earth's surface nearest the moon. Water reacts more dramatically to the pull and, thus, collects at the spot where it is nearest the moon. Another bulge forms at the same time on the opposite side and produces a smaller high tide there, Fig. 11-18.

The sun also has some effect on the tides. Its gravitational pull is not as strong, however, because of the greater distance to the sun.

Fig. 11-17. The level of large bodies of water rises and falls daily because of the gravitational pull of the moon and sun. Effect of large tidal range is apparent from these photos taken of a docked ship at high and low tide. (Nova Scotia Dept. Government Services.)

TIDAL RANGE

Tidal range is the difference between low tide and high tide. The greatest ranges occur when the gravitational pull of the sun and the moon are working

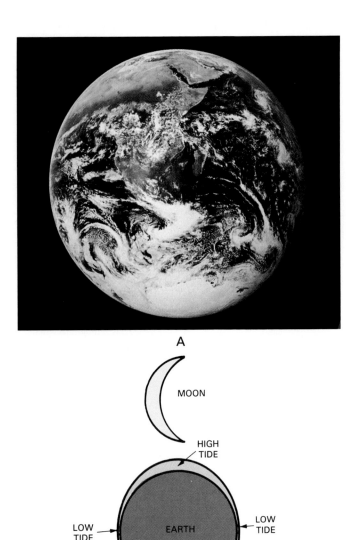

A

Fig. 11-18. A—View of earth from space shows its vast bodies of water. All are affected by the moon's gravitational pull. B—Diagram simplifies explanation of tides. If the earth were uniformly covered with water, this is how gravitational attraction would pull at the water to create high tides.

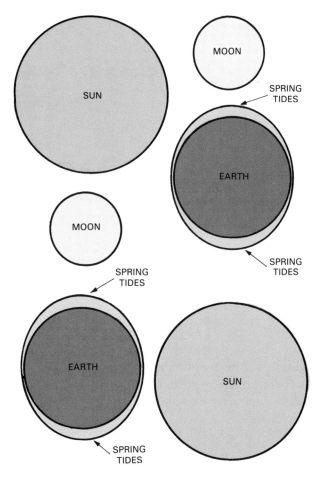

Fig. 11-19. Gravitational pull of sun and earth work together to cause spring (extra high) tides. Left. Sun and moon aligned on same side of earth. Right. Sun and moon on opposite sides of the earth work against each other's gravitational pull to produce neap tides.

together. The sun is on the opposite side of the earth from the moon, Fig. 11-19. At other times the two tend to counteract each other. The higher tides are called **spring tides** and the lower tides, **neap tides**.

FALLING AND FLOWING WATER

Falling and flowing water is the action of rivers moving water from higher ground to lower ground. Some of the great rivers of the world may flow for hundreds or thousands of miles before emptying into oceans. It is the action of the sun, working as a giant pump, that keeps the rivers flowing. It draws up moisture from the oceans and land and releases it over the land. This replenishes the flow of the rivers in a continuous cycle. The flow represents a huge supply of kinetic energy which can be tapped by humankind to do work. See Fig. 11-20.

STORED ENERGY

Some forms of kinetic energy are the result of planned storage. There are several types:
- Gravitational storage. The best examples of this are "grandfather" clocks and the water towers, Fig. 11-21, that provide the water pressure to feed running water into homes and factories. Another exple is the building of dams to store water for generating electricity.
- Storage by inertia (or **inertial storage**). This is the storing of energy in flywheels. Inertia is the tendency of a moving or spinning body to continue moving or spinning in the same direction. Such a flywheel can be used to drive electrical generators and propel vehicles or machines such as lawn mowers, Fig. 11-22.

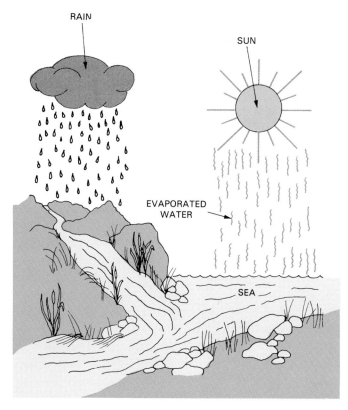

Fig. 11-20. The sun is like a giant pump. It draws up water vapor into the atmosphere in huge quantities. Released as rain, it replenishes the flow of the world's rivers.

Fig. 11-21. Planned gravitational storage. Energy of water held at a great height, causes the water to flow in homes when faucets are turned on.

- Storage of air under pressure. The air, being under pressure, can be used to drive generators. Compressed air is used to support transportation vehicles through pneumatic tires and air shocks, Fig. 11-23.
- Storage by deformation. The simplest example of this kind of storage is the winding up of a spring to drive the mechanisms in a clock. The spring tries to return to its original unwound state. As it unwinds it causes the clock wheels to spin and slowly turn

Fig. 11-22. Flywheels, because of their weight, will keep spinning and supplying energy long after the force spinning them has ceased operation. This one is part of an old steam engine used to generate electric power.

Fig. 11-23. Storage of compressed air in tires or air shocks supports all types of land transportation vehicles. (Ford Motor Co. and Goodyear Tire & Rubber Co.)

the hands. Another example is a toy which moves under power of a rubber band. See Fig. 11-24.

CONVERTING KINETIC ENERGY

Before any form of energy can do useful work some kind of change must be made in it. Kinetic energy

Fig. 11-24. Storage of energy is possible through deformation. A spool toy is given power of motion by the unwinding of the stretched, twisted rubber band.

comes closest to being ready to move loads. Thus, wind can be put to work directly by attaching a sail to a movable object such as a boat. A kite is like a sail. It is kept in the air by the blowing of the wind. So is a hang glider, Fig. 11-25. A river will carry logs to the sawmill or move a boat downstream. Windmills use the force of the wind to turn a wheel. The wheel drives gears and a crank that moves a rod up and down to pump water.

In most situations, however, kinetic energy found in nature is transformed (changed) into another form of energy. One of the biggest reasons for changing the kinetic energy is to make it possible to store it or move it to a location where it can be used.

The perfect example of this is the wheel or turbine of a wind machine connected to an electric generator. The electricity can be sent out to distant places over wire conductors. It can also be stored in batteries.

Wind energy can also be stored as a chemical. This is done through electrolysis. Electricity produced by the wind generator separates the hydrogen from water. The hydrogen can be stored or transported in containers and used anywhere as a fuel.

CONVERTING WIND ENERGY

As you learned in Chapter 2, wind has been used for centuries as a source of power. It pumped water for irrigation, moved ships, and ground flour from grain. In the 17th century, the Netherlands became the world's most industrialized nation simply by using wind power to drive trading ships, run mills, and pump water from lands below sea level. See Fig. 11-26.

Interest in windmills and sailing ships waned after the steam engine was invented. However, interest in wind power picked up again when Americans began to settle the Great Plains and the West. By 1900 the windmill industry represented a capital investment of more than $4 million. Twenty years later wind power had become a major source of electrical power not only in the rural U.S. but in Europe as well. In the 1930s the Rural Electrification Administration (REA) brought convenient electric power to the U.S. farms. By 1950, small-scale windmills were little used.

WIND GENERATORS

Interest in large wind generators never died. Between 1935 and 1955 a number of large units were built. One of the largest and best known, the 1.25 megawatt (MW) Smith-Putnam generator was built in the mountains

Fig. 11-25. Hang gliders depend on wind energy to stay airborne and support the weight of a person.

Fig. 11-26. This reproduction of an ancient Dutch mill is located in a park at Holland, Michigan. The vanes of the mill wheel have a framework designed to be covered with sailcloth.

of central Vermont. Palmer Cosslett Putnam, an engineer who lived on Cape Cod, first conceived the idea of a large generator which would feed electric power into the network of electric power lines normally fed by the fossil fueled electrical power station. He organized a few scientists and engineers to help him. This group designed a wind generator and found a site for testing it. The site chosen was Grandpa's Knob, a 2000 ft. peak in the Green Mountains.

The generator's two stainless steel blades had a 175 ft. (53 m) span and sat atop a 110 ft. (33.5 m) tower. It weighed 250 tons and could generate 1250 kilowatts (kW) of electricity. This was enough for a small town. The generator was completed in August, 1941. Testing, begun in October of that year, was interrupted frequently because of mechanical failures. In March of 1945, one of the blades fell off. It has not operated since. Though its performance was a disappointment to its designers and builders, it was an important effort by American scientists and engineers to harness wind power.

MORE RECENT DEVELOPMENTS

In 1973, following the oil crisis, the federal government began to sponsor wind generator projects. The first was started at the NASA Lewis Research Center by the National Science Foundation. A wind turbine was developed. The test machine was named "Mod-O." It had a two-blade wind rotor and a gear box which stepped up the speed. The rotor operated at 40 rpm and the gearbox drove the generator at 1800 rpm. In an 18 mph wind it could generate about 100 kW of electricity. It began operation at the NASA Plum Brook site in September, 1975. Its purpose was to gather engineering information that could be used in a large-scale wind energy program.

A larger wind turbine, Mod-OA, was developed at Clayton, New Mexico in 1977. It generates 200 kW of electricity at 40 rpm, double the power of Mod-O.

In 1976 the federal government added a third wind generator to their test program. It was the Mod-1 which was designed to produce 2000 kW of electric power at a wind speed of 25 mph (40 km/h). This machine, located at Boone, North Carolina, began operating in 1979, but was later shut down. In 1983 it was sold to a private company which planned to convert the generator for use with a water or steam turbine.

MOD-2 WIND TURBINE

Another experimental program in wind power is the Mod-2. Built by Boeing for the Department of Energy, it is a downwind design which can generate 2.5 MW

of electricity. Three of the units are in operation at Goodnoe Hills near Goldendale, Washington. See Fig. 11-27 and Fig. 11-28.

The Mod-2 is controlled completely by a microprocessor (a small computer) in the generator housing, Fig. 11-29. Power from the Mod-2 system is fed into the regional power grid (power lines) of the Bonneville Power Administration.

The units begin to generate electricity in 14 mph (about 22 km/h) winds. At winds of 27.5 mph (44 km/h) they reach their rated output of 2.5 MW. In stronger winds the blade tips feather (turn slightly to spill the wind). At winds of more than 45 mph the units shut down.

At wind speeds of 14 miles an hour a single Mod-2 machine can generate about 10 million kWh (kilowatt hours) of electricity a year. This would supply electricity for about 700 homes.

Fig. 11-27. One of three Mod-2 wind turbines are located in southern Washington state. This model weighs 20 percent less than Mod-1 but generates twice as much electricity. (Boeing Engineering and Construction Co.)

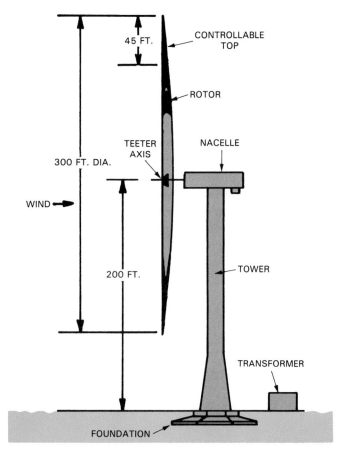

Fig. 11-28. Mod-2 dimensions. Its tower is tubular steel. The rotor blades are also steel.

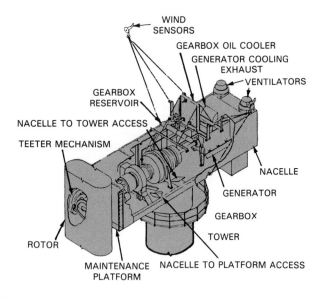

Fig. 11-29. Cutaway of a nacelle (housing) for Mod-2 shows gears and other mechanisms which control the electrical output. Not shown is the microcomputer control.

The goal of the Mod programs was to show that large wind machines are practical. Also, researchers wanted to determine the costs for going into commercial production. Similar testing was underway for small wind machines.

SMALL GENERATOR PROJECTS

Rocky Flats, Colorado, has become a test site for small wind generators. The first unit tested there had a 4.1 kW generating capacity. It was used for 2 1/2 months to generate power for a deep well irrigation pump. It was exposed to winds as high as 85 mph (137 km/h).

Other machines have been tested at the site. All have developed defects and failed. Problems have included broken tail vanes, bearing failures, and broken blades.

The tests were originally set up and sponsored by the Energy Research and Development Administration (ERDA). It contracted with a private company to develop the Rocky Flats test site. ERDA had identified a number of needs for wind power devices in three power ranges:

- A 1 kW direct current unit for remote locations and where climate is severe.
- An 8 kW alternating current unit for individual home and farm use.
- A 40 kW alternating current unit for powering irrigation pumps and to provide electricity for remote communities.

TYPES OF WIND MACHINES

There are two general classifications of wind machines:
- Horizontal axis.
- Vertical axis.

HORIZONTAL AXIS MACHINES

In a horizontal axis machine the rotor shaft is horizontal and the rotor or propeller revolves in a vertical plane (path). The farm windmill is a familiar example of this type. The rotor may have one, two, three, or many blades.

Single-bladed rotors, Fig. 11-30, require a counterweight on the opposite side of the rotor shaft

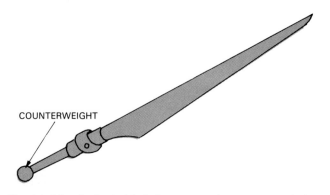

Fig. 11-30. A single bladed rotor requires a counterbalance to avoid destructive vibrations.

to stop vibration. This type is not practical where ice could build up on the rotor and throw it off balance. Two-bladed rotors, shown earlier in Fig. 11-27, are most popular because they are strong and less expensive to build. A three-blade rotor, Fig. 11-31, spreads stress more evenly, especially when the wind changes direction.

The bicycle multi-blade, Fig. 11-32, and the sail wing, Fig. 11-33, are newer rotor designs. The bicycle rotor has many narrow blades held in tension between the hub and a solid metal rim. It is light and strong.

The sail-wing rotor uses a metal tube for the leading edge of the blade. Short bars extend at right angles from the tube to form the rotor blades. Cable is stretched between the tip and the root to form the trailing edge. The blade is then covered with a cloth sleeve.

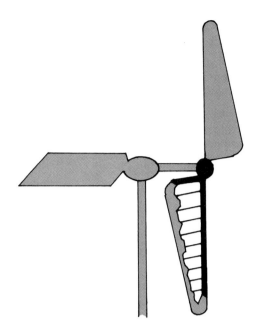

Fig. 11-33. Sail-wing rotor has a pipe and cable frame that can be covered with cloth sleeves.

Horizontal axis wind machines have two other characteristics that separate them into types:
• **Downwind** type, Fig. 11-34.
• **Upwind** type, Fig. 11-35.

DOWNWIND ROTORS

In downwind types, the rotor is located behind the tower and rotor housing. That is, the wind blows over the tower and housing before striking the rotor. This

Fig. 11-31. A three-bladed rotor spreads stress on the hub more evenly. (Niagara Mohawk Power Corp.)

Fig. 11-32. Bicycle multi-bladed rotor. Blades are attached to a steel hoop around the outer perimeter.

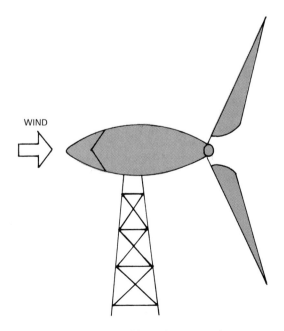

WIND

Fig. 11-34. Downwind machines do not need vanes to swing them into the wind.

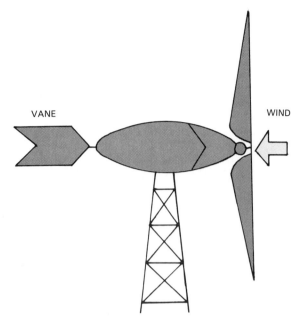

Fig. 11-35. Upwind machine. Rotor faces into the wind with the aid of a vane.

type has no vane. (This is the fan shaped metal piece keeping the rotor into the wind.) Downwind designs are preferred for larger wind machines.

UPWIND ROTORS

Upwind rotors are designed to head into the wind using a tail vane. A second mechanism pulls the rotor out of the wind when it is blowing too hard or when the machine is shut down.

VERTICAL AXIS WIND MACHINES

In vertical axis wind machines the rotor spins on a vertical spindle. No vanes are needed since the rotor can pick up the wind from any direction. Different kinds are shown in Fig. 11-36.

The Savonius or "S" rotor was patented in 1929 and resembles the ancient Persian windmills used to lift water from wells. It is well suited today to pumping

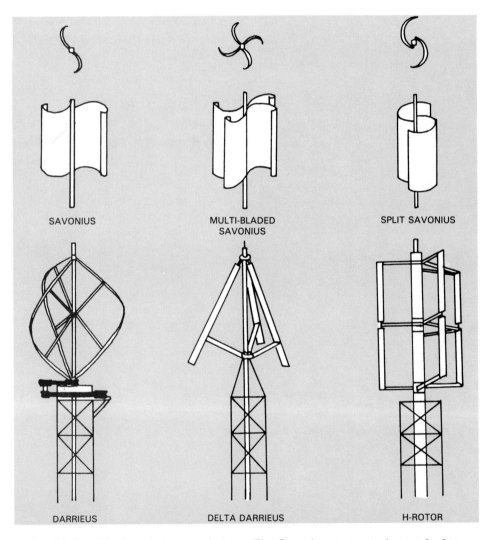

Fig. 11-36. Vertical shaft rotor designs. The Savonius types are best suited to pumping water and providing ventilation for buildings.

water and ventilating buildings. It rotates too slowly to generate electrical power. Fig. 11-37 shows a student-developed model of a split-Savonius rotor.

A variation of the Savonius is the Helius rotor, Fig. 11-38. The vanes are shaped and mounted like the Savonius but they are skewed to wring more energy out of the wind.

The Darrieus rotor was patented in 1931. It and the delta Darrieus are vertical axis versions of propeller wind machines. They are usually not self-starting and must be spun by an electric motor when the wind speeds are high enough for them to operate.

The H-rotor has adjustable blades which flip twice during each revolution to take greater advantage of the wind. It runs more slowly than the Darrieus, but develops more power.

WIND MEASUREMENT

Taking wind measurements is necessary to find a good site for a wind generator. The location is important since a good prevailing wind source is needed for generating electric power. Sometimes this information is available from weather stations, airports, or military bases. Many universities, public schools, and private colleges have weather stations and can supply wind records of an area. If these sources of information are not available, a wind speed indicator can be used.

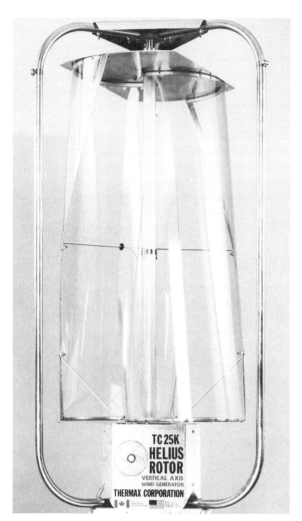

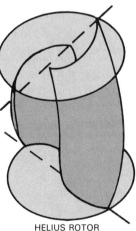

Fig. 11-38. Top. The Helius™ rotor is a variation of the Savonius. Bottom. Vanes are skewed to catch the wind better. (Thermax Corp.)

MEASURING INSTRUMENTS

Instruments which measure wind speed are called **wind speed indicators**, or **anemometers**, Fig. 11-39. A recording anemometer will keep a record of the wind speed over a period of time.

Fig. 11-37. A mock-up of a split-Savonius rotor. Refer to Fig. 11-36 for the basic design. (Barry David)

Fig. 11-39. This anemometer (right) measures wind velocity coming off Lake Michigan at a Coast Guard station.

EFFECT OF HEIGHT ON WIND SPEED

The surface of the earth and high structures, such as trees and buildings, affect the amount of wind available. Wind speeds at low heights are greatest over water and on level plains. Ground level winds are very weak in cities because of the many buildings.

COMPLETE WIND POWER SYSTEMS

The tower, rotor, and turbine or generator make up only a part of the wind power system. Generated electric power is carried away by conductors. Fig. 11-40 is a diagram of a complete, small-scale, wind power system.

A complete system will include the following additional components:

- Voltage regulator to keep voltage at a certain level.
- Storage batteries to contain the electrical energy until it is needed.
- An **inverter** which steps up voltage and converts direct current to alternating current.

While some wind generators produce alternating current, only direct current can be stored in batteries. Small inexpensive wind generators produce 12 V DC. Larger units produce 24, 36, 120, or even 240 V DC.

VOLTAGE REGULATORS

The purpose of the voltage regulator, as stated before, is to assure that a constant voltage output is supplied to the load. If the wind generator voltage is not above the voltage for which the regulator is set, a set of contacts in the regulator will close. This action allows current to flow through the circuit to the load. As the generator builds up voltage it will reach the rated voltage setting of the regulator. At that point

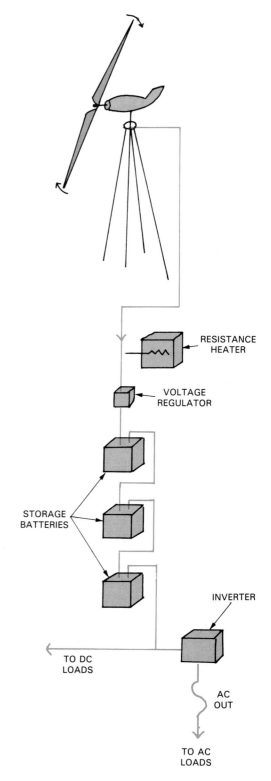

Fig. 11-40. A complete wind power system will require all these components. The inverter is needed to convert direct current to alternating current.

the contacts will separate so that current must go through a resistor. This cuts down the generator output. This action is repeated as often as needed to keep the generator from putting out too high a voltage for the circuit. See Fig. 11-41.

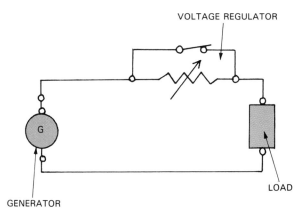

VOLTAGE REGULATOR

GENERATOR

G

LOAD

Fig. 11-41. This diagram is of a circuit containing a voltage regulator. When the voltage exceeds a certain level, the regulator's contacts open; then all voltage must go through the resistance in parallel with the regulator where it is reduced to a safe level.

BATTERIES

Excess electricity generated by the wind machine can be stored in a number of batteries which are connected together. A battery is made up of one or more cells. Each cell includes both positive and negative plates which are separated from each other. Positive plates are connected to the positive post of the battery and negative plates to the negative post. The plates are surrounded by electrolyte. Electrolyte is an acid and water mix which will conduct electricity.

A lead-acid battery, Fig. 11-42, has lead plates and a sulphuric acid-water electrolyte. An alkaline battery

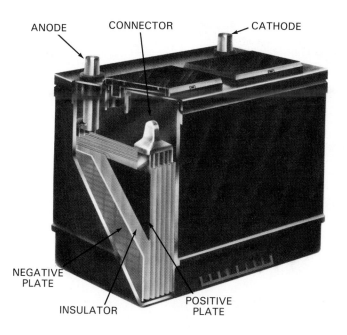

ANODE   CONNECTOR   CATHODE

NEGATIVE PLATE

INSULATOR   POSITIVE PLATE

Fig. 11-42. A lead-acid battery of the type used in automobiles is suitable for storing excess electricity from a wind power system.

has plates of nickel-iron and nickel dioxide. The electrolyte is a mix of potassium hydroxide and water.

When a lead-acid battery receives electric current a chemical change takes place in the battery. The water breaks down into hydrogen and oxygen ions. The hydrogen ions attract sulfate ions deposited on the plates when it is discharging. Sulphuric acid is formed.

When current is drawn from the battery the sulphuric acid breaks down into hydrogen and sulfate ions again. Hydrogen and oxygen ions combine and form water. The sulfate ions combine with lead and form a lead sulfate deposit on the plates.

INVERTERS

An inverter converts direct current electricity into alternating current. There are several kinds of inverters. A rotary inverter is a DC motor which drives an AC generator. An electronic inverter uses transistors and rectifiers to convert direct current.

When the excess current is to be fed into electric utility lines, a synchronous inverter is used. Then, when excess electric power is fed back into the utility's power lines, there is no need for batteries.

PUMPED STORAGE

Batteries are an expensive and limited way to store electrical energy. For power systems new storage methods are being used. One is pumped storage. In this method water is pumped to a high reservoir, Fig. 11-43. Later the water is released to power hydroelectric generators. This is done at Niagara Falls and other hydroelectric sites. A huge pumped storage facility is located near Ludington, Michigan on the Lake Michigan shore. The human made reservoir is 2 miles (about 3 km) long and nearly a mile wide. It holds 27 billion gallons of water at a head of 75 ft. (about 23 m). During off-peak hours, water is pumped into the reservoir. During peak demand periods, the water is released to generate the electric power needed. See Fig. 11-44.

Another possibility being explored is storage of energy as compressed air. Electric motors operating compressors would compress the air and store it in underground caverns. See Fig. 11-45. When power is needed, the compressed air would run turbines to regenerate electricity.

FUTURE OF WIND ENERGY

While wind is free, the equipment to change its energy into electricity is not. A well-designed wind generator can convert about half of the available energy

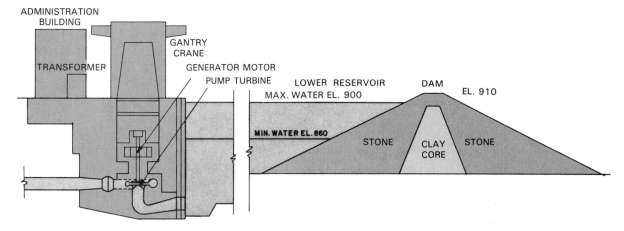

Fig. 11-43. A typical arrangement for pumped storage. During off-peak hours, when electric power demands are less, the pumping/generating station (left) pumps water to the storage reservoir (right). During peak-demand hours the stored water is released to generate additional electricity.

Fig. 11-44. This huge pumped storage facility is on the shore of Lake Michigan south of the city of Ludington. Top. A segment of the human-made lake showing the water intake. Bottom. The pumping station at the lake shore.

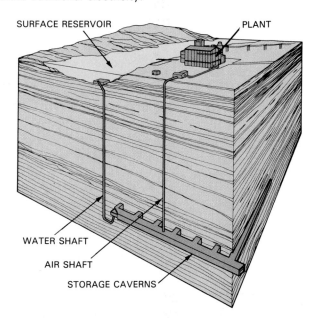

Fig. 11-45. Schematic of an air storage system. Excess electricity compresses air into hard-rock caverns. Water is used to keep air at a constant pressure. (EPRI Journal)

of the wind into electrical energy. There are other losses so that somewhere between 20 and 50 percent conversion is realized. Experts in this country are experimenting with the idea of large wind energy farms spread across the Great Plains and off our shores.

As energy requirements grow throughout the world, we cannot ignore the importance of wind energy. Since 1981, wind machines large enough to supply electrical power to utilities are being installed in greater numbers than ever before. There are now more than 1700 mW of rated capacity in the world. The U.S., alone has a capacity of 1400 mW. According to the latest estimates from the Annual Energy Review of the U.S. Energy Information Administration, wind provides less than 0.005 quad Btu or 5 trillion Btu of electricity annually to consumers. Fig. 11-46 graphs the growth of wind energy production in the U.S. from 1982 to 1989.

At the same time, while traditional energy costs have been rising, the costs of wind energy have been declining. Better technology, larger scale and more efficient manufacturing, and more experience with wind turbines have helped reduce costs. Fig. 11-47 compares costs of various energy sources.

ENVIRONMENTAL CONSIDERATIONS

Development of wind power systems has few undesirable by-products. There are no residues to cause smoke or rubbish. There are no health hazards. Still, there are some drawbacks. A single tower does not take up much space, but a giant "wind farm" with many towers and power lines would use up considerable land area. Furthermore, it may not be a pleasant sight.

Large wind generators with rotors fabricated of metals cause some annoyance. Residents near a large

wind generator in Boone, North Carolina complained of television interference and air disturbance when the unit was running.

On the other hand, land used for wind farming could be used for other purposes at the same time. Cattle could graze around the towers. Farming could be carried on as it is now under and around power line towers.

COSTS OF WIND POWER

Wind power systems have been slow in developing partly because of costs. However, it is thought that once systems are mass produced, wind generators will be able to produce electricity just as cheaply as fossil fueled or nuclear power stations. A major drawback of wind power is that the equipment is often idle for lack of wind. It is estimated that the system would be operating at full capacity only about 6 hours out of every 24. Coal and nuclear power systems can operate at capacity 50 to 60 percent of the time.

On the encouraging side, wind-powered systems have low maintenance and no fuel costs. Experts believe that the total cost of the electricity generated by the wind will be about 2 to 3.5 cents a kilowatt hour. This is close to the present average cost of electricity produced by conventional power plants. It is likely that wind power will be an important supplementary source of electric power.

HYDROELECTRIC POWER

As you learned from Chapter 2, one of the oldest methods of using a renewable energy resource is the waterwheel. Today when water energy is used, it is used almost entirely to produce electricity.

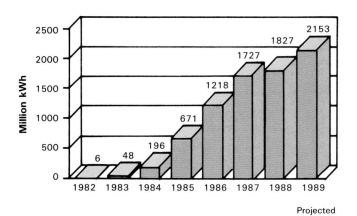

Fig. 11-46. Growth of wind energy production in the United States through 1989.
(Source: American Wind Energy Association)

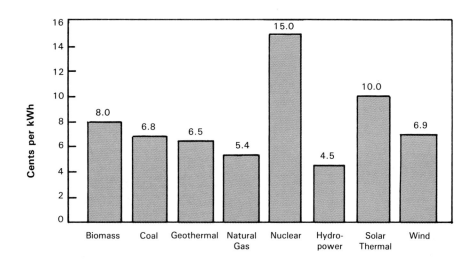

Fig. 11-47. Cost of wind-generated electricity compares favorably with those of traditional energy sources. (Source: Council of Renewable Energy Education, 1989)

There are several forms of hydro energy:
- Hydroelectric power derived from the natural force of falling water.
- Tidal power which comes from the force of the gravitational pull of the sun and the moon (discussed earlier).
- Wave power, caused by the wind driving the water toward the shore.

Nearly all power now taken from water energy is from hydroelectric plants at waterfalls or dams. The usual method is to dam up rivers where the terrain will produce a natural reservoir.

PERCENTAGE OF ENERGY PRODUCED BY HYDROELECTRIC

About 16 percent of all electric power used in the United States comes from hydroelectric installations. This is only 30 percent of what it could be if all resources were developed. It is thought that hydroelectric capacity could be doubled in time.

HOW HYDROELECTRIC POWER PLANTS ARE DESIGNED

Hydroelectric power plants must have a head of water at least 20 ft. (6 m). Head is the distance from the highest level of the dammed water to the point where it goes through the power producing turbine. Many low-head dams have a head of no more than 100 ft. (30 m). There are high-head hydroelectric plants with a water drop of 100 to 1000 ft. (30 to 300 m). See Fig. 11-48 and Fig. 11-49 for more information on design of dams.

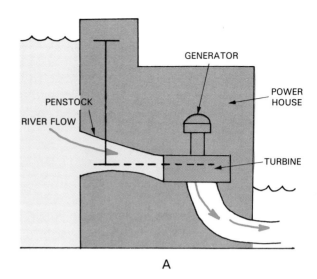

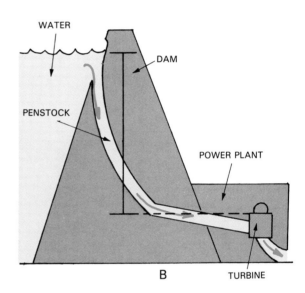

Fig. 11-49. Hydroelectric power plants operate with varying amounts of head. A—A typical low-head plant. B—Typical high-head plant.

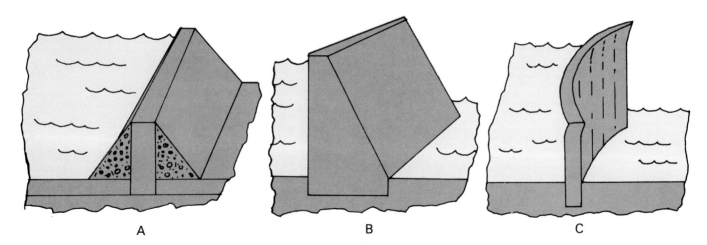

Fig. 11-48. Types of dams. A—Embankment dam uses weight of earth or rock fill. They are simplest and cheapest. B—Concrete gravity dams are massive and expensive, usually running across a broad valley. C—Arch dam is thin-walled and curved against the stream. The force of the stored water is transferred to the steep walls of the canyon and downward to the floor of the reservoir.

Water storage is another important factor in design. The amount of water storage needed depends on the electrical power demands on the power plant. This demand varies. Electrical loads are greatest during the day and gradually decrease in the afternoon, reaching a low point at night. Usage of electric power is also greater during June, July, and August because of air conditioning.

Water must be stored and released as demand for electricity increases and subsides. When demand is high, the gates of the dam are opened wider to allow more water to flow through to the turbine. More power is produced, as a result.

HYDROELECTRIC TECHNOLOGY

Fig. 11-50 is a simple schematic of a hydroelectric plant. The water flows from the dam down a large tube called a **penstock**. At the bottom of the penstock the fast moving water drives a **reaction turbine** or an **impulse turbine**. Reaction turbines work well with low-head hydroelectric plants.

OBSTACLES

There are several reasons why more hydroelectric power plants have not been built:
- The high cost of building dams to form a reservoir of water.
- Highways, railroads, power lines, and even towns must often be moved.
- Major equipment such as turbines, generators, and other power station equipment must be designed and built. This takes a great deal of time and money.
- Farm lands must often be abandoned behind the dam.
- Often, special groups oppose the destruction of other natural resources and wildlife caused by damming up rivers.

HARNESSING TIDAL WATERS

Rise and fall of tidal waters represents a source of energy which could be converted to useful power. Like wind energy, it is not a constant source. It would have to be captured and stored.

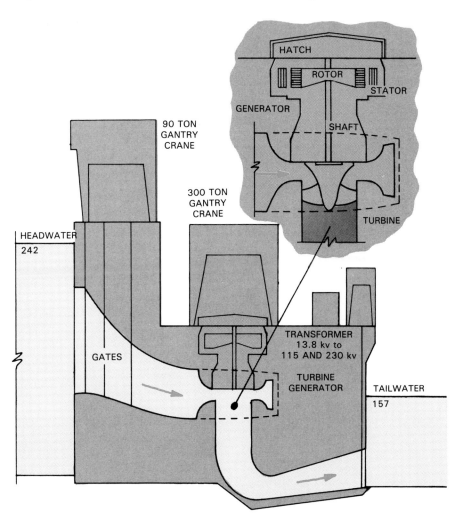

Fig. 11-50. Water flowing through a penstock and across the turbine creates the mechanical energy to drive the generator. (New York State Power Authority)

TIDAL ENERGY SITES

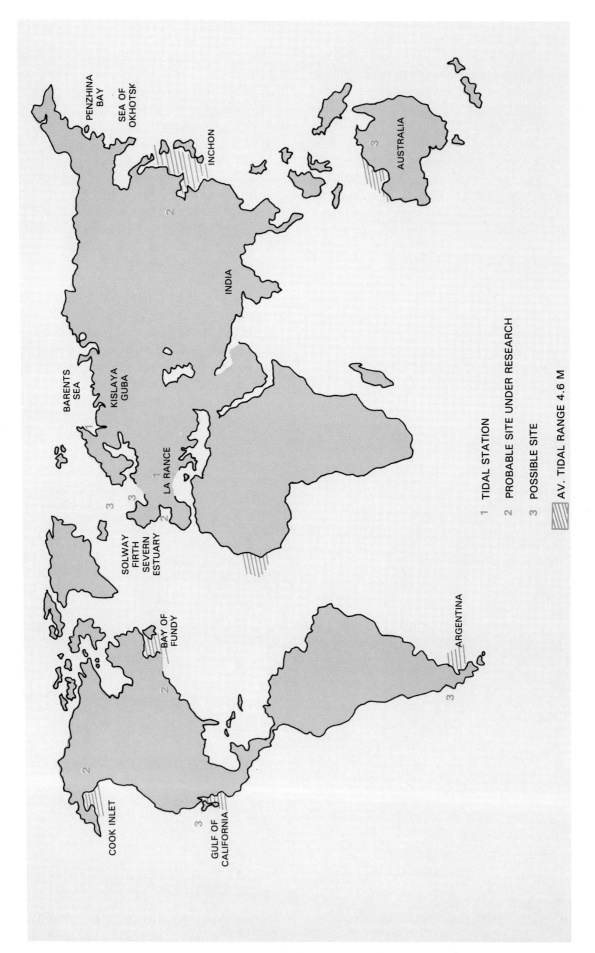

Fig. 11-51. Many potential tidal power sites in the world remain to be researched and developed. Rising costs of fossil fuels make them more attractive as sources of electric power.

TIDAL MILLS

Use of tidal power is not new. Hundreds of years ago tide waters were trapped at the mouths of rivers emptying into the sea. The stored energy was used to run tide mills. Dams were constructed to form a reservoir upriver. Incoming tide was allowed to flow through gates in the dam to flood the reservoir. At high tide the sluice gates were closed. Several hours later when the tide had fallen, the trapped water was released to turn waterwheels.

Between 1581 and 1822 one of the arches in the London Bridge had such a mill. It was used to pump water from the Thames River to the center of London.

A tidal power station operates much like the ancient tidal mills just described. Trapped tidal water flows through an opening in a dam to spin a turbine. Certain coastal areas of the earth would be well suited for trapping the tidal waters, Fig. 11-51.

The coastal feature best suited for a tidal power station is a narrow estuary (place where a river empties into the sea). The difference between high tide and low tide should be 30 ft. (about 10 m) or more. Such features are found at:
- The Bay of Fundy in eastern Canada.
- The Pacific coast of Alaska.
- The Severn and Solway estuaries in England.
- The Barents and Okhotsk Seas in Russia.
- Northwestern Australia.
- Northeastern Brazil.
- Southern Argentina.
- Western Africa.
- The eastern fringe of the West Indies.
- Northwestern France.

It is thought that tidal plants built on all these sites could produce nearly 2000 billion kilowatts of electricity every year. Tidal power stations need a plentiful supply of water to keep the generators producing power. Since the tide waters are available only for a short time everyday, the water must be captured by huge dams across the mouth of the estuary. It is very expensive to build such dams and to construct the turbines needed to turn the generators. The estuary must be able to hold a great deal of water behind the dam. Likewise the incoming tide must bring a great deal of water into the estuary.

EXISTING AND PLANNED TIDAL POWER STATIONS

The first tidal power station was opened at the La Rance River in northwestern France in 1962. See Fig. 11-52. A second station, Kislaya Guba, has since been constructed in Russia on the Barents Sea. A third is

Fig. 11-52. Tidal power dam at La Rance on the French coast. Its turbines can generate electrical power as water moves in either direction. (American Petroleum Institute)

in operation on the Bay of Fundy.

Turbines at the La Rance River tidal power station can generate electric power as the tidal waters move in either direction. The estuary forms a large natural reservoir. The tidal range (difference between low and high tide) can be as much as 28 ft. (8.5 m).

THE BAY OF FUNDY TIDAL POWER PROJECT

The Bay of Fundy tidal power project was designed not only to test the feasibility of tidal power in Fundy Bay but also to produce additional electrical power for Nova Scotia. A number of locations have been selected for future installations, Fig. 11-53. The first phase of construction was a pilot project, a generating station which has been built near Annapolis Royal on Hog's Island in northwest Nova Scotia. The tidal waters are held in a natural reservoir or basin formed by the mouth of the Annapolis River, Fig. 11-54. The project will test the suitability of a new turbine design. The unit is a modern adaptation of an axial flow turbine with a rim type generator. It is mounted horizontally to capture the energy of a horizontal flow of water. Fig. 11-55 is a cutaway drawing of the turbine and generator. The propeller consists of a hub and four blades. It rotates inside the huge generator stator. The turbine's operating head range is 4.6 ft. (1.4 m) to 22 ft. (6.8 m). It is capable of producing 50 GW of electric power annually.

The tidal range at Annapolis Royal and other points along the coast of Nova Scotia is as high as 28.5 ft. (8.7 m) during spring tides and as low as 14 ft. (4.4 m) during neap tides. The average range is 21 ft. (6.4 m).

ENERGY FROM WAVES

Ocean waves hold 10 times more energy than tides. Wave action in the Atlantic, Pacific, and Indian Oceans

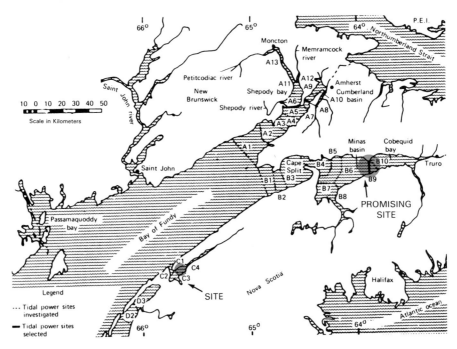

Fig. 11-53. Map of Fundy Bay area of Nova Scotia shows sites for potential development of tidal power stations. Site of the pilot project is near the mouth of the Annapolis River. The cite marked "B9" is considered the most promising for future development. (Tidal Power Corp.)

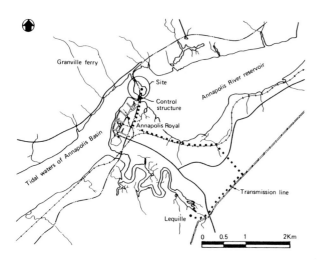

Fig. 11-54. Small area map shows more details of the Annapolis tidal generating station. The installation takes advantage of a barrage (like a levee) that had been installed to protect agricultural marshlands.

is capable of producing 40 to 70 kilowatts of electrical power for every meter of wave front. Great Britain is experimenting with devices which will trap the wave energy. See Fig. 11-56. While the cost of construction of the devices is not a problem, a system to carry the electrical power ashore would be expensive. The devices being built are for generating electricity. However, they could be used to power water electrolysis.

SUMMARY

Kinetic energy is the energy of matter in motion and is also known as mechanical energy. It can be made available to do work as a result of converting other forms of energy. However, kinetic energy is also available in nature. It occurs naturally in wind, waves, tides, and waterfalls.

The kinetic energy in nature is the result of the sun's heating effect on the earth. Uneven heating of the earth causes wind; solar heat also draws water into the atmosphere, setting up conditions for rain. The rain replenishes the water flow of rivers.

While the rotation of the earth has some effect on movement of air, most winds are the result of difference in air temperature. Warmed air rises. Wind is created as cooler air moves in to replace warmer air. There are patterns of wind direction that are peculiar to different parts of the world.

Winds blowing over large bodies of water create the energy in waves. This energy is dissipated (spent) as the waves reach a shore.

Tides are periodic changes of the water level in seas, oceans, and very large lakes. Rise and fall of the water occurs daily on a predictable schedule. It is the natural effect of the moon's gravitational pull on the earth. The sun also has some gravitational pull. However, it is not as strong because of the sun's greater distance from the earth.

Kinetic energy can be stored. Often water is pumped to a higher level and its kinetic energy used later. Hydroelectric plants sometimes pump water into a water storage lake to be used when electric power demands are high. Water is also stored in water towers to provide pressurized water supplies for communities and factories.

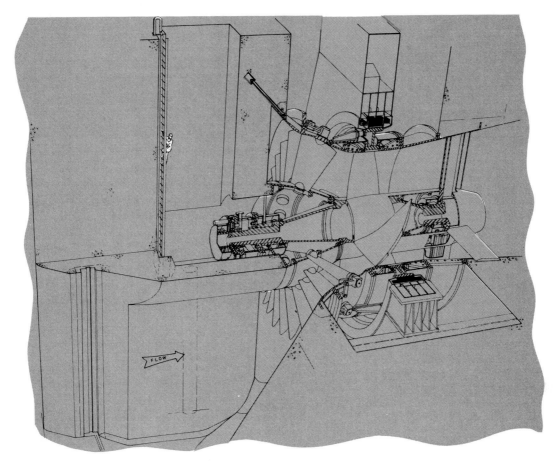

Fig. 11-55. A cutaway of one of the turbines in the tidal power station. It is designed to operate in either direction of water flow. Generator poles are arranged around the outside of the propeller blades. (Tidal Power Corp.)

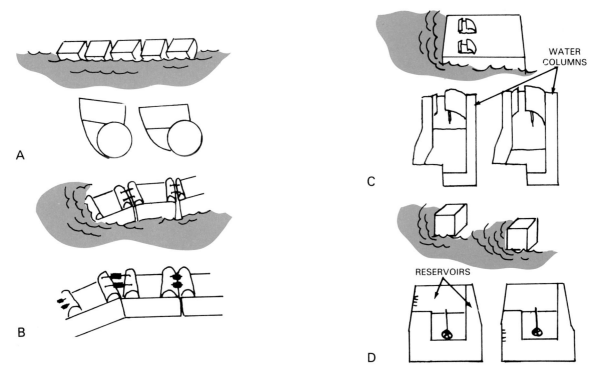

Fig. 11-56. Many different experiments are underway to capture the energy of waves. A—In an air pressure ring buoy, water action builds up air pressure that operates turbines in the buoys. B—Cockerell contouring raft. Waves operate hydraulic motors or pumps that pressurize a fluid. In turn, the fluid would operate a turbine. C—Oscillating water column in the foreground would create enough air pressure to drive an air turbine. D—Russell wave rectifier captures water in its reservoirs. The water would drive turbines.

Kinetic energy in nature can be converted to useful purposes through devices that convert the energy into electricity. Wind generators are one such device. Water turbines have been in use for hundreds of years at dams and waterfalls. Experimental devices are being designed to capture the energy in waves.

DO YOU KNOW THESE TERMS?

anemometers
cells
downwind type wind generator
foehn
horse latitudes
impulse turbine
inertial storage
inverter
local winds
mechanical energy
mistral
neap tides
penstock
planetary wind
reaction turbine
spring tides
tidal range
upwind type wind generator
wind speed indicators

SUGGESTED ACTIVITIES

1. Write to wind generator companies for literature. Report to your class on different sizes and designs of wind generators manufactured.
2. Select what you think are likely sites for wind generators in your community. Then, with the help of your instructor, take wind level readings with an anemometer. Be sure to secure permission from landowners before attempting to take readings.
3. Build a small, simple wind speed indicator and test it.
4. Build a model of an old windmill described in the text.
5. Build a small working model of a wind generator.

TEST YOUR KNOWLEDGE

1. List five natural kinetic energy forces.
2. Define wind and explain the role the uneven warming and cooling of the earth plays in causing it.
3. Some wind is created by the earth's rotation. True or False?
4. Total amount of wind energy available in the lower 48 states is equal to how much of our annual energy usage?
   a. Thirty percent.
   b. Two percent.
   c. Two times.
   d. Fourteen times.
   e. Forty times.
5. Name three properties of waves and define them.
6. What keeps rivers flowing?
7. Kinetic energy found in nature is usually converted to another form before use. The reason for this is (pick best answer or answers):
   a. It is not needed in its natural state.
   b. There is little for which kinetic energy can be used.
   c. Conversion makes it easier to store.
   d. Conversion makes it easier to transport to where it can be used.
8. A downwind generator needs a vane to keep it facing into the wind. True or False?
9. A _____ is a device for measuring wind speed.
10. What items make up a complete wind powered electrical system?
11. Hydroelectric power comes from the natural force of _____.
12. Explain the difference between a low-head and high-head dam.
13. What two natural features of a coast would make it a desirable place to build a tidal power station?
14. Ocean waves hold _____ times more energy than ocean tides.

12
CHAPTER

NUCLEAR ENERGY

The information given in this chapter will enable you to:
- Describe the basic principles of nuclear fission and nuclear fusion.
- Explain the process of extracting, refining, and fabricating a fuel assembly for a fission reactor.
- List and describe five basic types of nuclear reactors.
- Discuss the advantages and disadvantages of nuclear power.
- Discuss radiation and its effects on the human body.
- Describe current experimentation with fusion power systems.

Fig. 12-1. A nuclear power station in Pennsylvania. There are about 426 of them in operation throughout the world.

Present-day nuclear power plants, Fig. 12-1, use the heat from the fission of uranium fuel to heat water. (As you learned in Chapter 1, fission is the splitting of atoms of uranium.) The steam from the heated water drives a turbine which is connected to an electrical generator.

To understand the splitting of an atom, it is necessary to consider the atom and its parts. As we know, the atom is the smallest particle of any element that still retains all the chemical properties of that element.

We also know from a study of physics that all matter is made up of atoms. Water, wood, rocks, and even human bodies are composed of these tiny particles. So tiny are they that billions of them can rest on the tip of a pencil with room to spare.

Each atom is made up of even smaller particles called protons, neutrons, and electrons. Protons are positively charged; neutrons have no charge at all. Electrons are negatively charged. The protons and neutrons are held in the center of the atom which is called the nucleus.

The electrons, however, are in motion. They orbit the nucleus. See Fig. 12-2.

Every atom of an element has the same number of protons in the nucleus. However, there can be different numbers of neutrons in different atoms of the same element. These different forms of the same element are called isotopes, Fig. 12-3. Later, when we talk about different uranium isotopes, we will give them numbers to identify them.

An atom is mostly empty space. Most of the mass (weight) of an atom is in its nucleus. Electrons orbiting (moving in a more or less circular path) the nucleus are much lighter than protons and neutrons. Protons, for example can be at least 1800 times heavier than electrons.

The orbits of electrons are arranged in "shells." Each shell holds only a certain number of electrons. The shell

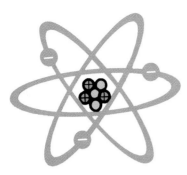

Fig. 12-2. An atom is a microscopic element made up of neutral and positively charged particles surrounded by orbiting negatively charged particles. Scientists have observed that electrons do not always have orbital paths; some are known to travel in a "figure eight" path! (Westinghouse Nuclear Energy Systems)

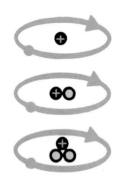

Fig. 12-3. Isotopes are atoms of the same element that have different numbers of neutrons in their nuclei.

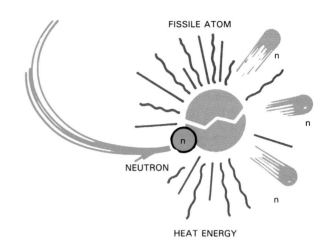

Fig. 12-4. Fission occurs when a uranium atom splits when struck by a fast-traveling free neutron. This action forms two new atoms along with tremendous amounts of heat. (Westinghouse Electric Corp., Advanced Power Systems Div.)

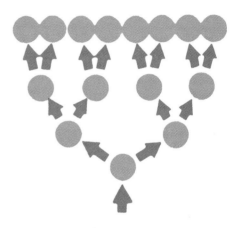

Fig. 12-5. A chain reaction occurs when neutrons from a split atom fly out, striking and splitting other atoms. (Westinghouse Nuclear Energy Systems)

closest to the nucleus may contain two electrons. The next can hold up to eight and the third one, up to 18. The fourth can contain up to 32. Each electron, regardless of how many are in the shell, has its own separate orbit.

At the beginning of nuclear fission, a neutron from one atom strikes another atom with great speed and causes the atom to split into two lighter atoms. See Fig. 12-4. If conditions are right, this sets up a chain reaction.

CHAIN REACTION

Nuclear power depends upon one reaction causing other reactions. Each neutron freed by fission can go on to split other uranium atoms so that the process continues, Fig. 12-5. This series of events is called a **chain reaction**. The process of splitting atoms causes a tremendous amount of heat.

FISSION

The chain reaction discussed earlier is started when a neutron "bullet" is fired at a U-235 atom. As the nucleus splits, it may shoot out two or three neutrons. Traveling at great speed, these neutrons, in turn, strike more U-235 atoms and split them. Once started, the splitting process will continue without any further help.

However, there is a need to control the number of reactions taking place at any one time. The flying neutrons may be controlled by **moderators.** These are any material that can slow down the neutrons and, thus, increase the chance of neutrons bumping into other U-235 atoms. Moderators are contained in **control rods.** They are loaded with neutron-absorbing elements such as boron or cadmium. The rods are movable so that they can be inserted into the fuel core when the number of reactions taking place at any one time need to be reduced; they are removed to increase the level of nuclear reaction. See Fig. 12-6.

All nuclear reactions take place in the pressure vessel of a strong, closed container called a **nuclear reactor.**

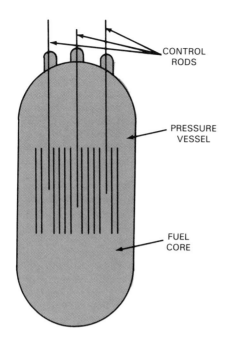

Fig. 12-6. Rods are used to slow down or stop the chain reaction of nuclear fission. The rods are lowered into the core where they can absorb some of the bombarding neutrons. (Westinghouse Nuclear Energy Systems)

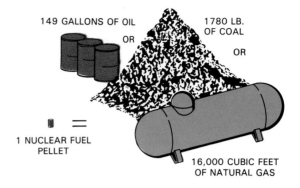

Fig. 12-7. A pellet of uranium fuel small enough to fit in a thimble contains as much heat as these amounts of fossil fuels.

HEAT OF FISSION

A very small amount of uranium fuel will produce a great deal of energy. It comes out of the chain reaction as heat.

The splitting of one atom releases about 190 MeV (million electron volts) of energy. (An electron volt is but a tiny fraction of a calorie. There are 2.6 x 1016 MeV in each calorie.)

CURRENT USAGE OF NUCLEAR POWER

Worldwide, there are about 426 nuclear reactors in operation producing 318,271 MWe (megawatts electric). This represents about 16 percent of the world's electrical energy. This is enough to supply all the electric power needs for some 159 million homes. See Fig. 12-7.

However, no source of power has generated as much controversy over its suitability. On the one hand, nuclear energy can produce a trillion times more power than kinetic forces such as wind and water and a million times more than chemical reactions such as combustion of fossil fuels. On the other hand, nuclear power activities generate radioactive waste and there is general concern about nuclear accidents such as were experienced in Chernobyl, Russia and Three Mile Island in the United States.

Because of these concerns, economic considerations, lower demand for electricity, and public opposition, the nuclear power industry in the U.S. has experienced

great difficulties. These circumstances and the uncertainties of governmental regulations have delayed construction as long as eight years and dramatically increased construction costs. As an example, the Shoreham (Long Island) nuclear power station in 1973 was to have cost $300 million to build. By the time it was completed in 1984, its cost had escalated to $5.5 billion! It has never been put into operation. Instead, it became the focus of a debate over which level of government—state or national—should control the location of nuclear power stations.

Even though the number of operating units and those planned or under construction reached an all-time high of 119 in 1990, this number is far below the total of 236 in 1975. Since then, many planned units have been canceled. No new orders have been placed since 1978. See Fig. 12-8 and Fig. 12-9.

Elsewhere, Switzerland and Germany have unofficially suspended reactor construction. In Sweden, a referendum (vote by the people) has called for ending nuclear power generation by 2010. The Russian Republics are arguing about the future of nuclear power in their countries.

Even so, there are powerful arguments for the continued development of nuclear power generation. Foremost of these is the growing need for electric power. Further, there is growing concern about the greenhouse effect caused by the burning of hydrocarbons. The use of more nuclear energy to produce power could help resolve these problems.

NUCLEAR FUELS

Fuel for nuclear power stations comes from uranium. **Uranium** is a silvery radioactive metallic element. It is thought to be one of the substances in volcanic ash which was spewed onto the earth's surface millions of years ago. Rains dissolved the uranium out of the ash and carried it back underground. There it hardened into an ore.

	REACTORS IN OPERATION	PERCENT OF ELECTRICITY GENERATION
NORTH & CENTRAL AMERICA		
Canada	18	15.6
Cuba	0	0
Mexico	1	—
U.S.	110	19.1
SOUTH AMERICA		
Argentina	2	11.4
Brazil	1	.7
EUROPE		
Belgium	7	60.8
Bulgaria	5	32.9
Czechoslovakia	8	27.6
East Germany	6	10.9
Finland	4	35.4
France	55	74.6
Hungary	4	49.8
Italy	2	—
Netherlands	2	5.4
Romania	0	0
Spain	10	38.4
Sweden	12	45.1
Switzerland	5	41.6
U.K.	39	21.7
West Germany	24	34.3
Yugoslavia	1	5.9
ASIA		
China	0	0
India	7	1.6
Iran	0	0
Japan	39	27.8
Pakistan	1	.2
South Korea	9	50.2
Taiwan	6	35.2
U.S.S.R	46	12.3
AFRICA		
South Africa	2	7.4
TOTALS	426	—

Fig. 12-8. Worldwide in 1989 this was the story of nuclear power usage. At that time 96 new reactors were under construction. (International Atomic Energy Agency, Vienna)

Since it is found near the earth's surface, like coal and oil, it can be mined in much the same way; that is, by open-pit mining or underground shaft mining. See Fig. 12-10.

A newer method of mining is also being used. Known as "in situ" (in place) leaching, it reverses the process which originally separated the uraniuim from the volcanic ash. Water is pumped into the ground, dissolving the uranium. It can then be pumped to the surface. Fig. 12-11 is a diagram of an in situ leaching operation.

URANIUM SUPPLIES

Like our fossil fuel supplies, uranium is exhaustible. Supplies will run out one day. Only one out of every 140 uranium atoms can cause fission. Thus, large amounts of uranium ore must be processed. A ton of uranium yields only about 14 lb. (about 6.5 kg) of fission fuel.

U.S. supplies of uranium could be as much as 4 million tons. This includes known and possible reserves. This reserve would be about 286,000 lb. (129 844 kg) of nuclear fuel. In 1989 there were 110 nuclear reactors in full operation in the U.S., 18 in Canada, and one in Mexico. Experts have estimated that even without reprocessing of spent fuels uranium supplies for the world's current number of reactors would last about 100 years.

However, there is a way to get more fuel from other atoms of uranium that will not go into fission in their natural state. Remember the definition of an isotope discussed earlier? You will recall we said that different atoms of the same element can have different numbers of neutrons in their nuclei. We call such atoms **isotopes.**

Table 101. Nuclear Generating Units, End of Year 1988-1990
(Number of Reactors)

Status	1988				1989				1990			
	Boiling Water Reactors	Pressurized Water Reactors	Other [1]	Total	Boiling Water Reactors	Pressurized Water Reactors	Other [2]	Total	Boiling Water Reactors	Pressurized Water Reactors	Other	Total
Operable [3]	37	70	1	108	38	72	0	110	38	73	0	111
In Startup [4]	1	2	0	3	0	1	0	1	0	0	0	0
Construction Permits Granted	3	10	0	12	2	8	0	10	1	7	0	8
Construction Permits Pending	0	0	0	0	0	0	0	0	0	0	0	0
On Order	0	0	0	0	0	0	0	0	0	0	0	0
Total	41	82	1	123	40	81	0	121	39	80	0	119

[1] Includes one gas-cooled reactor.
[2] High-temperature gas-cooled reactor.
[3] Units that have received a full-power license from the Nuclear Regulatory Commission, which includes the Hanford-N reactor for 1986 and 1987. Hanford-N, an unlicensed unit used for defense material production, was included in the operable category because power was produced as a by-product and sold commercially. The Hanford-N reactor was placed in a cold standby status by the U.S. Department of Energy in February 1988 and, consequently, is not included in the 1988 total. The Three Mile Island-2 reactor retains an operating license; however, there are no plans to resume operation of the unit, and it also is omitted from the 1988 total.
[4] Units that have received a low-power license from the Nuclear Regulatory Commission authorizing fuel loading and low-power testing.
Sources: Compiled by the Energy Information Administration from Nuclear Regulatory Commission sources.

Fig. 12-9. Construction of nuclear power generating units is at a standstill in the United States.

Fig. 12-10. Uranium mining in northern Saskatchewan, Canada. Left. An aerial view shows an open pit mine at Rabbit Lake. Right. Uranium-bearing rock is loaded onto a truck for transport to the mill. (Gulf Oil Co.)

One of these isotopes will produce fission if it is given another neutron for its nucleus. Scientists have discovered that this isotope, already present in nuclear fuel, will automatically absorb a stray neutron during nuclear fission. So the process of making new fuel, called **breeding** goes on all the time in a nuclear reactor.

To classify them, scientists have given certain isotopes of uranium names or numbers. The only isotope which is naturally capable of fission (being split) is called U-235. (The "U" stands for uranium.) Another isotope that can be modified to produce fission is labeled "U-238." When it gains a neutron it becomes Pu-239, another fissile isotope commonly known as **plutonium.** If the various uranium isotopes could be made fissile (able to go into fission), the supply of nuclear fuel could be extended to operate 1000 large nuclear reactors for about 30,000 years.

PROCESSING URANIUM

Uranium 235 makes up only 0.7 percent of natural uranium. The other 99.3 percent is U-238. Uranium ores, therefore, must be enriched through refining. Before it can be refined, uranium ore must be ground to the size of fine sand. Leaching with a water-chemical mixture then separates the uranium from other materials in the ore. This chemical treatment produces a mixture known as **yellow cake.** It may be in the form of a bright yellow slurry (soupy mix) or a powder.

Before it is used as a nuclear fuel, the yellow cake is converted into crystals of uranium hexafluoride. The crystals are enriched to increase the percentage of U-235 to about 3 or 4 percent.

FUEL FABRICATION

The enriched uranium hexafluoride crystals are then formed into rods or pellets. This is done by milling, pressing and sintering (baking in a furnace like pottery). Each pellet is about the size of a fingertip.

Pellets are packed into specially made metal tubes. Loaded tubes are welded shut and bundled together in metal frames. The frames maintain carefully engineered spacing between the rods. The bundled rods are called **fuel assemblies.** They will make up the core of a nuclear reactor. Other items in the core are the moderator material and the control rods. See Fig. 12-12.

NUCLEAR REACTORS

The nuclear reactor is to a nuclear power plant what a furnace is to your home. It contains the nuclear fission so that it does not destroy its surroundings. It also distributes the heat of fission so that the heat can be used to generate electric power. In this respect, it is again similar to the operation of the furnace which contains the burning fuel and circulates heated air or water to rooms in the house.

A steel pressure tank, called a **pressure vessel**, holds several thousand fuel rods which make up the fuel core (supply). The vessel, shown in Fig. 12-13, has walls 9 in. thick. Around the vessel is another structure called a "steel containment." It is made of leakproof steel plate. See Fig. 12-14. Around the vessel and the steel containment is a shield building. This building has walls of reinforced concrete three or more feet thick.

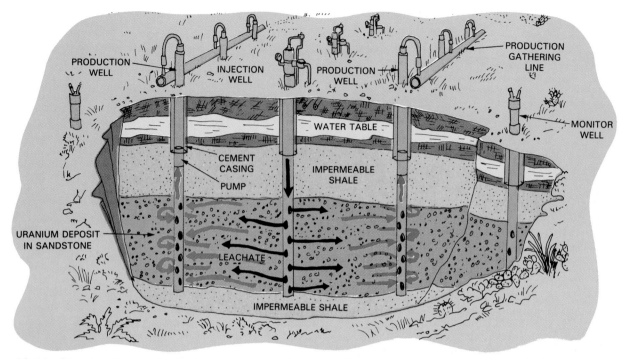

Fig. 12-11. Diagram of an in situ (in place) uranium leaching operation. A mixture of water and chemicals is pumped down injection wells to dissolve the ore. A producing well recovers the dissolved ore.   (Mobil Energy Minerals Div.)

Fig. 12-12. Quality control engineers inspect a newly built nuclear fuel assembly at a fabrication plant.   (Exxon Corp.)

REACTOR TYPES

Different types of reactors are being used or considered. They include:

- Light-water reactors (LWR). Seventy-five percent of the reactors in use today are of this type. In this category are two subtypes:
  a. Boiling-water reactors (BWR).
  b. Pressurized-water reactors (PWR).
- Heavy-water reactors (CANDU, an abbreviation of "Canadian Deuterium").
- High-temperature gas reactors (HTGR).
- Fast breeder reactors (FBR).
  a. Liquid metal fast breeder reactors (LMFBR).
  b. Gas-cooled fast breeder reactors (GCFBR).
- Integral fast reactor (IFR). This is a new design being tested at an experimental installation near Idaho Falls, Idaho.
- Several new-generation designs that are aimed at being simpler, safer, and less costly. These and the IFR could be the answer to future power needs.

LIGHT-WATER REACTORS

Light-water reactors use ordinary water as a moderator and coolant as well as a medium to transfer heat energy from the reactor to the turbine. The water flows around the uranium fuel assemblies carrying away the heat. The water turns to steam so that it can drive a turbine that powers a generator.

NUCLEAR REACTOR

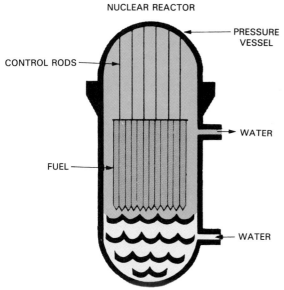

CONTROL RODS

PRESSURE VESSEL

WATER

FUEL

WATER

Fig. 12-13. A pressure vessel is that part of a nuclear reactor that holds the fuel core. Fission takes place inside the vessel. Top. A 900 ton vessel is lowered into a shielding wall at a nuclear generating station in New York state. (Niagara Mohawk Power Corp.) Bottom. Diagram of a typical pressure vessel. Water circulates through the vessel during fission to provide cooling and to transfer the heat needed to power the turbines. (Edison Electric Institute)

Fuel for these reactors must be enriched so that fission will take place. Enrichment involves increasing the U-235 concentration of the uranium from 0.7 percent to 3 percent. This enrichment is a difficult, expensive, and energy-consuming process.

In the United States, virtually all commercial nuclear power is generated by light-water reactors. About one-third of these are boiling-water reactors.

Fig. 12-14. A steel containment vessel such as pictured holds the pressure vessel.

BOILING-WATER REACTOR

Of the two types of light-water reactors, the boiling-water reactor (BWR) is the simplest in design. The water, in cooling the reactor, comes to a boil. Heat of fission brings the water to 545°F (285°C) under a pressure of 1000 psi (6894 kPa). The steam is piped directly to the turbine. A simple schematic of the BWR is shown in Fig. 12-15. Spent steam is cooled to a liquid and returns to the reactor where the cycle is repeated.

PRESSURIZED-WATER REACTOR

The pressurized-water reactor uses water under high pressure to cool the reactor and to absorb and transport the heat produced by fission. However, unlike the boiling-water reactor, the heated water or steam does not go directly to the turbine. It moves through its own piping system, called the **primary loop**, to a steam generator. The coolant water gives up its heat to water in a **secondary loop**. The radioactive water in the primary loop never mixes with water in the secondary loop. This separation prevents water in the secondary loop from becoming radioactive.

Heat, picked up in the steam generator, turns the water to steam that enters the turbine with great force.

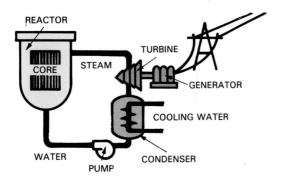

Fig. 12-15. A boiling-water reactor creates steam inside the reactor vessel. The steam is piped directly to the turbine. (Westinghouse Electric Corp., Advanced Power Systems Div.)

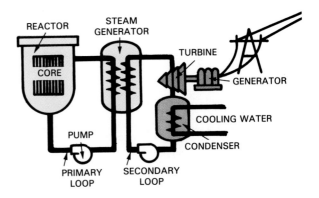

Fig. 12-16. A pressurized-water reactor uses two loops carrying water to transfer heat of fission from the core to the turbine. Cooled water is returned to repeat the cycle over and over again.   (Westinghouse Electric Corp.)

The turbine spins rapidly, providing power to a generator.

Water from the primary loop returns to the reactor after losing much of its heat to the steam generator. Likewise, steam in the secondary loop, after it passes through the turbine, is returned to the steam generator. However, it first passes through a condenser to cool it to a liquid. Fig. 12-16 is a simple schematic of a PWR.

The PWR operates with water at a pressure of 2250 psi (about 15,500 kPa) and a temperature of 600°F (315°C). Because of the high pressure in the primary loop, the coolant in this loop never boils. The boiling occurs in the secondary loop as the coolant picks up heat in the steam generator.

HEAVY-WATER REACTOR

Heavy water is ordinary water that has been enriched with deuterium. (**Deuterium**, also called heavy hydrogen, is an isotope of hydrogen which has twice the mass of an ordinary hydrogen atom.)

When used as a coolant and moderator in a nuclear reactor, heavy water does not absorb as many neutrons as light water. This being the case, the bombarding neutrons can travel farther before they slow down to the right level for controlled nuclear reaction. This allows use of fuel elements with lower amounts of U-235 in the fuel than the 3 or 4 percent required by light-water reactors. Some heavy-water reactors operate on natural uranium.

No commercial reactors of the heavy-water type have been constructed in the United States for the generation of electricity. Canada, France, Germany, Great Britain, and Sweden have them. See Fig. 12-17.

HIGH TEMPERATURE GAS-COOLED REACTOR (HTGR)

If you look at the diagram of the high temperature gas-cooled reactor in Fig. 12-18, you will not see much

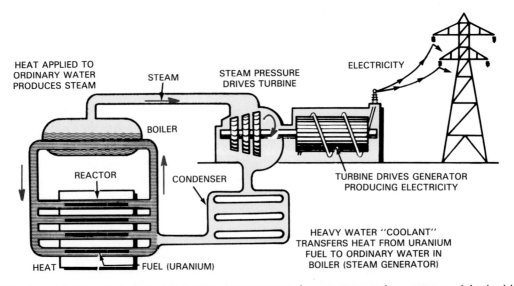

Fig. 12-17. Simplified sketch of the Canadian heavy water reactor. It uses as a coolant water enriched with deuterium, a hydrogen isotope.   (Nuclear Power in Canada/Canadian Nuclear Assn.)

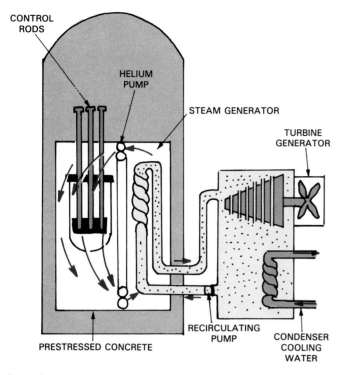

CONTROL RODS

HELIUM PUMP

STEAM GENERATOR

TURBINE GENERATOR

PRESTRESSED CONCRETE

RECIRCULATING PUMP

CONDENSER COOLING WATER

Fig. 12-18. The high temperature gas-cooled reactor (HTGR) uses helium gas to carry heat from the pressure vessel to the steam generator.

difference between it and the light-water reactor. They are constructed much the same except that the HTGR uses helium gas to carry the heat away from the reactor core to the generator. There are several advantages in the use of helium:

• Heat is transferred much better because the system can operate at higher temperatures and pressures.

• There is less danger from radiation since helium is an inert gas. (This means it does not react well with other chemical elements.)

Preparing the fuel

The U-235 and the thorium are formed into tiny particles called **microspheres**. One microsphere contains U-235 and the other contains thorium. Then the microspheres are coated with graphite and blended into a pellet. Each pellet is about 1/2 in. (12.5 mm) in diameter and 1 1/2 in. (38 mm) long. A single fuel element is made up of about 2000 fuel rods.

Controlling fission

As with the light-water reactor, the rate of fission in the HTGR must be controlled. Rods containing boron provide this control. Graphite (a material like the "lead" in a lead pencil) is used as the moderator. The graphite is machined into large blocks. Holes drilled into the blocks hold the fuel rods.

The HTGR uses two isotopes of uranium as fuel and natural thorium. U-235 is used to start the nuclear reac-

tion. During the nuclear reaction, thorium 232 (Th-232) is converted to U-233 which is fissionable. (Thorium is used because it is plentiful and less expensive than U-238.)

Like the pressurized-water reactor, the HTGR uses two separate heat transfer systems. Superheated helium in the first loop circulates around a steam generator inside the reactor vessel. Water flowing through the second loop turns to steam in the steam generator and is piped, under high pressure, to the turbine.

Pressure and temperature are very high. The helium reaches temperatures above 1425°F (775°C) and a pressure of 700 psi (4825 kPa). In the secondary loop, temperatures reach 1000°F (535°C) and a pressure of 2400 psi (16,545 kPa).

THE BREEDER REACTOR

The breeder reactor is usually called the fast breeder reactor because there is no moderator to slow down the neutrons that are creating fission. This design can produce more nuclear fuel than it uses while it is producing electricity. The process, as you learned earlier, is called breeding because it adds a neutron to U-238 and changes it to a type that can become part of a nuclear chain reaction, Fig. 12-19.

Use of the breeder reactor would produce fuel from useless isotopes making up over 99 percent of the mined uranium. This would cut down greatly the amount of fuel that must be mined, supporters of the breeder reactor program say.

Types of breeder reactors

There have been two types of breeder reactors under consideration. One is the liquid metal fast breeder reactor (LMFBR). The other, is called the gas-cooled fast breeder reactor (GCFBR). However, the latter appears to hold less interest among developers than the LMFBR.

Liquid metal breeder reactor

The liquid metal breeder reactor has been the mainstay of every breeder development program, including programs in the United States. Fig. 12-20 is a schematic of this reactor design.

The LMFBR uses liquid sodium as a coolant. (Sodium is a soft, waxy, silver white, metallic alkali.) Its value as a coolant lies in its ability to absorb more heat than water and helium. It melts at 210°F (98°C) and boils near 1600°F (880°C).

The primary loop carries the heated sodium to a heat exchanger inside the containment structure. This heat exchanger must be housed inside the building because the sodium becomes radioactive.

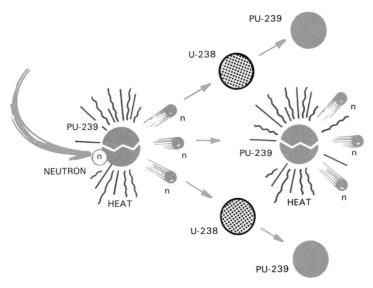

Fig. 12-19. Producing plutonium (Pu-239) from U-238. During fission U-238 isotopes each absorb an extra neutron and become plutonium isotopes. (Westinghouse Electric Corp., Advanced Power Systems Div.)

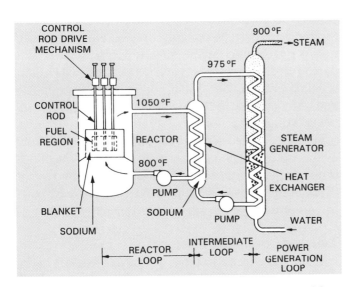

Fig. 12-20. A simple schematic of a typical liquid metal fast breeder reactor. It produces more new nuclear fuel than it uses. (Breeder Reactor Corp.)

A second sodium loop takes the heat from the first exchanger and delivers it to a second exchanger. A third loop receives water that is heated by the second exchanger and delivers it as steam to the turbine.

NUCLEAR PROPULSION SYSTEMS

Nuclear propulsion is the use of a nuclear reactor to power transportation vehicles. Nuclear fuel is required in very small amounts and does not need to be replaced often. The small nuclear reactor, by this fact, has some advantages for moving people and materials.

It is feasible for ships, aircraft, and rockets where cost is not as important as the need to go long distances and for long periods without refueling. Nuclear power

is not practical for small land vehicles because of the protective shielding needed.

NUCLEAR POWERED VESSELS

The U.S. Navy has several ships that are nuclear powered. Among them is the aircraft carrier, Dwight D. Eisenhower, Fig. 12-21.

The nuclear powering of ships requires a reactor similar to, but smaller than, the reactors used in electric power generation. Heated coolant (steam) drives a turbine connected directly to the propellers. Some of the steam is used to drive a generator which supplies electrical power.

Nuclear power for transportation was first used in submarines. The reactors are the pressurized light-water type. Molten sodium was tested in a submarine but was later replaced with a pressurized water reactor. Fig.

Fig. 12-21. The USS Dwight D. Eisenhower is a modern aircraft carrier that operates on nuclear power. (U.S. Navy)

12-22 shows a modern nuclear powered submarine and a schematic of its propulsion unit.

NUCLEAR AIRCRAFT ENGINES

Experimental models of two nuclear aircraft engines have been built:
• A direct-cycle engine.
• An indirect-cycle engine.

In the direct-cycle engine, compressed air is piped into the reactor core. It is heated and carried back to the engine. Here it is expelled to create thrust like that of a jet engine.

The operation of the indirect-cycle is similar. A liquid metal coolant enters the reactor core, picks up heat, and transfers it to compressed air outside the reactor. Then the heated compressed air is exhausted in the engine to drive the aircraft.

NUCLEAR POWERED ROCKETS

During the 1960s the United States developed a nuclear reactor suitable for propelling rockets. Known

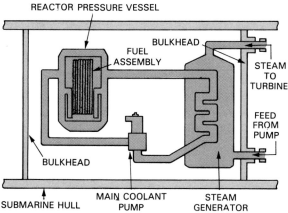

Fig. 12-22. Nuclear power for propulsion was first installed in submarines. Top. A starboard quarter view of the nuclear-powered submarine, Oklahoma City under way. Bottom. Schematic of a typical submarine nuclear power plant. (U.S. Navy)

as the Kiwi, it was a low-powered reactor using hydrogen gas as a propellant. In the 1990s there was discussion among scientists and environmentalists whether nuclear-powered space vehicles should be allowed.

BY-PRODUCTS OF FISSION

Fission has several by-products. Some are useful; others are not. As U-235 isotopes are splitting, fragments of materials are left over. One of the fragments is a heavy isotope whose name, plutonium, you have heard earlier in this chapter. You will recall that plutonium results when the isotope U-238 gains a neutron during a chain reaction. Other by-products include elements that are lighter in weight than uranium. A few by-products are gases. Some fragments are not radioactive while in others the radioactivity lasts only fractions of a second. Still others are radioactive for thousands of years.

Every nuclear reactor builds up these by-products during fission. After a while the by-products become so plentiful that the efficiency of the reactor is reduced. When this happens, the fuel rods containing the pellets of fuel must be replaced. This usually happens after about three years of operation. Every year about one-third of the fuel rods are replaced.

REPROCESSING AND STORAGE

Old rods removed from the reactor, in addition to having usable by-products, still contain about 30 percent of the original fuel. The large numbers of by-products would have stopped fission had the rods been allowed to remain longer in the reactor. Both the left-over U-235 and the plutonium by-product can be reclaimed and reprocessed into new fuel.

TEMPORARY STORAGE

The spent fuel rods are first stored in pools of water at the reactor because of their radioactivity. After four months, when the radioactivity has been reduced by more than 90 percent, the rods are safe to ship in special containers. See Fig. 12-23.

Again, the fuel is stored. The shipping casks are opened under water. Once more, the fuel rods are removed and stored in pools.

REPROCESSING

About 99 percent of the uranium and plutonium can be recovered from the spent fuel, Fig. 12-24. In the recovery process the rods are moved to concrete-walled

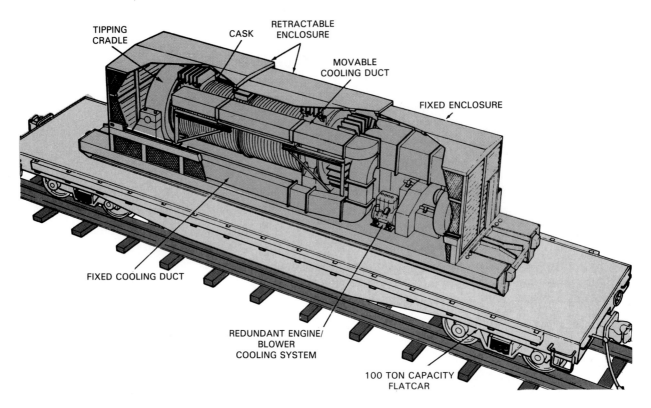

Fig. 12-23. Fuel removed from a nuclear reactor can be shipped in a spent-fuel cask shown here in cutaway. The shielding protects against radiation. The cask is mounted on a railcar. (Atomic Industrial Forum, Inc.)

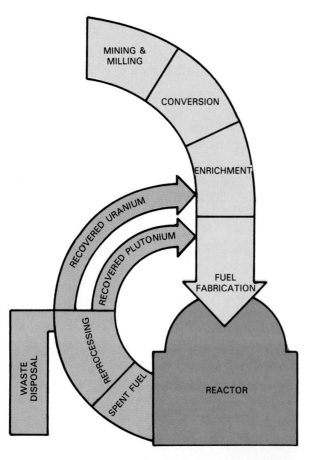

Fig. 12-24. Reprocessing of spent fuel recovers 30 percent of the original fuel load of U-235 as well as a large amount of plutonium converted during fission.

work areas. The work rooms, called "hot cells," have walls several feet thick to protect workers from radiation. Here the rods are cut into small pieces. The fuel must be handled entirely by remote controlled machinery.

Next, the pieces are treated to a nitric acid bath to leach out (dissolve) the spent fuel. Another chemical separates the plutonium and the other uranium isotopes from the acid solution.

Special plants are required for reprocessing of spent fuels. Depending on its capacity, each could service 30-50 commercial nuclear power plants of 1000 MWe. Probably few countries will have reprocessing plants until the end of the century.

WASTE TREATMENT AND STORAGE

Nuclear wastes are classified as either high level or low level. High level wastes are the fission products and fuel elements that have been removed from operating nuclear reactors. Among other things, these wastes contain quantities of strontium, cesium, and elements with half-lives of more than 25-30 years. Low-level wastes contain mostly short-lived radioactive materials whose radioactivity will likely disappear within 25-30 years. They usually come from contaminated solid waste. These wastes are produced in every state by medical centers, manufacturing facilities, universities, and other organizations licensed to use nuclear materials. Included in this waste stream are such things as machine parts, protective clothing, tools, test tubes, filters, and general scrap. Those resulting from operation of nuclear power generating facilities include particles lodged in filters, pumps, valves, or which collect on surfaces inside the plant. The waste is picked up by items such as filters, cloth, paper wipes, plastic shoe covers, tools and materials, resins, and other residue.

Waste containing plutonium, neptunium, and other radioactive materials with short half-lives, are considered low-level waste having low radioactivity.

Leftover waste must be safely stored so that it does not present a danger to health. (Since some wastes are found in liquid form, its volume can be reduced through evaporation.) The waste is then stored for up to five years in double-walled stainless steel tanks. Finally, the waste, still liquid, will be solidified and stored indefinitely.

It is estimated that the yearly volume of waste from one nuclear reactor will occupy 70 cu. ft. (1982 L), or a space roughly 4 ft. square and 4 1/2 ft. high. This solid waste would be contained in 12 stainless steel canisters 1 ft. (about 30 cm) in diameter and 10 ft. (about 3 m) long.

It is estimated that there are over 256 million acres (103 hectares) of underground salt formations in the United States. About 2000 acres would be needed as storage. Fig. 12-25 shows how an underground storage vault might look.

TRANSMUTATION

Scientists are exploring still another method of disposing of the radioactive waste through transmutation. This is a process which separates radioactive chemicals into materials with a shorter radioactive life.

THE FUTURE OF FISSION

Nuclear power stations now account for about 21 percent of the electric power generated in the United States and could eventually produce 45 percent or more. The prospect of having a fuel source, plutonium, which lasts 30,000 years is attractive and exciting. This would allow technology the time to develop other energy sources, especially solar, wind power, and, perhaps, nuclear fusion.

However, nuclear fission as a power source is, as was noted earlier, burdened with a certain amount of public opposition stemming from fear of radiation as a result of potential nuclear accidents and disposal of nuclear wastes.

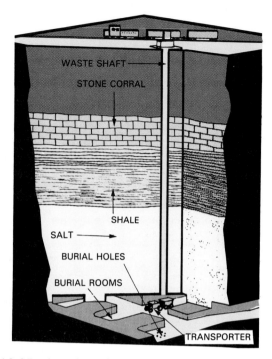

Fig. 12-25. A method of storing radioactive wastes is to bury them underground in salt formations. (Atomic Industrial Forum, Inc.)

Running concurrently with the fear of radiation is the fear that plutonium, while being transported, could fall into the "wrong hands." There is concern that, as more and more plutonium is transported, it could be seized by terrorists or blackmailers and used to produce bombs.

Experts in the nuclear power field have responded with facts and assurances aimed at reducing or dispelling these fears.

• The amount of radiation from the emissions of nuclear generating stations are insignificant when compared to the average radiation received from other sources. The average North American receives about 55 millirem of radiation a year from X-rays and medical procedures that use radioactive materials. Another 10 millirems per person are received from the mining of minerals, burning of fuels, building of houses, and other activities that redistribute natural radiation. Other contributory factors are natural radiation in the earth, fallout from past testing of nuclear weapons, and consumer products including color television sets and luminous-dial clocks. Altogether, natural and human-made radiation expose the average person to about 360 millirem a year. Even the Three Mile Island accident, the worst in the history of U.S. commercial reactor history caused an average exposure of only 1.5 millirem to people living within 50 miles (about 80 km) of the reactor.

- Safeguards used at nuclear power stations reduce, if they do not eliminate, the danger of accidental exposure to radiation. In fact, the operating stations have a good safety record. Further, designs that led to the disastrous meltdown at Chernobyl are not permitted in any other country. Built-in safety measures would stop fission long before there would be any danger of a meltdown.

On the downside, a concern still to be addressed is the potential for nuclear weapons development from hijacked plutonium or other fissile isotopes. Fissile isotopes such as U-235, U-233, and Pu-239, that are used to fuel commercial reactors are not sufficiently "enriched" to be suitable weapons material. It is also true that terrorist groups would not have the facilities to enrich the isotopes to weapons grade. However, the development of enrichment plants would put enriched uranium in the grasp of some nations. Another concern is that stored wastes will accidentally leak radiation into the water supply. Once there, it would be introduced into the entire food chain.

WHAT IS RADIATION?

Radiation in the nuclear sense, is ionizing radiation. It is the result of unstable atoms that want to become something else that is stable. As these atoms (also called radionuclides) break up they give off particles of energy. Such radiation is found naturally in soil, water, rocks, building materials, and food.

Radiation may take a number of forms, including:
- Beta rays. These are electrons or positrons (the counterparts of electrons).
- Alpha particles. These are helium nuclei.
- Electromagnetic radiation such as gamma rays and X rays.

These forms of radiation cause the formation of ion pairs in matter that they pass through. Ionization (gaining or losing an electron) as the radiation passes through tissue, deposits energy that changes the tissue in a certain way.

This energy entering tissue is called the absorbed dose. The amount of radiation absorbed per gram of body tissue is measured in a unit called a **rad**.(The name comes from **r**adiation **a**bsorbed **d**ose. This term is being replaced by a new international unit called a gray. One gray = 100 rad.) Any dose must be weighted for damage potential. This measurement is expressed as **rem** (for **r**oentgen **e**quivalent **m**an; a roentgen is the measure of X ray and gamma ray radiation given off by radioactive material). The international unit is the sievert. One sievert = 100 rem.

The average person in North America receives an annual dose of about 180 million millirems (mrem) from natural sources. A millirem is one-thousandth of a rem.

Two categories of effects result from radiation exposure: one is acute; the other is long term. Exposure to radiation causes cell damage that, depending on the dosage and rate, may show itself almost immediately or after a long period of dormancy. A dose of 10,000 rem will bring death quickly because of damage to the nervous system. At 300 rem delivered to the whole body, death will ensue in 50 percent of the cases. Dosages of from 100 to 300 rem will likely cause injury.

Smaller doses spread over a long period do not cause enough tissue damage to cause death or illness. However, the effects may show up years later. While there are a number of ill effects, two are of greatest concern: cancer and genetic damage. Cancer affects only the persons exposed; genetic damage affects succeeding generations.

(The nuclear meltdown and explosion at Chernobyl caused radioactivity of 15,000 to 20,000 roentgen per hour at the reactor core. As distance from the rubble of the ruined reactor increased, radioactivity declined in proportion to the square of the distance. Thus, a person on site at Chernobyl immediately after the accident would have absorbed 15,000 to 20,000 rem per hour of radiation. The permissible maximum exposure for power plant workers is 5 rem per year and 1.3 millirem in a 24 hour period.)

A number of national and international bodies, including the World Health Organization, have made the following recommendations:
- Persons working where exposed to radiation should receive no more than 5 rem per year of radiation in the course of their work.
- The general public should receive no more radiation than 0.5 rem per person per year.
- Large population groups should receive no more than 0.17 rem per year.

These limits are on exposures from human-made sources. Exposure from medical treatment are not included.

Public opposition to nuclear power is based upon the fear of the hazards of radiation. In response to this expressed fear, proponents of nuclear power, among them scientists, point out that since 1970 the radiation from all commercial nuclear energy activities averaged 0.01 mrem per person. Moreover, they say, no person in North America, whether working in the nuclear power industry or living near an installation, have ever been exposed to radiation levels above the annual dose limits established by international safety standards.

Science and technology are working to build in ever better safeguards against accidental or malicious discharge of radiation materials into the air or water.

Eventually people must weigh these safeguards against the risks and such factors as:
• The need for electrical energy.
• The side effects of fossil fueled power generation.
• The likelihood of alternate sources being able to generate enough electric power.

Then they must decide on the appropriate energy source or sources.

NUCLEAR FUSION

Another source of nuclear energy is called **fusion.** Unlike fission which is the splitting of atoms, fusion is the combining of atoms. This is how the sun produces radiation while using hydrogen as a fuel.

Getting two atoms of any kind of matter to fuse is very difficult since the nucleus of the one atom repels the nucleus of another atom. You can think of the nuclei as similar poles of a magnet. If you try to push the positive poles together the magnets will resist. Nuclei are positively charged with protons. Therefore one nucleus will have to be hurled at another with great force to cause them to fuse. The more protons the nucleus has, the more it resists fusion. It is therefore necessary to use nuclei with the smallest number of protons. See Fig. 12-26.

HYDROGEN ISOTOPE

Two isotopes of hydrogen are the most suitable fuel for fusion. One is the isotope **deuterium** (D) also known as "heavy hydrogen." The other is the isotope **tritium** (T). It can be extracted (separated) from water and from lithium, a soft, silver white metal. Though more rare than deuterium, it is abundant enough that supplies from lithium alone would provide sufficient energy for 50,000 years.

Heavy hydrogen is in plentiful supply on the earth since it is present in water. For every 6500 atoms of ordinary hydrogen in water there is one atom of deuterium. It is calculated that the earth's oceans, rivers, and lakes hold about 10 trillion tons of deuterium. Between six and eight trillion tons of this are recoverable.

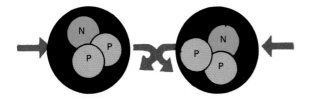

Fig. 12-26. Nuclei of atoms tend to repel each other. This makes fusion difficult to achieve. The fewer protons in the nuclei, the easier fusion becomes.

DEUTERIUM AS AN ENERGY SUPPLY

The amount of energy available from deuterium is staggering when compared to any other energy source. When two deuterium particles fuse, the new mass (weight) of the combined particles is less than the sum of their masses before the fusion. The lost mass is converted into energy. The burst of energy from even a tiny amount of deuterium is gigantic. A look at the four reactions in fusion indicates how much energy is released, Fig. 12-27.
• In the first two types of reactions, deuterium particles react with other deuterium particles:
  a. A deuterium particle fuses with another deuterium particle. This forms a tritium particle and releases one proton and 3.3 Mev (75 million Btu per gram) of energy.
  b. A deuterium particle fuses with another deuterium particle forming a helium particle and releasing a neutron and 4 Mev (92 million Btu per gram) of energy.
• In two additional reactions, deuterium forms energy while fusing with products of the first two reactions, Fig. 12-28.
  a. Deuterium fuses with tritium (T), forming a new isotope of helium (He4) while releasing a neutron and 17.6 Mev (313 million Btu per gram) of energy.
  b. Deuterium fuses with helium (He3) forming a new helium isotope and releasing a neutron.

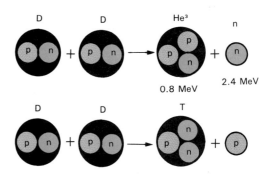

Fig. 12-27. The fusion of two nuclei of deuterium release from 2.4 to 3 million electron volts of energy. (U.S. Atomic Energy Commission)

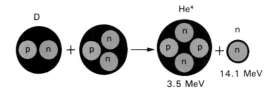

Fig. 12-28. When fusion takes place between deuterium and tritium, the release of energy equals more than 14 million electron volts.

Energy released equals 18.3 Mev (334 million Btu per gram) of energy.

By comparison, a ton of coal produces only 25 million Btu when burned. Each ton of deuterium produces 60 million times more energy. At present rates of energy consumption, the world could be using fusion energy for the next 50 billion years!

CREATING FUSION

As you already know, getting two particles of matter to fuse is not easy. Something must be done to get them to move with such force that they overcome their electrical repulsion. They must be hurled at each other with great force.

In early experiments, scientists used a **particle accelerator** to hurl deuterium particles against a stationary target made up of deuterium or tritium. This caused fusion but no chain reaction. Much more energy was consumed in causing the fusion than was released.

Later experiments proved that it is much more practical to use heat to speed up the motion of the particles. If a mixture of deuterium and tritium is heated to a temperature of 90,000,000 to 180,000,000°F (50,000,000 to 100,000,000°C), the particles will collide and fuse.

PLASMA

Matter subjected to such high temperatures as are needed for fusion will not remain in their normal state (solid, liquid, or gaseous). They turn into a fourth state called **plasma** or **ionized gas.** In this state the electrons leave the nuclei and form a mix of charged electrons and the stripped nuclei. These nuclei, called ions, are positively charged while the electrons are negatively charged.

Although plasma contains free positive ions and free negative electrons, the numbers of positive and negative electrical charges are in balance. The plasma as a whole, then, is electrically neutral.

CONFINEMENT

Maintaining high temperatures to create plasma is hard. It calls for special containers. Ordinary bottles, cans, or tanks will not do. The particles have to be contained in a vacuum where they will not touch any matter which will take away their heat. There are two problems then:
- Finding a way to contain the mixture. It must be confined for about one second.
- Suspending the mixture in a vacuum where it can be kept free of any foreign substance and out of con-

tact with the sides of the container.

Two ways of containing and protecting the mixture are:
- Magnetic confinement.
- Inertial confinement.

MAGNETIC CONFINEMENT

Since the plasma must literally float in a vacuum, a method is needed to suspend it in its containing vessel. If it were allowed to touch the sides, the plasma would cool off so rapidly that fusion would stop altogether. This is done by surrounding it with a magnetic field which presses inward on the plasma keeping it away from the vessel sides. This magnetic action is often referred to as a "magnetic bottle."

Experiments are underway using three different systems of magnetic confinement.

The "doughnut" chamber
The toroidal-shaped (doughnut-like) chamber is hollow. The plasma travels around in this vacuum chamber, Fig. 12-29. The Russian version, known as "Tokamak," has been the most successful. Most U.S. experiments are being done with this type. See Fig. 12-30.

Magnetic mirrors
The magnetic mirror system, Fig. 12-31, uses open-ended tubes rather than closed tubes to contain the plasma. The system is based on the fact that the plasma cannot pass through a magnetic field. The tubes have field coils wrapped around them. When a current passes through the coils a magnetic field is set up. The coils are arranged so that the magnetic field is strongest at the open ends. The stronger field of magnetism is called the "mirror." It turns back the plasma toward the center of the tube where the magnetic field is weaker. The plasma will loop back and forth trapped between the magnetic mirrors, Fig. 12-32.

Magnetic pinch device
In a magnetic pinch device, the torus shaped vessel is used. The interior is filled with plasma that is compressed or "pinched" by rapid compression of the magnetic field. This "squeezing" is produced by increasing the strength of the field and forcing the plasma toward the center of the tube. See Fig. 12-33.

Heating the plasma
Heating of the plasma can be done in one of three ways:
- Electric currents induced into the plasma. The plasma heats up like the heating element of an elec-

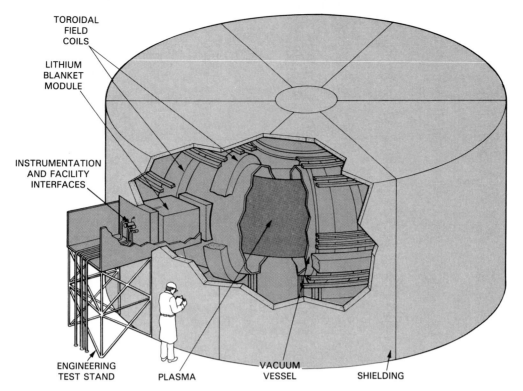

TOROIDAL FIELD COILS

LITHIUM BLANKET MODULE

INSTRUMENTATION AND FACILITY INTERFACES

ENGINEERING TEST STAND

PLASMA

VACUUM VESSEL

SHIELDING

Fig. 12-29. Cutaway of the Princeton University fusion test reactor. Hot plasma is suspended in the chamber by magnetism.   (EPRI Journal)

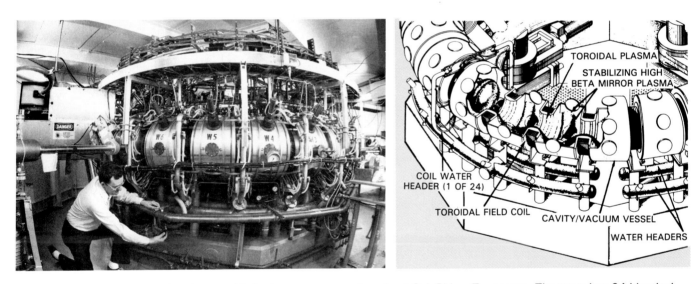

TOROIDAL PLASMA

STABILIZING HIGH BETA MIRROR PLASMA

COIL WATER HEADER (1 OF 24)

TOROIDAL FIELD COIL   CAVITY/VACUUM VESSEL

WATER HEADERS

Fig. 12-30. Left. Fusion chamber used in fusion energy experiments at Oak Ridge, Tennessee. The torus has 24 identical cavities. Right. Cutaway of the chamber.   (Union Carbide Corp., Nuclear Div., Oak Ridge National Laboratory)

tric toaster. Toroidal vessels rely, to some extent, on this kind of heating.

- A nuclear beam injected into the plasma. The beam is made up of very energetic deuterons (deuterium isotopes). They are fired into the plasma inside confinement vessels.
- Compression of the plasma. This causes the plasma to heat up enough to start the fusion reaction. This technique can be used in the magnetic pinch confinement vessels. Pinching heats up the plasma while confining it.

INERTIAL CONFINEMENT

The inertial system does not use magnetism to confine the plasma. Rather, the fusion fuel, a mixture of deuterium and tritium, is frozen into small, round droplets less than 1 mm (about 1/25 in.) in diameter. One of the droplets is located in the middle of a vacuum chamber. Then it is hit from all sides by an energy source. This source can be a laser beam or a beam of charged nuclear particles. The particles may be ions or electrons. (An ion is either positively or negatively

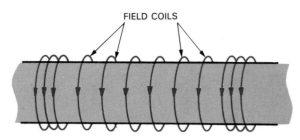

Fig. 12-31. Magnetic mirror system. Field coil circling the plasma chamber sets up a magnetic "girdle" to contain the plasma.

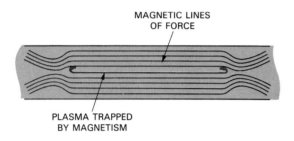

Fig. 12-32. Strong magnetic lines of force at the ends of the chamber trap the plasma, turning it back to the center of the chamber. (Union Carbide Corp., Nuclear Div., Oak Ridge National Laboratory)

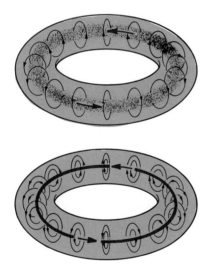

Fig. 12-33. How plasma is "pinched" by magnetism. Top. At first, plasma takes up much space in the toroidal (ring shaped) tube. Bottom. However, when heavy current flows through the coil surrounding the tube, the magnetic lines of force squeeze the plasma into a smaller space and make the plasma more dense. (Union Carbide Corp.)

charged as a result of having gained or lost one or more electrons.)

LASER BEAM BOMBARDMENT

A powerful laser beam may be focused with great accuracy on the pellet. A system of mirrors breaks the beam into several smaller beams that hit the pellet from all sides.

Speed is of the essence. The bombardment must take place in a split second. This heats the fusion fuel rapidly without expanding it. The outside layer of the pellet vaporizes and rushes outward. The rapid vaporization creates a force against the remainder of the pellet causing it to implode (fall inward with crushing force). The great density and heat ignites the fusion reaction. The pellet explodes as shown in Fig. 12-34 and fusion takes place. Most of the energy is carried by the released neutrons.

FUSION ENERGY AND THE ENVIRONMENT

Successful fusion reaction does not threaten the environment to the same extent as fission reaction. Some writers contend that fusion could be as nonpolluting as any energy we have except solar. The advantages of fusion energy as an environmentally acceptable energy are:

• The danger of a nuclear accident is sharply reduced. Fusion reactors do not contain the huge amounts of radioactive material characteristic of fission reactors.
• Danger of meltdown would be greatly reduced, if not altogether eliminated. If any part of the system fails, the plasma is destroyed and the reaction stops.
• Storage of waste material from spent fuel is less of a problem for two reasons. First, the toxic nature of the waste is not so long-lived as fission by-products. Second, the amount of waste is small compared to fission.
• Since plutonium is not a by-product of fusion, the danger of nuclear materials falling into unfriendly or irresponsible hands is removed.
• Gathering the fusion fuel does not destroy the en-

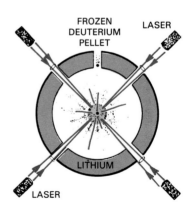

Fig. 12-34. One way of creating fusion. Laser beams can be directed at a frozen pellet of deuterium inside a vacuum chamber. The laser beams cause violent compression of the deuterium pellet resulting in the release of heat and neutrons. A layer of lithium recovers the heat and breeds tritium. (Nuclear Power in Canada/Canadian Nuclear Assn.)

vironment. Removing one hydrogen atom in 6500 from sea water will have no measurable effect on the water.

SUMMARY

Nuclear power plants use heat derived from nuclear fission to generate electric power. Worldwide, there are over 400 nuclear reactors in operation producing more than 300,000 megawatts—enough to supply all the electric power for 159 million homes. Still, in the United States, nuclear power activities encounter considerable public opposition due to the fear of radiation and its effects on life and health. Nuclear power engineers, nuclear power experts, and many scientists attest to the safety of nuclear energy and offer facts and statistics to back up their claims. Still, the fears and opposition persist.

The fuel for nuclear reactors comes from certain isotopes of uranium, specifically U-235 and Pu-239. The mined uranium ores must be refined and enriched. The resulting fuel is called yellow cake.

Fuel is contained in a reactor. Fission takes place in a strong container called a pressure vessel. There are a number of designs of nuclear reactors. They are: Light-water reactors, Boiling-water reactors, Pressurized-water reactors, Heavy-water reactors, High-temperature gas reactors, Fast breeder reactors, and Integral fast reactors. Smaller reactors are used in certain types of vehicles as a source of propulsion energy.

Spent fuel can be stored and reprocessed into new fuel. For now, North American nuclear power stations have been storing spent fuel on site. Eventually satisfactory storage off site will have to be arranged. Proposed sites have met with public opposition.

Another source of heat energy is fusion. Efforts to develop this source continue although it has not been possible to sustain a chain reaction without using more energy than created. Experts involved in fusion development believe that fusion may be commercially available within the next 20-25 years.

DO YOU KNOW THESE TERMS?

breeding
chain reaction
control rods
deuterium
fuel assemblies
fusion
ionized gas
isotopes
microspheres
moderators
nuclear reactor
particle accelerator
plasma
plutonium
pressure vessel
primary loop
rad
radiation
rem
secondary loop
tritium
uranium
yellow cake

SUGGESTED ACTIVITIES

1. As a class, research the texts and other literature on nuclear energy and develop a written report. State your conclusions, if any, about the future of nuclear power in North America.
2. Research your library or resource center on the subject of fusion energy. Develop a report on whether the United States should continue research and development of a commercial fusion reactor.
3. Construct a model of one of the following:
   a. Containment vessel for a nuclear reactor.
   b. Nuclear fission chain reaction.
   c. Underground storage vault for nuclear waste.
   d. Fuel assembly for a nuclear reactor.

TEST YOUR KNOWLEDGE

1. Different forms of the same element are called _____.
2. Describe, in simple terms, how nuclear fission takes place.
3. Heat of fission is due to (select the correct answer):
   a. Friction between neutrons.
   b. Some atoms which burn up.
   c. Conversion of matter into energy.
   d. The speed at which the freed neutrons are traveling.
4. _____ _____, a mixture of oxides of uranium, may be in the form of a bright yellow slurry or a powder.
5. Name six basic types of nuclear reactors.
6. The _____ reactor will actually produce more fuel than it uses to achieve fission and chain reactions.
7. A fuel element is a bundle of hollow rods containing pellets of uranium fuel. True or False?
8. Plutonium is an atom (or isotope) of uranium

which has gained a _____.

9. Explain how nuclear power can be an advantage for propelling vehicles.

10. Nuclear fusion (select correct answer):
   a. Is the combining of two atoms into one.
   b. Is easiest using two isotopes of hydrogen.
   c. Is an attractive energy source because the supply of fuel is virtually unlimited.
   d. Is already commercially feasible.
   e. All of the above.

11. To create fusion, the deuterium must be heated to temperatures in excess of (select correct answer):
   a. 5000°C.
   b. 10,000°C.
   c. 25,000°C.
   d. 50,000°C.
   e. 100,000°C.

12. List the three ways in which fusion fuel (plasma) may be heated.

13. List four advantages of fusion energy.

13
CHAPTER

BIOMASS ENERGY

The information given in this chapter will enable you to:
- *Define biomass and list its various forms.*
- *List and describe processes by which biomass is converted to energy.*
- *List and describe fuels produced through biomass conversion.*
- *Describe the densification process for producing solid biomass fuel.*
- *Discuss the potential for a biomass fuel industry.*
- *Discuss problems connected with growing and collecting biomass materials.*

Biomass is live organic material such as trees, farm crops, seaweed, or algae, Fig. 13-1. It is also wastes such as manure, trash, sewage, and garbage. All biomass is a form of indirect solar energy. It represents material that once grew and thus captured sunlight. The process by which plants store energy is called **photosynthesis**.

The term, **biomass**, has become a common part of our energy language. It comes from the words biology and mass. **Biology** is a study of living organisms, which includes plant life. **Mass** refers to the body of organic matter available in a region, country, or in the whole world.

HOW BIOMASS IS PRODUCED

As you learned in Chapter 1, plant life is a form of potential energy. The energy is stored in the cells of plants.

The plants use radiant energy to convert carbon dioxide, water, and nutrients from the soil into sugar and oxygen. Certain **enzymes** within the leaves of the plant

Fig. 13-1. Biomass resources on earth include all that grows. Forests are important sources of wood for heating and lumber products for construction. Croplands produce fibers that are used as animal feeds, alcohol, and other energy materials.

help make the process work. (An enzyme is a type of protein which causes chemical action to take place within plant and animal cells.) Fig. 13-2 illustrates the process of photosynthesis in a plant.

Carbon is an important element found in the air and in all kinds of organisms living and dead, Fig. 13-3. The amount of carbon stored every year through photosynthesis is about 10 times the world's total annual energy use. Only a half percent of this carbon is

LOCATION	QUANTITIES (in billions of tons)
Atmosphere	700
Dead organic matter land	700
Land plants	450
Marine zooplankton	5
Marine phytoplankton	more than 5
Dead organic matter marine	3000
Fossil fuels	10,000
Sedimentary rock	20,000,000
Sea surface layers	500
Deep sea	34,500

Fig. 13-3. Carbon is distributed throughout the world in plant and animal life as well as in the atmosphere.

consumed as food or used to produce fibers for manufactured products. The rest is available as a source of largely untapped energy.

BIOFUEL USAGE

Wood and other biomass sources provides close to 5 percent of the total energy consumed in the United States. Energy from wood totaled about 2.5 quadrillion Btu in 1989, the last reported year. Energy derived from other biofuels, such as agricultural wastes, solid municipal wastes, and alcohol fuels amounted to 0.4 quadrillion Btu.

Many states have begun to increase the percentage of biomass in their energy mix. Hawaii is an example. Fifty percent of its energy is obtained from renewable resources, half of that amount from biomass. Maine generates 23 percent of its electric power from biomass. Nationwide, electrical power generation with biomass has grown from 200 megawatts in 1979 to more than 7300 megawatts in 1989.

Internationally, biofuels are becoming more important as an energy source. Currently, biomass use for energy accounts for 12 percent of all energy production. In many developing countries biomass is 50 percent of their energy totals. The European Economic Community (EEC) encourages the increased use of renewable resources. Sweden plans to cultivate fast-growing hardwood hybrids as fuel for future power plants.

CONVERTING BIOMASS ENERGY

Before it can be used as any kind of energy or food source, biomass must be converted to a more usable form. Plant life is converted in several ways:
• Metabolic conversion. This involves consumption (eating) of plants (or other food) by animals, Fig.

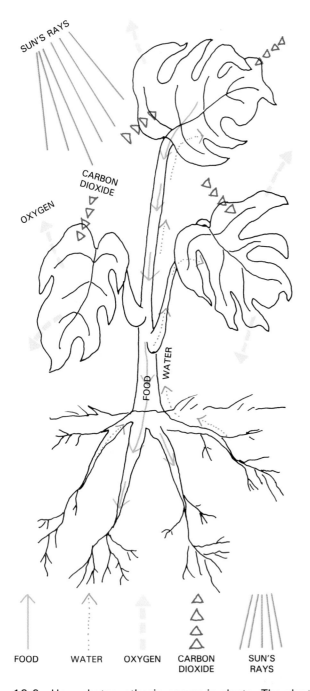

Fig. 13-2. How photosynthesis occurs in plants. The plant takes in sunlight, water, carbon dioxide, and nutrients (from the soil). The plant combines the carbon with hydrogen from the water and stores it as plant material. Oxygen is released to the air as photosynthesis is taking place.

13-4, or humans. The metabolism (digestion) of the animal and human body is such that it reverses the process of photosynthesis. Heat, at a very low level, oxygen, and enzyme action break down the plant cells to create food for energy and growth. The by-product is carbon dioxide and fecal wastes.

- Thermochemical conversion. This is an industrial process which depends upon higher levels of heat to burn the biomass or to convert it to a combustible gas or liquid.
- Biological conversion. This, too, is an industrial process. It uses enzymes, fungi (molds), or micro-organisms (bacteria) to cause a chemical change resulting in a gaseous or liquid fuel.

BIOMASS FUELS

Biomass conversion yields several types of fuel:
- Biogas (also called synthetic natural gas) which is chiefly methane, a flammable gaseous form of hydrocarbon.
- Ethanol and methanol, volatile, flammable liquids.
- Petroleum.
- Solid fuels such as char and chips. These include charcoal made from wood and the combustible residue remaining after the destructive distillation of coal.

THERMOCHEMICAL SYSTEMS

Thermochemical processes include:
- Direct combustion.
- Pyrolysis.
- Gasification.
- Liquefaction.

DIRECT COMBUSTION

This is the simplest and best developed of all the processes. Dry wastes and residues from forests, farms, and cities are burned directly to produce steam, electricity, or heat for industries, utilities, and homes, Fig. 13-5.

The forest products industry, for many years, has used biomass successfully to produce process steam and electricity for lumbering operations. According to estimates, forest wastes contribute 1.1 quadrillion Btu or about 45 percent of the energy used by the industry. Other industries producing textiles and paper products can also use biomass to provide heat for their manufacturing processes.

Home heating is also possible with a wood burning stove or furnace. In a well designed and well insulated home, wood may well provide all the heat needed.

Direct burning of biomass has been limited because of the problems it presents. As a solid fuel it is quite bulky and difficult to transport long distances. Further, it causes some air pollution because of the gases and solids given off during burning. Therefore, it is more desirable to convert it to combustible gases through other thermal processes.

Fig. 13-4. Animals, as well as humans, reverse the photosynthesis process. At low heat levels, their bodies use oxygen and enzymes to break down plant life into food.

Fig. 13-5. Some communities collect garbage and separate out the organic waste to use as a boiler fuel. (Union Electric Co.)

Gas fuels are clean burning and easy to distribute. Gases also convert easily to liquids or chemicals.

BIOCONVERSION BY PYROLYSIS

Pyrolysis is a breakdown of matter such as biomass by heat. Burning takes place in the absence of oxygen. This process was developed by a research and development company in California. First the organic waste is separated from inorganic materials (glass, metals, and other noncombustibles). Then the organic materials are shredded, dried, and fed into an air-tight furnace. It is heated to a temperature of $932°F$ ($500°C$) to produce oil, "char" (charcoal), and a low grade gas fuel. Some of the char and the gas fuel are recycled to provide heating for continuing the pyrolysis. Each ton of refuse produces one barrel of oil. Fig. 13-6 is a diagram of the process.

Coal, as well as biomass, may be converted by pyrolysis. The char is known as **coke**.

ECONOMICS OF PYROLYSIS

Use of pyrolysis to process biomass into usable fuels will be possible in some locations but not in others. Some of the organic waste in the United States is produced in small quantities at many sites. Collecting it in large enough quantities to feed a processing plant would be difficult.

However, enough of it is accumulated at large factory operations, big cities, cattle feed lots, and sawmills to produce 170 million barrels of oil a year. It has been estimated that a 2000 ton-per-day plant could process refuse waste at $5 per ton and handle the waste output from a city of 50,000.

GASIFICATION

Pyrolysis is also used in gasification processes. The waste is transformed into fuel with heat plus air or oxygen. See Fig. 13-7 for gasification processes and their products.

AIR GASIFICATION

A second type of gasification is known as air gasification. It uses small amounts of air and steam to convert char to gas. The process produces a low energy gas (LEG) which is diluted with nitrogen from the air. Though not suitable for pipeline distribution, it can be burned by boilers now using oil or natural gas. It is good enough also to power internal combustion engines in vehicles or in power stations.

OXYGEN GASIFICATION

Like air gasification, oxygen gasification is a simple process. It is carried on in an oxygen atmosphere which produces a medium energy gas made up of carbon monoxide (CO) and hydrogen (H_2). It makes a good fuel and can be used as a feedstock to make methanol (a form of alcohol), ammonia, H_2, or gasoline. This type of gas is also known as "synthesis gas" or "syngas." Fig. 13-8 shows three types of gasifiers.

Many industries produce wastes that would provide the biomass needed for gasification. Among them are sawmills, processors of animal feeds, food processors, and processors of raw materials for the textile industries.

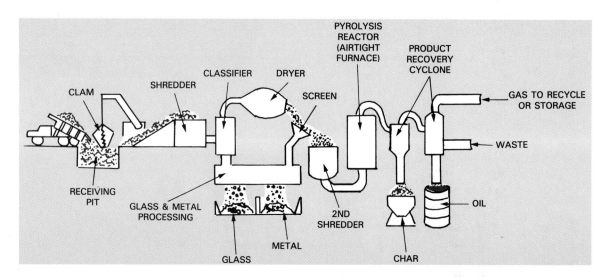

Fig. 13-6. The pyrolysis process at a glance. Organic waste is shredded and heated to $932°F$ in an airtight furnace. Oil and combustible solids are the resulting energy products.

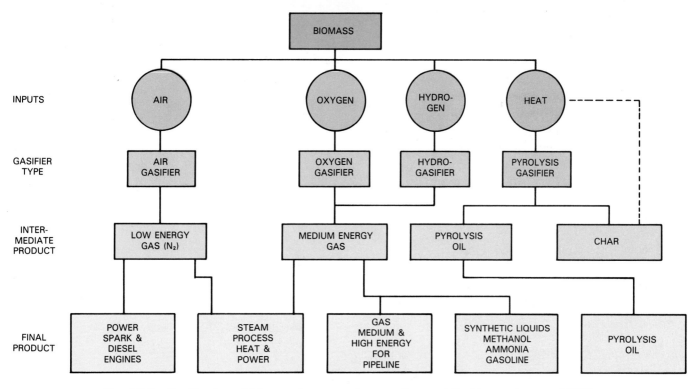

Fig. 13-7. Gasification of biomass has many intermediate and final by-products.   (SERI)

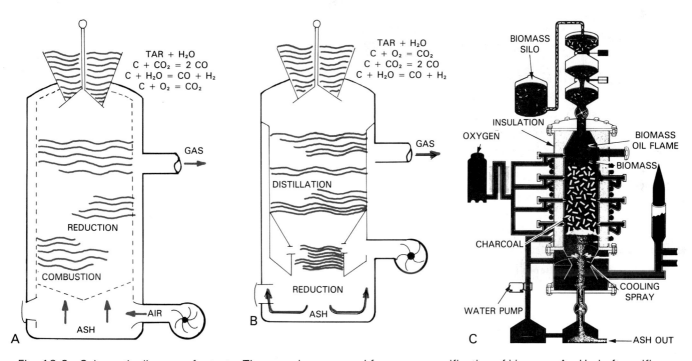

Fig. 13-8. Schematic diagram of retorts. These are burners used for oxygen gasification of biomass. A—Updraft gasifier. B—Downdraft gasifier. C—Stratified downdraft gasifier.

Waste materials that can easily be gasified include: sawdust, corncobs, fruit pits, nut shells, and rice hulls. Cotton gins also produce waste which could be gasified.

LIQUEFACTION

Liquefaction is also called **hydrogenation**. It is a process used to produce a gaseous or liquid fuel. Generally,

liquefaction takes place in reaction with hydrogen or a mixture of hydrogen and carbon monoxide.

Basically, the process has been used to extract oil or gas from coal. However, it has also been used successfully in experiments to make the same fuels from organic wastes.

The organic waste is placed in a reaction vessel and heated to 716°F (380°C) in a mixture of carbon monoxide and steam under pressure. Within 20 minutes, nearly half of the waste becomes fuel.

About 99 percent of the carbon content of the waste can be converted to fuel oil. A ton of dry waste will produce about 1.25 barrels of oil, Fig. 13-9. Animal wastes have a heat value of 15,000,000 Btu per ton. This is a little less than half the heat value of petroleum.

The fuel extracted by this process contains more oxygen than crude oil and needs much refining unless it is to be used only for direct combustion.

COGENERATION

Biomass conversion processes can be used by industry to produce electricity and process heat at the same time, Fig. 13-10. (Process heat is the heat used by a manufacturing operation to process raw material into a product. For example, a sugar processing plant

Fig. 13-9. A ton of dry organic waste will produce about a barrel of oil by the hydrogenation process.

may use its own wastes to produce electricity for its plant machinery and steam for heating and evaporating processes.)

There have been pilot projects in the past to separate and incinerate the combustibles in garbage. The heat captured in steam was used to generate electricity that was fed into the local electrical power grid.

BIOLOGICAL CONVERSION

Biological conversion produces either liquid or gaseous fuels. It is accomplished through **anaerobic digestion**.

ANAEROBIC DIGESTION

Anaerobic digestion is the controlled decay of organic matter without oxygen. Waste materials such as sewage, manure, paper, seaweed, and algae can be converted to produce methane gas.

The organisms which do the work are known as **anaerobes**. These cells are able to maintain their life processes without air or free oxygen. Anaerobes can oxidize organic matter without using free oxygen. However, they do obtain oxygen by breaking up molecules containing oxygen. They then use this oxygen to produce chemical change in the oxidized matter.

The process is simple and cheap. Bacteria break down the waste at low temperatures—between 59°F and 122°F (15°C and 50°C) at atmospheric pressure. Two types of bacteria are involved. One breaks down the organic solids into organic acids, hydrogen, and carbon dioxide. The second feeds on the waste products of the first types and produces methane-rich biogas as its own by-product. See Fig. 13-11.

Fig. 13-10. Spent steam from the turbine of a power plant can be piped to a factory and used as process heat.

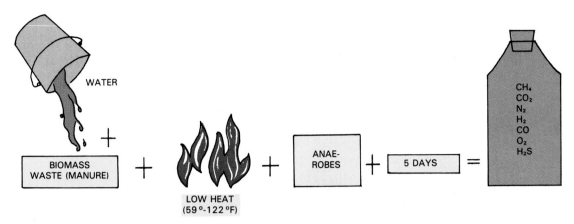

Fig. 13-11. In anaerobic digestion, low heat—between 59°F (15°C) and 122°F (50°C)—starts the process. In about five days it can produce methane-rich biogas from waste slurry.

One of the foremost countries in development of anaerobic digestion is India. Some years ago, an Indian scientist, Ram Bux Singh, built many units which he called biogas plants. The residue of the process is a rich organic plant food which can be returned to the soil.

Rising costs of natural gas are making the construction of methane digesters at sewage disposal plants a more attractive alternative to purchasing gas. Small-scale digesters are being used on farms particularly in Europe and Asia.

Many U.S. farms have large enough biomass supplies to make small-scale anaerobic digestion possible. An estimated 20 percent of current U.S. natural gas requirements could be met with anaerobic digestion.

FERMENTATION

Fermentation is an anaerobic process for producing alcohol. The process depends on the action of enzymes on the starches and sugars in fruits, grains, and other biomass forms. It is a primitive form of respiration. The enzymes which carry on the process are secretions (yeasts) of microorganisms found in nature.

Alcohols are an important biofuel which can be blended with gasoline or burned alone. There are several types. The one produced by fermentation of grains, fruits, and other biomass is called **ethanol** or ethyl alcohol.

Fermentation produces a mixture of water and alcohol. Since water boils at 212°F (100°C) and alcohol around 173°F (78°C), the two are separated by distillation. The mixture is heated to a point where the alcohol boils off. The vapor is caught and condensed.

The formula for pure ethanol is CH_3CH_2OH. It is a colorless, sweet-smelling liquid that burns with a hot, smokeless, pale blue flame.

With the growing scarcity and cost of importing gasoline, alcohol has long been seriously considered as an alternate fuel. Some oil companies are blending it with gasoline at 10 parts of alcohol to 90 parts of gasoline. This blend is sold as "gasohol."

ANAEROBIC DIGESTERS

A digester is basically an insulated, airtight container where gases from decaying organic wastes are trapped. Anaerobes break down the waste into a methane fuel.

The gas produced by anaerobic digestion is a mixture of seven different gases. It is mainly methane and carbon dioxide, however. Usually it is called either **biogas** or **methane**. See Fig. 13-12.

Digester types
There are two basic types of digesters:
- Batch load. This digester is filled with a soupy mix of organic waste called **slurry**. Then the digester is sealed. It will not be emptied and refilled until the raw materials have stopped producing gas. Fig. 13-13 shows a simple batch load design.
- Continuous load. This type will accept a small amount of fresh slurry continuously. A small amount

Makeup of Biogas

SYMBOL	GAS	PERCENTAGE
CH_4	Methane	54-70
CO_2	Carbon dioxide	27-45
N_2	Nitrogen	.5-3
H_2	Hydrogen	1-10
CO	Carbon monoxide	0.1
O_2	Oxygen	0.1
H_2S	Hydrogen sulfide	trace

Fig. 13-12. Biogas is made up of seven different gases. Usually the carbon dioxide is removed by other processes.

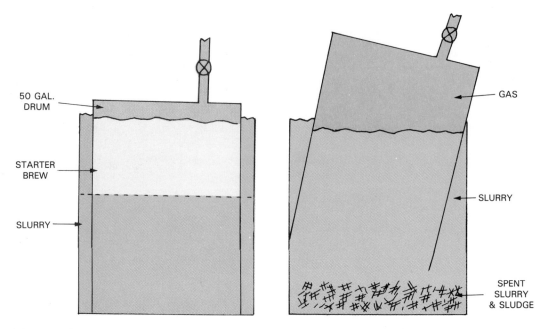

Fig. 13-13. Diagram of a simple batch digester. When anaerobic action is finished, the container must be dumped and refilled. Left. A freshly loaded digester. Right. A working digester. Spent solids can be used as fertilizer.

can be added daily. The digester will produce gas as long as it is being fed.

There are two types of continuous load digesters. The first type, shown in Fig. 13-14, is a **vertical-mixing digester**. The materials are digested in vertical chambers. Spent slurry is lighter and will flow out over the top. Units like the one in Fig. 13-14, right, have two chambers. As the spent slurry overflows into the second chamber it continues to digest. Finally, it is drawn out the effluent pipe.

A second type designed for continuous load is known as a **displacement digester**. It is a long horizontal cylinder (often made from oil drums welded end-to-end). Fig. 13-15 is a simple sketch of a displacement digester. By the time the spent slurry arrives at the opposite end of the digester it has separated into distinct layers. See Fig. 13-16.

Starting from the top, the working digester contains:
• Biogas.
• Scum. This is a mixture of fibers and solids separated

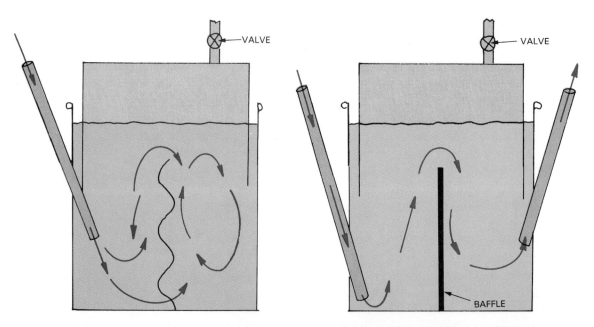

Fig. 13-14. Small amounts of slurry can be added daily to continuous load digesters through a pipe system. Diagrams show two types.

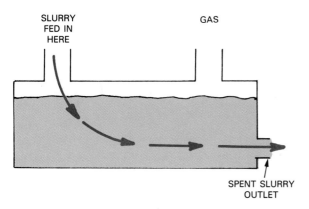

Fig. 13-15. Diagram of a displacement digester. As waste is added, it moves along, is digested, and leaves by the outlet when spent.

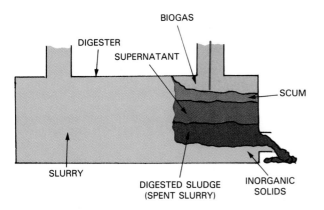

Fig. 13-16. Biogas and by-products of digestion separate into layers when the slurry is spent.

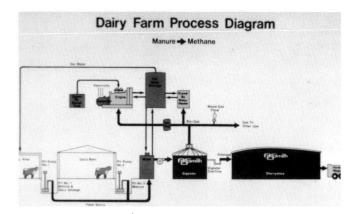

Fig. 13-17. A prototype of a farm digester. Top. A diagram of the system. Bottom. A farm installation. (Harvestore Systems; A.O. Smith Harvestore Products, Inc.)

from the organic waste. If not present in large amounts, the scum insulates the slurry and helps the digestion process. However, scum is usually a problem if there is too much of it since it can cause the digester to stop working.

- **Supernatant.** This is the spent liquid of the slurry. Its value as a fertilizer is equal to that of the sludge since it contains dissolved solids.
- Sludge. These are solids contained in the original waste. They have been reduced to around 40 percent of their original volume. Sludge is used for fertilizer.
- Sand and inorganic matter.

Anaerobic digesters are designed to be used on farms where large quantities of manure are available. The methane produced can be used as a fuel for stoves, agricultural dryers, or internal combustion engines. See Fig. 13-17.

WASTE-TO-ENERGY SYSTEMS

The day may not be far off when many American cities will be able to reduce their trash disposal prob-

lems and reap energy rewards by burning their refuse. Municipal waste can be shredded, mixed with other fuel, such as coal or oil, and burned. The heat created will generate electricity or provide heat for buildings. European cities have been harvesting the energy from trash and garbage for many years. Trash and garbage incineration requires special incinerators equipped with moving grates. Some North American cities, such as Chicago, the Norfolk (Virginia) Naval Base, and several Canadian cities, already have incineration programs.

ST. LOUIS WASTE BURNING PROGRAM

The city of St. Louis, Missouri, working with an electric utility company, built an experimental plant in the 1970s. The refuse, Fig. 13-18, was first shredded in a hammer mill. Massive metal blades of the mill, driven by a 1250 hp motor reduced the trash to particles less than 1 1/2 in. in diameter. See Fig. 13-19. Next, magnetic separators removed ferrous metals which were stored until they could be sold to a steel mill, Fig. 13-20.

Because the refuse still contained heavy material that would not burn, it was passed over heavy jets of air to remove light, burnable material, Fig. 13-21. Conveyors carried away the combustible material to the power plant boilers. Here it was mixed with coal and

Fig. 13-18. An experiment in using trash as fuel for electric power had garbage trucks bringing trash to a processing plant for shredding. The same plant separated combustibles from metals. (Union Electric Co.)

Fig. 13-20. Separated metals were moved to storage by conveyors. Later, the metal was hauled to a steel mill. (Union Electric Co.)

Fig. 13-19. After shredding, the trash looked like this. It contained bits of paper, cloth, and other fibrous material.

Fig. 13-21. An air classifier, using jets of high-pressure air, separated lighter combustibles from noncombustibles like glass before the combustibles were fed into the boilers of the power station.

fed into the boiler through pneumatic tubes. Fig. 13-22 is a diagram showing the entire process.

CROP RESIDUE AND DRYING GRAIN

Farmers have been experimenting with the use of crop residues as a fuel for drying grain. A manufacturer of drying bins has been testing a biomass furnace. The furnace attaches to the drying bin and can burn nearly any form of combustible material, Fig. 13-23. A logical source of fuel is corn crop residue. According to the manufacturer, a 95 bushels per acre corn crop will yield 5500 lb. of cobs, leaves, and stalks per acre. This is equal to the energy of 320 gal. of propane.

The furnace is designed to heat air that is then blown into the grain bins. Through use of a heat exchanger, combustion flames never contact the grain. (A heat exchanger is a device which collects heat at one point and

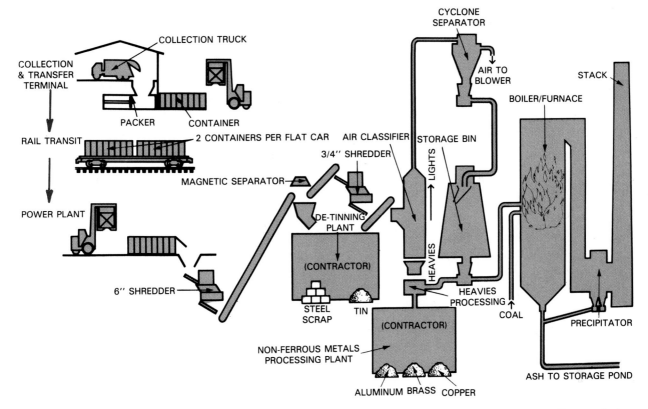

Fig. 13-22. This simplified diagram shows the entire St. Louis waste recovery plant. (Union Electric Co.)

Fig. 13-23. Agricultural biomass, consisting of material such as straw, corncobs, and cornstalks can be used to dry stored grain. Here, bales of straw are being fed into a biomass furnace.

delivers it someplace else without mixing the two air sources.)

A seed corn company is using corncobs from its own croplands to dry its seed. The cobs are fed into a gasifier chamber where they are burned with a limited oxygen supply. This creates combustible gases that provide a fuel for the seed dryer.

CORNCOB GASIFICATION RESEARCH

Iowa State University at Ames has conducted research on corncob gasification. It is believed that gas engines will run on this fuel if carburetors are modified. Their research centers around compact corncob gasifiers that mount on tractors. Crop residue would produce all the energy the tractor needs.

DENSIFYING BIOMASS

Because biomass is difficult to collect, store, ship, and use, efforts are being made to compress it. When so treated, it is called **densified biomass fuel** or DBF.

Before it is converted, the raw biomass undergoes some changes. First the noncombustibles must be separated. Then the remaining combustible materials are shredded and dried.

Five methods of biomass densification are now in commercial use:

- Pelletizing. Pellets are formed by hard steel dies full of holes 1/8 to 1/2 in. (3 to 13 mm) in diameter, Fig. 13-24. The die rotates against rollers that force biomass through the holes under great force (10,000 psi or 7 kg/mm). The extruded (pushed through dies) material comes out as pellets.
- Cubing. This is a kind of pelletizing in which the dies

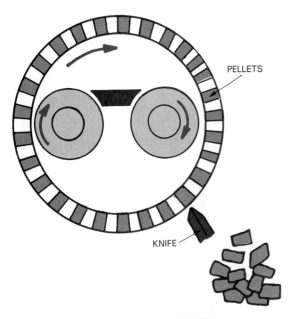

create larger pellets or cubes 1 to 2 in. (25 to 50 mm) across. See Fig. 13-25. This method is currently used to pelletize straw and paper. However, it could be used to pelletize a variety of other materials.

- Briquetting. The biomass is pressed between rollers that have depressions in their surfaces. They produce shapes like barbecue charcoal briquettes.

- Extrusion. This method depends upon an auger such as is used in a food grinder. The auger forces the biomass through a large die. The biomass particles are 1 to 4 in. (25 to 102 mm) in diameter. Pitch or paraffin is often added to the mix to bind the particles together and provide more heat content. Supermarkets sell them as fireplace logs. See Fig. 13-26.

Fig. 13-24. Pelletizing of biomass reduces its bulk. The diagram at the top shows a pelletizing drum with biomass being forced through dies under high pressure. (SERI) Middle illustration shows a sample of a pelletized wood/peat mixture. Bottom. Pelletized wood particles. (Guaranty Fuels, Inc.)

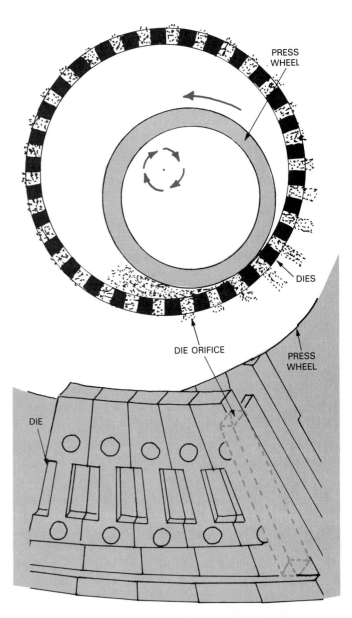

Fig. 13-25. A cubing process produces larger pellets. Top. A press wheel and dies compress and shape the cubes. Bottom. An enlarged section of the die. Phantom lines show die openings through which biomass is pushed. (Papakube Corp.)

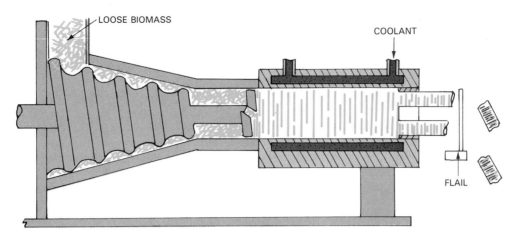

Fig. 13-26. Taiga extrusion process. The auger mechanism compresses the biomass and forces it through a huge die. The material emerges as chunks or logs. A flail cuts them to length.

- Rolling-compressing. In this process, fibrous material, such as straw or hay, is fed continuously into four rollers, Fig. 13-27. The rollers, being power driven and under heavy spring tension, squeeze and wrap the biomass into densely packed cylinders. As the rollers turn and compress, the rolled material is moving out of the rollers. This action produces a continuous cylinder of compacted material 5 to 7 in. (13 to 18 cm) in diameter. It is cut into cylinders 4 to 12 in. (10 to 30.5 cm) long, Fig. 13-28.

Solid waste densification is an attractive process for city refuse disposal departments. It produces easily stored energy and solves waste disposal problems.

POWDERED BIOMASS FUEL

In another process presently under development, a dense powder, called Ecofuel 11, is produced out of waste. The waste is shredded and milled under hydrochloric or sulfuric acid. The resulting powder is fed into suspension-fired boilers.

Fig. 13-28. This hay sample has been wrapped into a densely packed cylinder.

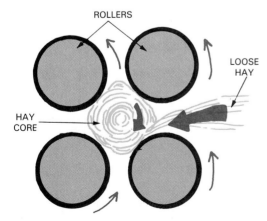

Fig. 13-27. Another method of compressing biomass is to wrap it between rollers at high pressure. (Agropack)

USES OF DBF

Densified biomass fuel (DBF) does not produce sulfur pollution like coal. The ash residue can be used as a fertilizer.

DBF can be used in factories to produce process heat. It can be burned directly in modified gas and oil boilers or it can be fed into a gasifier for conversion to a fuel gas. Other possible uses of DBF are:
- Fueling of residential, commercial, or industrial furnaces.
- Fueling of airtight wood stoves.

Biomass Energy 233

- As a fuel for external combustion engines such as the Brayton and Stirling engines.
- Fuel for fireplaces and outdoor grills.
- Feedstock for pyrolysis of oil and high density charcoal.
- Feedstock for gasification plants.

Fig. 13-29 shows a flow chart of biomass from its source, through processing, and final use. A typical biomass compaction plant is pictured in Fig. 13-30.

BIOMASS SOURCES

Organic matter making up the biomass could come from different sources. Included in these sources are:
- Wastes and residues from cities and factories.
- Wastes and residues from forests and farms.
- Energy farming to grow plants in soil and fresh water.
- Energy farming in salt water.

Wastes and residues are a valuable resource which, today, are too often thrown away. The forest and food processing industries have tons of waste which could be recycled to produce energy. The same is true of American cities which generate so much garbage and other solid wastes that their disposal is a national problem. See Fig. 13-31 for a partial list of wastes and residues.

TO GROW OR TO SALVAGE

The four biomass sources just listed represent two very different approaches to supplying biomass for

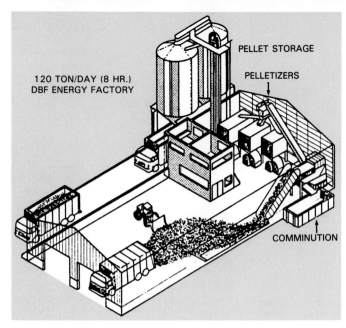

Fig. 13-30. Diagram shows a biomass compaction plant that turns biomass into pellets for fuel. (Papakube Corp.)

energy. The last two sources (energy farming) would require a new industry that would grow plant life solely for the purpose of producing energy. The other two would redirect the disposal of often unused materials. It would require the setting up of new systems for collecting and converting the wastes and residues.

Either approach has its own set of problems. These will be discussed later.

BIOMASS SUPPLIES

Annually the U.S. generates about a billion tons of wastes from all sources. Not all of this is useful as fuel.

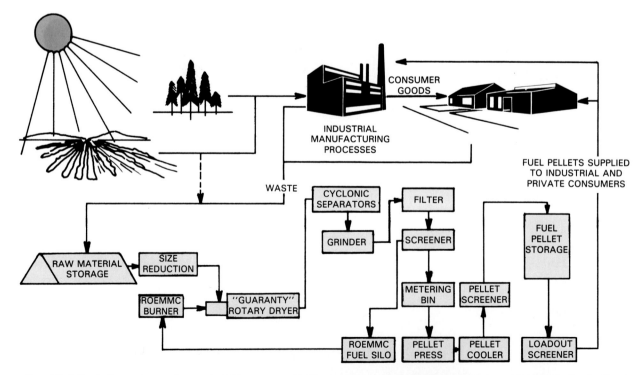

Fig. 13-29. A flow chart of a typical biomass cycle from source to final consumption. (Guaranty Fuels, Inc.)

WASTES	RESIDUES
BY-PRODUCTS OF MANUFACTURING PROCESS	UNUSABLE "SCRAP" FROM HARVEST OF RAW MATERIALS FROM FORESTS AND FARMS
FOREST Sawdust Bark Paper pulp Wood shavings Scrap lumber Wood dust Paper	FOREST Noncommerical timber Diseased trees Slash
AGRICULTURAL & FOOD PROCESSING Corncobs Fruit pits Manures Rice hulls Sugarcane bagasse Nut shells	AGRICULTURAL Corn stalks Oat straw Wheat straw Barley straw Rice straw
MUNICIPAL Sewage Trash Garbage	

Fig. 13-31. Biomass materials such as these used to be wasted. Proper recycling methods can turn them into useful energy.

Some of it is not organic and will not burn. Much of it is too widely scattered to be worth collecting.

Mixed with municipal organic wastes are many inorganic materials. These wastes include:
- Minerals. Among these are iron, aluminum, brass, copper, and tin.
- Manufactured noncombustibles. Glass, stone, concrete, and bricks are the most common of these.

Many of these materials can be separated and recycled. Others can be used as clean landfill or building aggregate.

Recycling of metals represents energy saving also. For example, only 5 percent as much energy is needed to recycle aluminum as to produce it from raw ores.

Cities of the United States create huge quantities of organic waste in the form of trash and garbage, Fig. 13-32. Until now, little of this material has been recycled. Consequently it has posed a serious problem of disposal. Reports are seen with growing frequency that communities are running out of landfill sites for disposal of garbage and trash. The costs of trash collection and disposal are considerable. In 1964 it cost $26 per ton, but in 1984 is nearer to $50 per ton. These costs will continue to rise.

The energy that can be generated from municipal wastes would be roughly equal to 150 million barrels of oil per year. This would represent $7.2 billion in energy savings if oil were selling at $48 a barrel.

Material	NATIONAL TOTALS (million tons)	AVERAGE HOUSEHOLD ACCUMULATIONS (pounds)	ANNUAL HOUSEHOLD ENERGY SAVINGS EQUIVALENT	
			RECYCLING (gal. of gasoline)	BURNING (gal. of gasoline)
PAPER Newspaper Other Separable Nonseparable	7.0 5.3 10.9	233 173 370	11.2 or 8.3 or —	10.2 7.6 16.3
GLASS	10.8	368	3.7	—
FERROUS METALS	6.3	210	10.1	—
ALUMINUM	.6	20	16.0	—
OTHER NONFERROUS	.2	7	—	—
PLASTICS Separable Nonseparable	1.0 2.3	33 77	4.9 or —	2.9 3.4
RUBBER & LEATHER	1.0	33	—	1.4
TEXTILES	1.0	33	—	1.4
WOOD	.6	20	—	.9
FOOD	17.8	593	—	26.1
MISC. INORGANIC	1.0	33	—	—
Subtotal	65.8	2203	54.2	70.2
BULKY MATERIALS	6.4	213	—	—
YARD WASTES	18.1	600	—	19.2
Total	90.3	3016	54.2	89.4

Fig. 13-32. Cities create mountains of trash and garbage.

ENERGY FARMS

Scientists have been experimenting with large-scale production of green plants for fuel since the early 1950s. Experiments have been underway to increase the crop yield of certain plants, Fig. 13-33.

An energy farm is operated solely to produce energy. It is one of the alternatives that have been considered to replace dwindling fossil fuels.

However, it is doubtful that such farms could be relied upon at present for several reasons:

- Biomass farming would require large amounts of energy. This is true of most processes for converting biomass energy to other forms of fuel and chemical feedstocks.
- High quality, fertile soil is needed for biomass farming. This would take some of our best farmlands out of food production.
- Energy needs would compete with rising demands for food to feed a growing world population.

On the other hand there are 220 million acres of pasture and range land capable of growing biomass crops. In addition, 160 million acres of forest lands might be used for biocrops. Even so, biomass farming of these lands would make them unavailable for growing feed and fiber products.

AQUACULTURE

Algae, hyacinth, and kelp are three plant types being studied for possible use in biomass aquaculture. While harvesting may be difficult, there is no problem of using precious water. In addition, yields are greater in water than on land.

For more than 30 years, algae has been the subject of experiments in fast-growing biomass farming. This single cell organism which is grown in either oceans or fresh water ponds is known all over the world.

A small-scale pilot plant was built more than 30 years ago to see if it might be feasible to produce the algae as a cheap source of biomass. Special nutrients and carbon dioxide were fed into pools set up on a rooftop to grow the algae. The test program produced 75 to 80 lb. of dried algae. While the project proved the feasibility the cost was relatively high. Production costs were about $0.25 a pound compared to $0.10 per pound for traditional growing methods. In spite of its rapid growth rate, algae does not, at present, appear to be a suitable alternative for meeting future energy needs.

KELP FARMING

Dr. Wheeler J. North, a biological oceanographer at California Institute of Technology tested kelp growing in the sea during the 1980s. Kelp grows at the rate of up to 2 ft. a day. Once harvested kelp could be loaded into barges and allowed to decompose, producing methane gas.

It was estimated that the kelp could supply the energy needs of 300 people at current U.S. levels. At the same time it could easily yield enough food for the needs of 3000 to 5000 people.

Since the oceans represent an area of more than 80 million square miles, kelp farming could conceivably support a population of from 20 to 200 billion people.

ENERGY FROM PHYTOPLANKTON

Phytoplankton is a name given to a huge number of plants and animals that drift near the surface in large bodies of water. Algae is one of them. They occur in both fresh and salt waters. Some are moved through the water by tides and currents. The animals, called nekton, move about independently of the tides. An estimated 10 billion tons of plankton are produced in the sea every year, making it the largest form of sea life and vegetation.

Dr. William Von Arx, noted marine scientist, speculates about the energy potential of plankton. He says that if the plankton were harvested for power alone, and converted to methane it would produce 10 times the present annual power needs.

COST OF BIOMASS FARMING

At present, producing biomass energy on energy farms would be expensive. The cost of such energy would be several times that of crude oil and coal. Sharp increases in the cost of fossil fuels and nuclear energy would not make biomass energy less costly by comparison since the production of biomass, its transport, and processing requires a great deal of energy in itself. Two things would have to happen to make energy farming competitive.

- The productivity of energy farming would have to increase greatly.

BIOMASS TYPE	YIELD	
	(tons per acre)	(metric tons/km²)
Algae	12 to 30	0.044 to 0.108
Sunflower	10 to 20	0.036 to 0.072
Sugar Cane	12 to 50	0.044 to 0.108
Eucalyptus	8 to 30	0.029 to 0.108
Sorghum	8 to 30	0.029 to 0.108

Fig. 13-33. Some plants, such as those listed above, produce large yields of biomass. They may be grown in the future as a source of energy.

- The efficiency of the conversion processes would have to be vastly improved.

The better the land on which the biomass is grown the better the yield. It is estimated that to produce 1 percent of U.S. energy needs would require 10 million acres of very good quality farmland. Producing the same amount of biomass on poorer land would require more acres, possibly as much as 30 or 40 million.

ADVANTAGES OF BIOMASS ENERGY FARMING

There are a number of reasons for considering the use of biomass energy. For one thing, its use creates less pollution and health risk than either nuclear power or fossil fuel power. Unlike fossil fuels, biofuels are renewable and duplicate rapidly the whole process which formed the fossil fuels.

Lower sulfur and ash content in biomass would reduce the environmental pollution from combustion processes that produce heat energy. The sulfur content of coal is 2500 percent higher than in biomass.

Biomass stores easily. Energy systems using solar energy, for example, require separate and expensive storage facilities.

DRAWBACKS OF BIOMASS ENERGY

Biomass, however, does have several disadvantages which make it less feasible for energy:
- It is inefficient as a converter of solar energy. Photosynthesis generally converts less than 1 percent of the incoming solar energy to chemical energy. Even under the most favorable conditions efficiency does not exceed 4 percent.
- Biomass fuels are spread out over a wide area and are expensive to collect for processing.
- Being perishable and bulky, the raw materials present problems of transport and storage until they can be used.
- The energy content of a ton of biomass is much lower than equal weight of fossil fuels or nuclear fuel. Dried biomass has less than half the heat content of oil, for example.

SUMMARY

Biomass, which includes all live organic material, living or once living, is a form of indirect solar energy. The process of collecting the solar energy and then storing it in leaves and fibrous materials is called photosynthesis. This process involves the conversion of carbon dioxide, water, and certain nutrients (from the soil) into sugar and oxygen.

Biomass can be a useful source of energy if converted to a more usable form by metabolic, thermochemical, or biological conversion processes. In metabolic conversion, humans and animals digest the biomass. Heat, oxygen, and enzyme action break down the plant cells to create food for energy and growth. Thermochemical conversion is an industrial process that either burns the biomass or uses heat to convert it to a combustible gas or liquid. Also an industrial process, biological conversion employs enzymes, fungi, and bacteria to cause a chemical change. The result is a gaseous or liquid fuel.

Biomass yields such fuels as biogas, ethanol or methanol, petroleum, and solid fuels such as char and chips.

Thermochemical conversion systems include direct combustion (burning), pyrolysis, gasification, and liquefaction. Biological conversion is accomplished by a process known as anaerobic digestion or fermentation.

Biomass energy has not been used to any extent for the reason that it is not well understood. Furthermore, collecting the biomass is difficult while it has only about one-third the energy of a like amount of oil.

Many communities have operated pilot plants to burn garbage and trash, using the heat for generation of electric power or other purposes. On farms, crop residues have been used for drying grain. There is also a small industry producing alcohol to be mixed with gasoline, thus stretching petroleum supplies.

DO YOU KNOW THESE TERMS?

anaerobes
anerobic digestion
biology
biogas
biomass
coke
densified biomass fuel
displacement digester
enzymes
ethanol
hydrogenation
mass
methane
photosynthesis
slurry
supernatant
vertical-mixing digester

SUGGESTED ACTIVITIES

1. Take part in a debate with other members of your class on one of the following topics:

a. Should surplus grain be used for production of alternate fuels or should it be used to feed the hungry people of the world?

b. Should large cities and large companies be required to separate the organic wastes they generate and recycle them as energy?

c. Should surplus crop acreages be turned to production of energy crops that can be converted to alcohol or methane gas?

2. Keep a record for a week of the organic wastes thrown out by one household. Report on:

a. Your calculation of how much recyclable waste is disposed of in a year.

b. Figure the amount of energy in this waste in terms of gallons of gasoline. Then, using the current cost of gasoline, determine the dollar value of the energy in the waste. Use the chart in Fig. 13-33 to make your calculations.

3. Construct a small methane digester and demonstrate it.

TEST YOUR KNOWLEDGE

1. What is biomass?

2. The process by which plants capture the energy of the sun and store it is called _____.

3. List the three methods by which biomass energy is converted to other forms of energy.

4. List at least three of the biomass fuels.

5. Burning biomass to produce heat is called:
   a. Direct combustion.
   b. Pyrolysis.
   c. Liquefaction.
   d. Gasification.

6. A ton of dry waste will produce how much energy?
   a. Equivalent of one barrel of oil.
   b. Equivalent of 25 barrels of oil.
   c. Equivalent of 125 barrels of oil.

7. Anaerobic digestion is one of the forms of _____ conversion.

8. Alcohol can be made from fruit. True or False?

9. List the two types of methane digester.

10. List three sources of biomass.

14
CHAPTER

THERMAL ENERGY

The information given in this chapter will enable you to:
- *Describe the types of thermal energy which are found naturally on earth.*
- *Describe methods for tapping and using heat stored in the earth's crust.*
- *Discuss the amount of heat energy available in the earth.*
- *List and describe methods of tapping and using ocean thermal energy and geothermal energy.*

Heat is stored naturally in the earth. One source of this heat, solar radiation, is collected and held in the top few feet of the earth's soil and in its oceans, Fig.

14-1. Another heat source is trapped deep within the earth. The sources are known as:
- Ground thermal energy.
- Ocean thermal energy.
- Geothermal energy.

GROUND THERMAL ENERGY

Both ground thermal and geothermal energy come from under ground. While coming from the same source there is a difference in these two. The way in which the energy was created is different and the way in which it is captured and used is also much different.

Ground thermal energy is partly the result of solar radiation storage and partly the result of the heat locked in the earth when it was first formed. Some of the sun's rays are captured by the rock and soil on the earth's surface.

Fig. 14-1. The earth's surface, both soil and water, absorbs and stores solar energy in the form of heat.

EARTH AS A SHOCK ABSORBER

Unlike air, which loses its heat energy rapidly into the upper atmosphere, the earth gives up its heat energy much more slowly. At the same time, the ground absorbs heat at a slower rate than does the air.

As a result, the earth stores heat better than air and acts as a temperature "shock absorber." In one study, temperature was taken 9 ft. below the surface and at the surface over a period of a year. The purpose of the study was to compare temperature differences as the seasons changed. The coldest ground-level temperature came in February. The coldest temperature 9 ft. down, however, came in late May. The hottest ground-level temperature was recorded in August. Below-ground temperature reached its highest reading in November.

This shock-absorbing quality has a decided impact on the soil temperature in North America. In most of the United States, the underground temperature, except very near the surface, varies from 40°F (4.4°C) to 70°F (21°C) between summer and winter. Canadian temperatures are only a few degrees lower. This is considerably higher than the winter air temperature in most of North America.

USING EARTH'S STORED HEAT

Until the mid 20th century, humankind made little use of the earth's heat. True, we did put down foundations for buildings deeply enough to get below the frost line. (This is the depth at which the ground does not get cold enough to freeze in cold weather.) However, use of this heat energy was not common until we began to construct earth homes and machines that could collect the heat energy and transfer it into dwellings.

Earth protected homes

An earth protected home is one which uses the natural heat of the earth as protection against both excessive cold and excessive heat.

An earth home may be entirely in the ground, partially in the ground, or merely sheltered by earth. See Fig. 14-2. Because of the "shock absorbing" effect

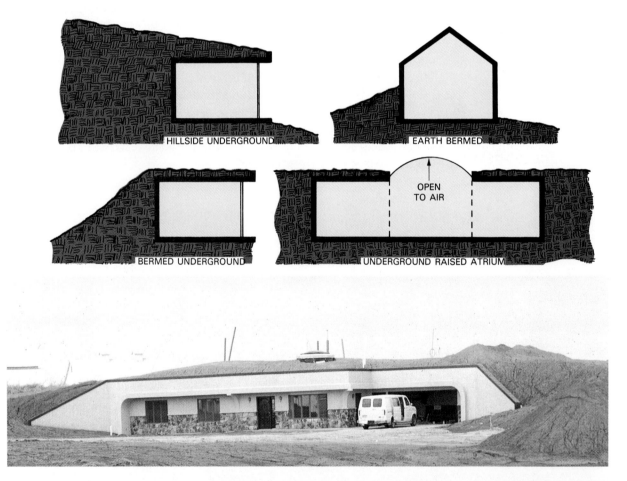

Fig. 14-2. Only a few feet below the surface, earth temperatures are more temperate the year round. Home builders often take advantage of this condition, building dwellings entirely or partially underground. Top. Types of earth sheltered homes and their names. Bottom. This earth sheltered home is located in a midwestern state where summer temperatures can reach over 100°F (38°C), while winter temperatures are sometimes below 0°F (−17°C).

described earlier, the temperature of earth homes, even without artificial heating or cooling, is more even the year round. Construction costs for earth protected homes are higher (from 30 to 49 percent).

Heat pumps

Heat pumps, Fig. 14-3, are a type of machine that can collect even low-level heat from one place and transfer it to another location. It uses gases or liquids called refrigerants and a pair of coils. Put simply, heat pumps are air conditioners that can either heat or cool space. They are often teamed up with sources of thermal energy to provide comfortable temperatures in living and working space. The working principles are explained in Chapter 16, Conservation of Energy.

Since the heat pump is designed to extract low-level heat from its surroundings, it is the perfect machine for drawing the heat from beneath the ground. This is accomplished by placing the outdoor coil of the heat pump in the ground, down a well, or deep below the surface of a pond or lake. In cold climates these may be the only practical ways of using the heat pump. In even in the coldest climates, the temperature 4 to 6 ft. below the surface does not vary much from one season to the next. Fig. 14-4 is a schematic for a heat pump system with its outdoor coil located down a well.

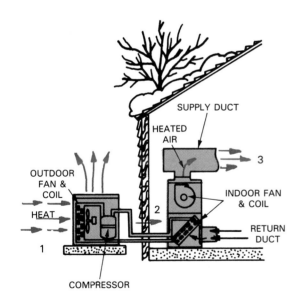

1—HEAT IS EXTRACTED FROM AIR BY OUTDOOR COIL.

2—REFRIGERANT GAS CARRIES HEAT TO INDOOR UNIT.

3—CIRCULATING INDOOR AIR PICKS UP HEAT AND CARRIES IT THROUGHOUT HOME.

Fig. 14-3. A heat pump extracts heat from one location and delivers it to another. Its winter mode is shown here. (Commonwealth Edison)

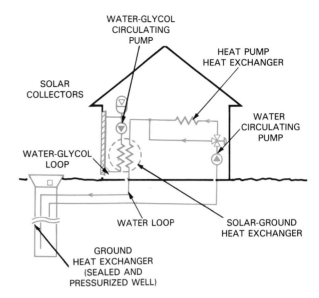

Fig. 14-4. Schematic shows a heat pump designed to draw heat from a well during winter. In summer the process reverses and heat that is drawn from the house is delivered to the well.

OCEAN THERMAL ENERGY

Ocean thermal energy conversion (OTEC) is a method of using stored heat from ocean surfaces in and near the equator as an energy source. A Rankine cycle heat engine floating on the ocean surface uses the heat to generate electricity. It is simply a giant heat pump harnessed to an electrical generator.

The heat collected by the Rankine cycle heat engine drives a turbine that is connected to a generator. Whereas a fossil-fueled or nuclear power plant uses steam to drive its turbines, OTEC turbines are driven by a fluid with a much lower boiling point than water.

HOW OTEC WORKS

The operation of the OTEC system can be explained simply. Warm water from the ocean surface is drawn into the power plant, Fig. 14-5. As the water passes through an evaporator it loses a quantity of its heat to a fluid such as ammonia, propane, freon, or some other type of refrigerant. The low-level heat causes the fluid to boil. This turns the fluid to a gas. As it becomes a gas it expands. This creates pressure great enough to drive the turbine. After the heated gas has given up its energy to the turbine, its pressure drops. Next, it passes into a condenser. Cold water from deep in the ocean passes around the condenser. The gas transfers some of its remaining heat to the cold water.

The cooled gas returns to a liquid state. It moves along to the pump and is ready to repeat the Rankine cycle. Fig. 14-6 is a cutaway of a typical OTEC power station.

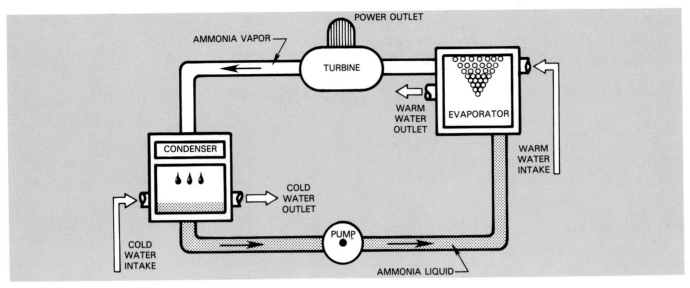

Fig. 14-5. This simplified schematic drawing shows how electricity is generated from heat stored in oceans. Seawater heats up ammonia in the system. The ammonia turns from a liquid to a gas as a result and its pressure drives a turbine. The turbine drives a generator. (SERI)

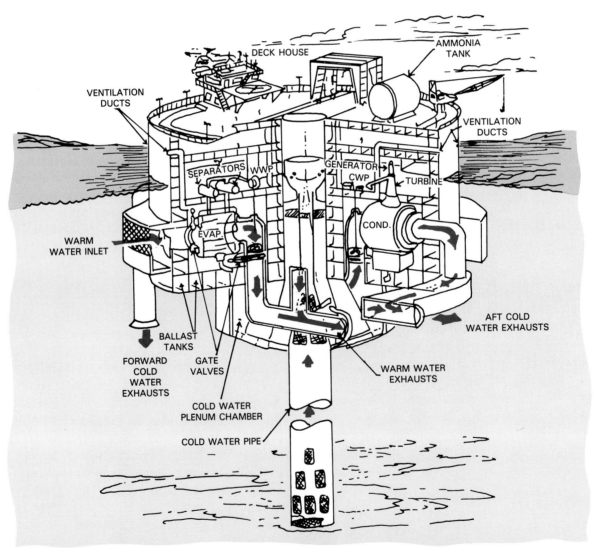

Fig. 14-6. Cutaway of an ocean thermal energy conversion plant. It uses volatile fluids to generate electricity with the ocean's heat energy. The plant is designed to float in the ocean in tropical areas. Cables (not shown) carry generated electricity to shore. (Lockheed Missiles and Space Co., Inc.)

The electric power generated by the power station could be carried to shore through power cables and fed into any electrical power system. Excess electric power could also be used to convert seawater to hydrogen. The hydrogen could then be piped ashore or to a ship and stored for future use.

Another use for the electric power generated is for the production of ammonia, using hydrogen from the water and nitrogen from the air. At present, ammonia is obtained from natural gas which may soon be in short supply. Also, the ammonia can easily be changed back into hydrogen and nitrogen. As such, it could be used as an energy source for fuel cells. See Fig. 14-7. (The fuel cell is discussed in Chapter 16.)

ENERGY SOURCE FOR OTEC

As we know, the sun bathes the earth each year with 18,000 times as much energy as humans can use. Seventy percent of the earth's surface being water, a great deal of the energy is absorbed and stored in upper layers of the oceans.

A huge band of water 10 degrees on either side of the equator acts as a massive solar storage tank. Temperature of the surface water is a constant 80°F (27°C). Meanwhile, 3000 ft. below the surface, currents of 40°F (4°C) water flow from the North and South poles. The difference in temperature is what makes the OTEC power plant possible. See Fig. 14-8.

A great advantage of the OTEC plant is this ready source of stored heat. Other advantages are:
• It costs nothing.
• It does not need to be transported.
• It is constantly being replenished, and, generally speaking, is nonpolluting.

TESTING THE OTEC

No large OTEC plant has yet been built. However, Hawaii and two private companies have entered a partnership agreement to develop the first Rankine closed-cycle, self-sustaining OTEC system. It will be a miniature plant capable of generating 50,000 W of electric power, using ammonia as a working fluid.

The federal government has been working on programs to develop and test OTEC components and subsystems. It is also planning storage and delivery systems. The first operating plant will generate about 25 MW of power.

At most sites, the cold water would be brought up from depths of 2000 to 3000 ft. (610 to 914 m) with a pipe diameter of 100 ft. (20 m). This size would serve

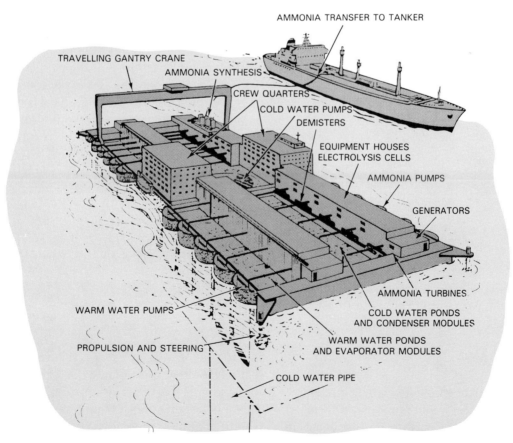

Fig. 14-7. Electric power, generated by an OTEC unit can also be used to extract ammonia or hydrogen from seawater. This sketch is of an ammonia plant ship designed by the Applied Physics Laboratory at Johns Hopkins University. (SERI)

Thermal Resources

Hawaiian Islands

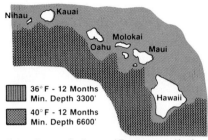

Puerto Rico

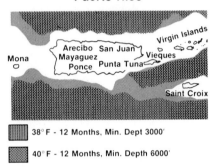

Gulf of Mexico

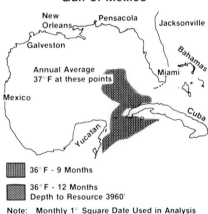

Fig. 14-8. These are prime locations for setting up OTEC power plants.

a system with a generating capacity of 250 MW. The flow through such a pipe would equal the flow of the Missouri River past Omaha, Nebraska!

OTEC'S PROBLEMS

An OTEC power plant would have to be quite large to house both machinery and an operating crew. Dimensions would vary according to advances in design; however, such a plant might be close to 400 ft. or about 122 m in diameter. Only living quarters and space for small equipment would be above the surface. The main structure, including buoyancy tanks would be underwater. Parts of the plant would extend downward 500 to 600 ft. (150 to 180 m).

Problems of such a structure might include:
• Corrosion due to saltwater.
• Damage or instability due to stormy seas.
• Clogging of inlets, ducts, and heat exchangers with algae and marine life.
• Possible damage to environment. Discharge of cooling waters could endanger sea life accustomed to colder temperatures.

These problems pose challenges to the designers. Special materials would have to be developed to deal with saltwater corrosion. Special attention would have to be given to strong mooring systems that would secure the plant against turbulent seas. Some of the problems could be solved by a land-based OTEC plant.

Several companies have received grants to develop 40 MW facilities in Hawaii. These facilities could either be based ashore or on an artificial island.

GEOTHERMAL ENERGY

Geothermal energy is heat energy that comes from deep within the earth. The earth is made up of several layers. A thin crust, known as the **lithosphere**, makes up the outer layer. This layer varies in thickness. Under land masses it is 50 to 62 miles (80 to 100 km) thick. Under the oceans it is much thinner. Rock temperature at the base of this crust is as high as 2000°F (1100°C). Beneath the lithosphere is a layer composed of rock that is molten at its upper edge. This layer is called the **mantle**. It extends about 1800 miles (about 2900 km) downward. Temperatures in the mantle range from 2000°F (1100°C) to around 6700°F (3700°C). The entire mantle would be molten were it not for the tremendous pressure. (Molten rock is known as **magma**.)

The earth's iron **core** is roughly 2175 miles (3500 km) in diameter. Its temperature goes as high as 7770°F (4300°C). The outer shell is molten while the interior is solid.

ORIGIN OF EARTH'S HEAT

The origin of the earth's heat is not certain. According to the most likely theory, the earth was originally entirely molten. It is thought that earth's core and most of its mantle still contains much of the earth's original heat. It became trapped there after the cooling and solidification (hardening) of the crust and upper mantle. The heat still found in the crust and upper mantle could be the result of slow decay of radioactive elements. Some of these elements are plentiful in the crust. See Fig. 14-9.

Another theory suggests that radioactivity alone generated the intense heat in the earth's interior. This heat could not leak out to the atmosphere but heated its surroundings. Eventually, the heat reached the melting point of iron. Once molten, the iron would have sunk to the center of the earth. This formed the iron core the planet has today. Gravitational energy of the iron sinking to the core of the earth would have released so much heat that the Earth turned molten. Lighter rocks then floated to the surface where they cooled to form the Earth's crust.

USING GEOTHERMAL ENERGY

From readings taken in mines and wells, it is known that the earth's temperature near the surface increases slightly less than 2°F for every 100 ft. of depth. At 400 miles down the temperature could be about 4532°F (2500°C). As you can see, this heat is too deep beneath the earth to be tapped as an energy source.

TAPPING GEOTHERMAL ENERGY

Theoretically, geothermal energy can be collected and used by drilling down to it from anywhere. However, in most parts of the earth this is not possible or practical. The hot interior mass lies too deep within the earth to reach. Even if we had the technology, drilling would be too costly.

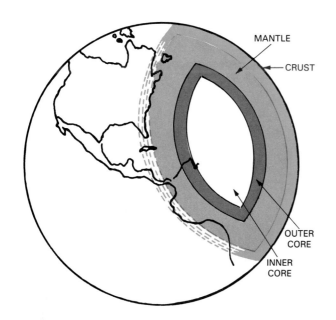

Fig. 14-9. Cutaway of earth shows source of geothermal energy. Temperature at core is estimated to be about 18,000°F (9982°C).

Fortunately, as just mentioned, the magma sometimes escapes from its underground prison and works its way closer to the surface. This has occurred in a few spots around the world, Fig. 14-10. Geothermal sites are found in areas where the earth's crust has shifted, producing fissures through which the molten materials travel. Volcanoes are a more violent form of geothermal energy working its way to the earth's surface.

GEOTHERMAL SOURCES

Reserves include all the geothermal energy known to exist that is economically recoverable with existing technology. Resources are the total amount of geothermal energy including estimates for deposits which have not yet been discovered. Ultimately the resources of geothermal energy may be 10 to 200 times the present reserves.

The actual amount of energy available and reachable from geothermal reservoirs under the earth is not certain. It has been said that if all the geothermal sources were tapped for energy, it would still take 41 billion years to drop the earth's interior temperature 1°C (about 2°F).

One estimate places U.S. resources at 6×10^{24} calories. However, not all of this energy can be recovered. The heat must be concentrated in reservoirs where it can be stored over a long period of time.

Current estimates of reserves are 60,000 MW of electrical power. This means that all of the recoverable energy from geothermal sources would generate 60,000 times 1 million watts of electric power annually. This is enough electricity to meet the needs of 60 million homes.

LOCATION OF U.S. RESOURCES

Most of the known U.S. geothermal reserves are in the western part of the country. See Fig. 14-11. California, Nevada, and Oregon have the most. Alaska, Idaho, Montana, New Mexico, Utah, and Washington also have large amounts. Probably, the greatest reserves will always be in the western regions.

The Electric Power Research Institute has estimated that up to 6 million kilowatts of geothermal energy could be generated in the United States by the year 2000 if development moved forward.

FINDING GEOTHERMAL ENERGY

Often, geothermal sites are identified by the steam rising through cracks or the presence of hot springs. Without these signs the presence of geothermal energy

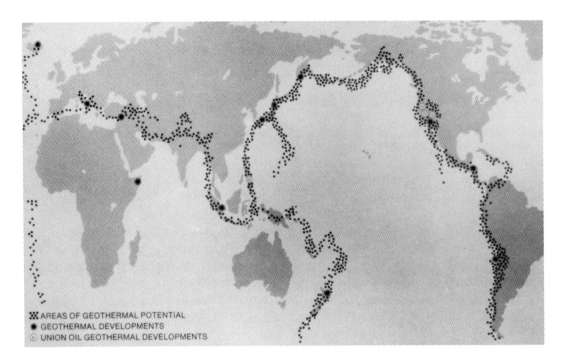

Fig. 14-10. Likely spots where geothermal energy might be tapped are to be found where the earth's crust shifts. Thus, volcanic activity is usually connected with geothermal activity. (Union Oil Co.)

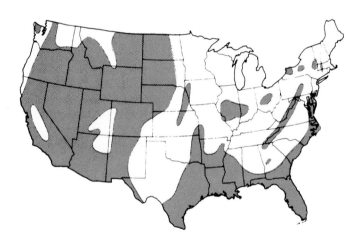

Fig. 14-11. This map of the lower 48 states shows shaded areas where geothermal energy might be found near enough to the surface to tap. (U.S. Dept. of Energy)

is hard to detect. About 1000 hot springs have been mapped in the lower 48 states.

Many different techniques were borrowed from the oil and gas industry for finding "hot spots." These include:

- Studying geologic maps and satellite photos which will show evidence of fault lines in the earth.
- Measurement of how shock waves and electricity travel through the earth.
- Drilling of shallow holes to take temperature and heat flow measurements.
- Drilling deep exploratory wells in promising areas. These will indicate if the resource is there and if it

contains enough heat to make further development worthwhile.

TYPES OF GEOTHERMAL HEAT

Geothermal heat is found in three natural conditions:
- Steam.
- Hot water.
- Dry rock.

As magma works its way through cracks in the earth, it heats up the layers of rock above. If the rocks are water soaked, hot water or steam is created. If the steam is close enough to the surface, hot springs, geysers, or **fumaroles** (wisps of steam) may occur.

GEOTHERMAL WELLS

Wells are drilled into geothermal fields to reach the hot water or steam, Fig. 14-12. The pressurized steam or hot water is piped to its point of use.

In some localities geothermal heat is collected to heat buildings. Very often it is used to generate electric power.

PRODUCING ELECTRICITY FROM GEOTHERMAL SOURCES

Geothermal steam that is to be used for generating power must be "cleaned" first to remove impurities and dissolved minerals. Then it is piped to a power station turbine.

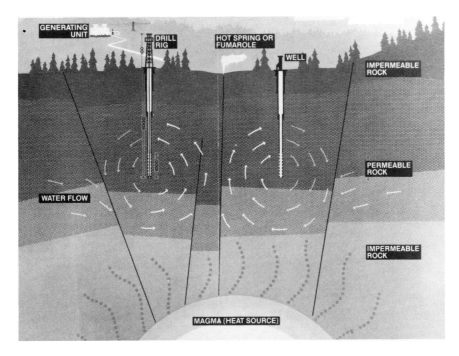

Fig. 14-12. Artist's conception of a geothermal well drilling operation. Arrows represent flow of hot water and dotted lines represent heat radiation through the impermeable rock.

In some geothermal operations only hot water is produced. Two methods are employed to make the water usable for power generation.

- The flash method. In this process, when the hot water is brought to the surface its pressure is reduced in special equipment. Some of the water "flashes" into steam. The remaining hot water is returned to the reservoir from which it came through an injection well.

- The vapor-turbine method. In this process, hot water is passed through a heat exchanger. Its heat is used to vaporize **isobutane** (a hydrocarbon liquid that vaporizes at low temperature). This is very similar to the exchangers used in ocean thermal conversion systems. The vaporized isobutane is under great pressure as it is piped away to drive the turbine. See Fig. 14-13. Once it has given up its usable heat, the water is returned to the underground reservoir. Fig. 14-14 is an aerial view of a geothermal power station.

SPACE HEATING

It is often very efficient and economical to use geothermal energy as a direct source of heat. Some localities have long been using it to heat homes. Examples are Klamath Falls, Oregon, and Boise, Idaho. For 80 years, Boise has been using geothermal energy to heat houses. The heating system is now being expanded to include city and state public buildings. Klamath Falls is installing a similar system. The city

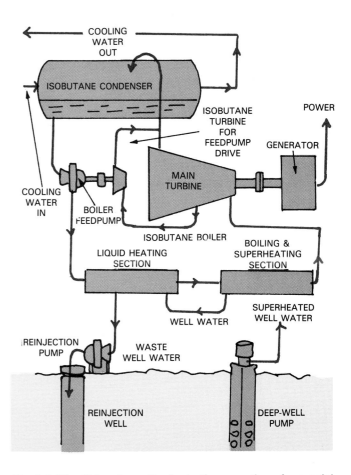

Fig. 14-13. This schematic charts the operation of a special type of heat engine used to tap the energy in hot water from a geothermal well. It is called a vapor-turbine cycle because the refrigerant material vaporizes as it draws heat from the hot water. The pressure of the vapor provides the force to drive a turbine.

Fig. 14-14. A fumarole (steam vent) and a geothermal power station. Steam is piped from the fumarole in the background to the power station in the foreground.

TEMPERATURE		USE
F°	C°	
356	180	Evaporation of concentrated solutions Refrigeration by ammonia absorption Digestion in pulp manufacturing
338	170	Heavy water by Hydrogen Sulfide Process Drying diatomaceous earth
320	160	Drying of fish meal Drying of timber
284	140	Drying farm products Canning of food
266	130	Evaporation in sugar refining Extraction of salt by evaporation
248	120	Distilling water Refrigeration by medium temperatures
230	110	Drying & curing light aggregate cement slabs
212	100	Drying organic materials (seaweed & grass) and vegetables Washing and drying wood
194	90	Drying of stock fish
176	80	Space heating Greenhouse space heat
158	70	Low temperature refrigeration
140	60	Heating buildings, housing animals Heating greenhouses
122	50	Mushroom growing Therapeutic baths
86	30	Swimming pools, biodegradation, fermentation
68	20	Fish farming

Fig. 14-15. Geothermal energy can be tapped for a number of processes needing the application of different levels of heat.

of Rykjavick, Iceland, heats between 70 and 90 percent of the community's homes with a geothermal system.

OTHER USES OF GEOTHERMAL ENERGY

Besides generation of electricity and heating of homes there are other uses for geothermal heat:
• Agriculture and horticulture.
  Heating of greenhouses.
  Hay drying.
  Animal husbandry.
  Fish farming.
  Hatcheries.
• Process heat.
  Paper mills.
  Drying and seasoning of lumber.
  Salt recovery from seawater.
  Brewing and distillation.
Fig. 14-15 is a chart of various uses of thermal energy and the temperatures required.

ARTIFICIAL GEOTHERMAL SYSTEMS

Hot dry rock contains vast quantities of geothermal heat that could not be tapped up to now. Nevertheless,

it has the potential of being the largest source of geothermal energy. One method, hydrofracturing, borrowed from the petroleum industry, shows promise for capturing great quantities of heat.

In hydrofracturing, cold water, under high pressure, is forced down a well into the dense, hot rock. The water fractures the dense rock. More pressurized water is forced through the cracks to extract the heat. See Fig. 14-16.

Another extraction method, still in the experimental stage, is the creating of large artificial cracks by setting off underground explosions. Little is known about the possible effects of the blast waves on surface structures or the costs of explosive fracturing.

ENVIRONMENTAL EFFECTS

Geothermal energy has many advantages from the standpoint of environment. For one thing, the production of the energy takes place right where it is used. There are no mining, processing, or storage activities. Moreover, it produces no solid wastes or atmospheric pollution of a significant nature.

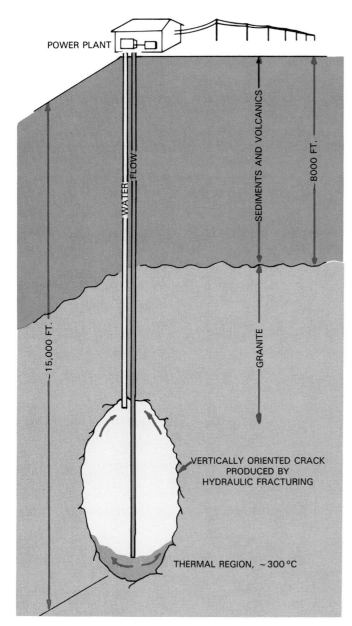

Fig. 14-16. Dry, superheated rock beds near enough to the surface can be converted into geothermal wells. It is possible to fracture the rock by pumping pressurized water into it. Continuing to pump water through the fractured rock "harvests" the heat energy. (Union Oil Co.)

This is not to say that there are no environmental problems, however. Within the immediate area of the power plant gaseous discharge such as hydrogen sulfides can be disagreeable. Other problems are subsidence (sinking) of soil over the well.

Even so, the problems caused by geothermal energy extraction are not as damaging to the environment as problems connected with power plants using fossil or nuclear fuel. Most of the pollution can be controlled or eliminated.

FUTURE OF GEOTHERMAL ENERGY

Geothermal energy will never be able to contribute great amounts of electric power. In 1990 it provided 0.2 quadrillion Btu toward the generation of electric power. This represents a decrease for the third consecutive year. Its largest contributions will be in the South and Southwest for heating of homes and factories.

SUMMARY

Heat is stored naturally in the earth from two sources: solar radiation and what is thought to be the earth's original heat trapped in the interior as the surface cooled.

Ground thermal energy is partly the result of solar radiation and partly the result of the earth's own interior heat.

The earth is slower to heat than air; likewise, it is slower to lose or transfer heat that is stored in it. Thus, it acts as a giant shock absorber. It tempers (evens out) the extremes of heat and cold from season to season. We make use of this principle in conditioning the air in modern dwellings. Homes are dug into the earth or bermed. Heat pumps also can draw warmth from the earth by having their coils buried in the ground or installed in a well.

Another giant thermal storage system is available in tropical oceans. Huge turbines are under development driven by Rankine engines. These can be floated on the ocean surface. They will create electrical power using the temperature differential in the water.

Heat deep from the earth is known as geothermal energy. It, too, is a source of thermal energy that can be tapped where it is near enough to the surface to be reached with current technology. When this heat contacts surface water geysers are the result. This superheated water is being tapped to generate electric power and to heat homes. If all sites could be tapped, they could generate 60,000 MW (megawatts) of electricity. This is enough electricity to supply 60 million homes. Artificial geothermal wells have been considered. This could be accomplished by drilling deep into the rock and fracturing it with water or explosives. Then water could be pumped into the wells, heated and utilized. This process is know as hydrofracturing.

DO YOU KNOW THESE TERMS?

core
fumaroles
ground thermal energy
heat pumps
isobutane
lithosphere
magma
mantle

SUGGESTED ACTIVITIES

1. Using two thermometers, compare the temperature a few inches under the ground with air temperature on any given day. Take the temperature readings at different times of the day. For example, first reading, 7 a.m.; second reading, 12 noon; third reading, one-half hour after sundown. Suggestion: use a darkroom or pool thermometer to take the ground temperature. Dig a hole with a spade, insert the thermometer and carefully replace the dirt around the thermometer. Discuss your temperature findings with your class.
2. Construct a cutaway of a geothermal well. Refer to Fig. 14-12 and Fig. 14-16.
3. Research construction of earth sheltered homes and build a scale model of one. Arrange with your instructor to display it at school.

TEST YOUR KNOWLEDGE

1. _____ energy is the result of heat locked in the earth when it was first formed.
2. Why is the earth referred to as a "shock absorber"?
3. A _____ is a device that will draw the heat from the earth and use it to heat buildings.
4. What does the acronym, OTEC, stand for?
5. A big advantage of the OTEC system is that the fuel costs nothing. True or False?
6. Geothermal heat comes from (pick best answer):
   a. Heat of the sun stored in the earth.
   b. Tremendous pressure of rocks in the earth's mantle.
   c. Molten matter in the mantle and core of the earth.
7. At a depth of 400 miles the temperature of the earth could be:
   a. About 2500°F (1371°C).
   b. About 3600°F (1982°C).
   c. About 4500°F (2482°C).
   d. About 10,000°F (5538°C).
8. Since geothermal hot water cannot be used directly to power turbines, two methods must be employed to use it. Name and describe them.
9. Discuss the potential of geothermal energy as a source for generating electrical power.

15
CHAPTER

CONSERVATION OF ENERGY

The information given in this chapter will enable you to:
- Define energy conservation and understand why it is important as a part of a national energy program.
- Review the various sectors of society that use energy and indicate how energy might be saved in each.
- Suggest the importance of each individual becoming active in energy conservation.

Conservation of energy has two meanings. In the study of physics, it means that the amount of energy in a system remains unchanged even though there are changes in its ability to do work. In our everyday usage, **conservation of energy** means to save any energy source by using it efficiently and not wasting it.

Chapter 1 stated that you cannot destroy matter. The amount of matter in the world always remains the same. However, as energy changes from one form to another it eventually enters a state where it is not able to do any more work. This is the principle known as entropy.

For example, when a gallon of gasoline is burned in an internal combustion engine, about 28 percent of the heat energy is changed to mechanical energy. The rest becomes random heat energy, Fig. 15-1. It is no longer useful as an energy source. A small amount may be used to heat the passenger compartment of the automobile, but most is lost to the atmosphere.

WHY CONSERVE ENERGY?

Not until the early 1970s did people come to realize that fossil fuels could soon be in short supply. At the same time, before energy costs began to rise, it was

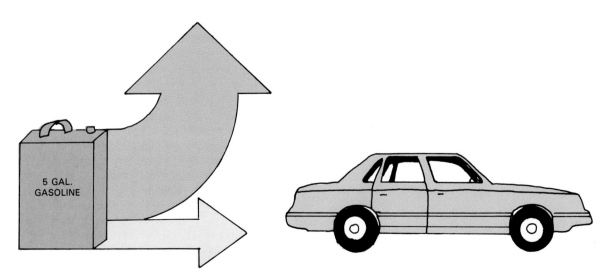

5 GAL. GASOLINE

Fig. 15-1. Roughly 28 percent of the energy in gasoline is used to propel an automobile. The rest is passed off to the atmosphere as low grade heat that can do no more useful work. A small amount is used to heat the interior of the vehicle.

not always economical to conserve energy. It was cheaper, for example, to buy fuel to heat homes than it was to install heavier insulation, Fig. 15-2.

Today, conditions have changed. Fuel is becoming more and more expensive and conservation more attractive. We must find ways to save energy. We know that eventually there will be a shortage of fossil fuels. The purposes of conservation are fivefold:

- To reduce the cost of energy to individuals and to industry.
- To save precious supplies of fossil fuels so they will be available to future generations.
- Conservation of petroleum stock's feedstock (raw material) for medicines and health products. When petroleum is gone, substitutes may be difficult to find.
- The reduction of dependence upon imported oil. The oil embargo of 1973 demonstrated the unreliability of imported energy in times of war or foreign political unrest.

If the foregoing reasons are not enough to move us to conserve energy resources, then we should consider one more fact. We may well be visited by severe energy shortages in our own lifetime if we continue to use up energy at the rate we have in the past.

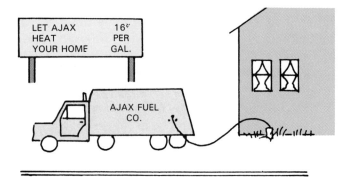

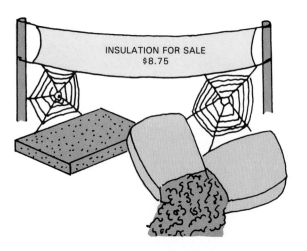

Fig. 15-2. There was a time when it was cheaper to buy fuel than to buy insulation.

HOW ENERGY IS LOST

It is an unfortunate fact that 50 percent of all energy available is lost as we convert it to useful forms. As we know from Chapter 1, some of this loss cannot be helped. It is a natural result of the laws of conversion. Still, there are ways to prevent some of the loss through correcting wasteful practices and through improving the efficiency of our conversion methods. For example, furnaces manufactured just a few years ago are able to use about 30 or 50 percent of the heat energy in fuel. Now, design improvements produce furnaces that are 80-96 percent efficient.

Improvement of design is a job for engineers and scientists, of course. However, elimination of waste requires the help and cooperation of every person who consumes energy.

In 1977 the President of the United States emphasized the need for conservation. Some of the specific goals he called for included the following:
- Reducing the annual growth in energy demand.
- Reducing gasoline consumption.
- Improving insulation of homes. This was to involve upgrading the insulation of existing homes and drafting of higher insulation standards for new construction.

ENERGY-USING SECTORS

In Chapter 4 we defined the energy-using sectors in our society. Let us review them:
- Residential/commercial. These include houses, apartment buildings, schools, office buildings, and stores. These buildings consume gas, oil, and coal to warm them in winter and cool them in summer. They also contain lighting, appliances, and equipment which consume electricity.
- Transportation. In this sector are all types of vehicles—automobiles, trucks, buses, trains, airplanes—which move people and goods. Transportation also takes in pipelines for moving gas, petroleum, and water from one place to another.
- Industrial sector. This sector takes in all businesses organized for producing goods and materials. Essentially these involve mining and manufacturing. Such enterprises use fuels and electricity for space heat, process heat, lighting, and operation of tools and machines.

Examples include:
  a. A steel mill which uses vast quantities of heat to melt ores and scrap until they are molten.
  b. A factory which produces clothing.

Total consumption of energy in all energy-using sectors in 1990 amounted to 81.44 quadrillion Btu, according to the U.S. Department of Energy. Per capita consumption was 327 million Btu. See Fig. 15-3.

Generation of electric power takes a large part of our available energy. This power is distributed among and is used by every one of the sectors, Fig. 15-4.

CONSERVATION MEASURES

There are ways of saving energy in each of these sectors. Many, if not all, of the methods require the help of the whole community if they are to be effective. Individual action is required in all of the following:
- Choosing and purchasing energy-efficient housing, appliances, and automobiles.
- Providing better maintenance to retain or improve energy efficiency of housing and machines.
- Developing habits which will conserve energy. Individual habits can affect all sectors.
- Becoming active in collecting recyclable materials.

CONSERVATION RESIDENTIAL/ COMMERCIAL

A major portion of the energy used in buildings goes for heat. About 75 percent of the energy is spent on heating space and water. The rest is for lighting, air conditioning (cooling), and operation of a variety of appliances.

INEFFICIENT HOME ENERGY USE

Homes may not be efficient users of energy for many reasons, Fig. 15-5:

ENERGY CONSUMED BY ELECTRIC UTILITIES*

YEAR	AMOUNT (in quadrillion Btu)
1985	26.48
1986	26.64
1987	27.55
1988	28.63
1989	29.30
1990	29.58

*Includes hydroelectric.

Fig. 15-4. Thirty-six percent of all energy used in 1990 was consumed for production of electricity.

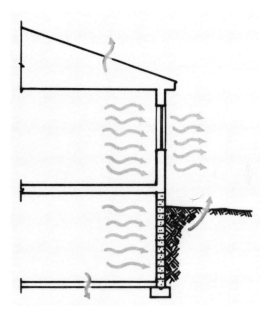

Fig. 15-5. Inadequate insulation in structures allows heat conduction through construction materials. Cracks around windows and doors allow heat to enter or escape through convection. (Portland Cement Co.)

				Consumption per Capita			
				Total Energy		End-Use Energy[1]	
Year	Total Energy Consumption (quadrillion Btu)	End-Use Energy Consumption[1] (quadrillion Btu)	Population[2] (million)	Quantity (million Btu)	Change from Previous Year (percent)[3]	Quantity (million Btu)	Change from Previous Year (percent)[3]
1979	78.90	61.84	224.6	351	-0.3	275	-0.4
1980	75.96	58.60	226.5	335	-4.6	259	-5.8
1981	73.99	56.56	229.6	322	-3.9	246	-5.0
1982	70.85	53.70	232.0	305	-5.3	231	-6.1
1983	70.52	52.91	234.3	301	-1.3	226	-2.2
1984	74.10	55.92	236.5	313	4.0	236	4.4
1985	73.95	55.39	238.7	310	-1.0	232	-1.7
1986	74.24	55.68	241.1	308	-0.6	231	-0.4
1987	76.84	57.68	243.4	316	2.6	237	2.6
1988	80.20	60.37	245.8	326	3.2	246	3.8
1989	81.35	61.08	248.2	328	0.6	246	0.0
1990[4]	81.44	61.10	248.7	327	-0.3	246	0.0

[1]End-use energy consumption is total energy consumption less losses incurred in the generation, transmimssion, and distribution of electricity, less power plant electricity use and unaccounted for electrical system energy losses. (See Glossary).
[2]Resident population of the 50 States and the District of Columbia estimated for July 1 of each year, except for the April 1 census count in 1950, 1960, 1970, and 1980.
[3]Percent change calculated from data prior to rounding.
[4]Previous-year data may have been revised. Current-year data are preliminary and may be revised in future publicationa.
Sources: **Total Energy Consumption:** Table 3. **End-Use Energy Consumption:** Tables 3 and 92. **Population:** • 1949—Bureau of the Census, *Current Population Reports,* "Population Estimates and Projections," Series P-25 No. 802, may 1979. • 1950 through 1980—Bureau of the Census, *Current Population Reports,* "Population Estimates and Projections," Series P-25, No. 990, July 1986. • 1981 and forward—unpublished data consistent with the Bureau of the Census Press Release CB90-204, December 1990. **Consumption per Capita:** Calculated by Energy Information Administration.

Fig. 15-3. While our total energy usage increased by about 6 quad, per capita consumption has gone down 24 million Btu per year since 1979. (Energy Information Administration)

- Structures are not tight enough. Cracks around windows, doors, and foundations allow air to move in and out freely.
- Too little insulation. Heat loss or entry through the walls, ceilings, and windows is too high. Furnaces and air conditioners must work harder to maintain comfort levels; thus, they must use more energy.
- Electrical appliances, especially older ones, use electricity inefficiently.
- Uninsulated hot water pipes allow too much heat to escape to the atmosphere.
- Older furnaces and boilers do not deliver enough of the heat value in their fuel. Some units have an efficiency rating of less than 35 percent.
- Dwellings are often larger than they need to be. Too much energy is used to heat them.
- Living quarters are often kept too warm in the winter and too cool in the summer.

It has been suggested that remedying these problems would save as much as 50 percent on residential/commercial energy usage.

HOW HEAT IS LOST

We know from our study of the nature of heat (Chapter 1) that it will always move from a warmer substance or warmer surroundings to those that are cooler. It tends to do this until all the surroundings have the same temperature.

For example, a building which is heated to 70°F (21°C) will lose heat to the outside when the temperature outside is below 69°F (20°C) or lower. It will gain heat from the outside if the outside temperature is above the inside temperature. This will happen in spite of barriers such as doors, windows, walls, and roofs.

Buildings gain or lose heat primarily through:
- Conduction.
- Convection.

When heat moves through materials by increased molecular activity, it is called **conduction**. Loss of heat or gain of heat by a building through conduction is also called exfiltration and infiltration. It is a slower process than convection.

Some materials such as stone or brick are good conductors of heat. Other building materials conduct heat poorly and are known as insulators.

Convection is air in motion. Currents of air can carry heat through cracks around doors, windows, and other openings in a building. Desirable heat can also be exhausted by such mechanisms as fans. Convection also carries useful heat up chimneys. These same openings can also carry unwanted heat into a building during the cooling season.

CHIMNEY EFFECT

One problem involving air leaks is the "chimney effect." Warm air rises and leaks out through cracks or openings at upper levels or through ceilings. Cold air leaks in at lower levels, Fig. 15-6. The cracks act like a chimney drawing in air at a low level and letting it out at an upper level.

Conserving energy in housing is closely related to the nature of heat. Conservation starts with measures that will make movement of heat in or out of a building more difficult.

These measures are:
- Use of building materials that resist passage of heat through them. (You can tell the difference in materials on a cold day by simply touching a wood door and then the glass in a window.) Glass does not resist the passage of heat as well as the wood. A material which is highly effective at reducing heat passage is called **insulation**. You will generally find that good heat insulators are able to trap air easily.

Generally speaking, building materials must be strong to support the loads placed upon them. The very qualities which make them good at supporting loads also make them poor insulators. Therefore, builders use strong materials to build a frame. This frame must support insulation, siding, wall coverings and more.
- Seal points in a building's frame where convection can draw warm air in or out of the structure.

R AND U RATINGS

To recognize energy efficiency in building, you should understand the U and R values of construction materials. An **R value** is the value given a material's resistance to passage of heat. A high R value means the material is a good insulator.

A **U value** is the measure of a material's ability to conduct heat. A high U rating means that the material conducts heat easily. A material with a high U rating is not a good insulator. As you can see, U and R values are the opposite of each other.

Until the 1970s, when energy costs began to rise rapidly, little attention was paid to R or U values and insulation. Energy was so cheap. Insulation was considered expensive. However, with higher energy costs, it is important to be familiar with insulating standards for housing and insulating qualities of different construction materials, Fig. 15-7.

DEGREE DAY

Recommendations for R values are set by the Federal Housing Administration which issues the FHA Minimum Property Standards. See Fig. 15-8. These

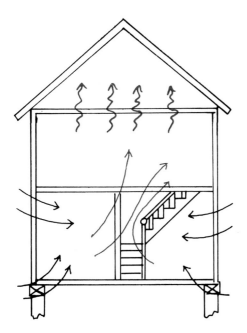

Fig. 15-6. The chimney effect occurs in a buildng when air enters through cracks and leaves through openings at a higher level or through a poorly insulated ceiling. The effect robs heat from the building's rooms.

MATERIAL	THERMAL RESISTANCE (R)
Still surface air	0.68
Air space	0.97
Face brick	0.39
Wood siding	0.98
1 in. wood floor	0.98
Concrete block	1.11
1/2 in. wood siding	0.85
Gypsum wallboard, 3/8"	0.32
Urethane insulation	9.0
Building paper	0.06
Linoleum or tile	0.05
Asphalt shingles	0.95
Plywood	0.95
Single pane, glass	0.9
Double glazing, 1/2 in. space	1.54
Triple glazing, 1/2 in. space	2.79
Storm windows	1.79

Fig. 15-7. Insulating (R) values of common building materials. These values can be used in computing the thermal resistance of different parts of a structure.

Normal Number of Degree-Days Per year

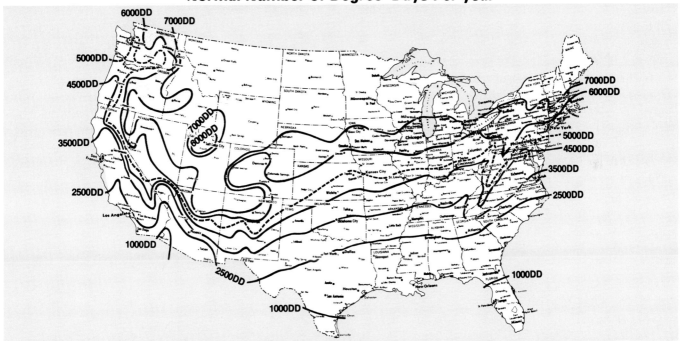

This map is reasonably accurate for most parts of the United States but is necessarily highly generalized, and consequently not too accurate in mountainous regions, particularly in the Rockies.

RECOMMENDED INSULATION R VALUES

DEGREE DAYS	CEILINGS	WALLS	FLOORS
Above 7000	38	17	19
6001-7000	38 (30F)	17 (12F)	19 (11F)
4501-6000	30	17 (12F)	19 (11F)
3501-4500	30	12 (17E)	11 (19E)
2501-3000	22 (30E)	11 (17E)	0 (11E)
1001-2500	19 (22E)	11 (12E)	0
1000 and under	19	11	0

Fossil fuel (gas & oil) = F
Electric resistance heat = E
Heat pumps with electric heat = H
R Values apply to all unless otherwise shown.

These insulation "R" values are recommended to meet the FHA Minimum Property Standards for thermal resistance. The R value shown for walls can be met with any combination of sheathing and cavity wall insulation, or blanket insulation alone. For additional information, refer to Owens-Corning Pub. No. 5-BL-9293-A.

Fig. 15-8. Insulation requirements for housing depend on the severity of the winters and the intensity of the summer heat. (Owens-Corning Fiberglas Corp.)

standards are based on the severity of the winters and the heat of the summer in a particular locality. This severity is measured in degree days. A **degree day** is a measure of how much temperature varies above or below a standard temperature of 65°F (18°C). It is assumed that at this temperature neither heating nor cooling is needed.

This is how degree days are found:

1. Average the high and low temperature for a 24-hour period.
2. Subtract this temperature from the standard temperature (65° F or 18°C). The answer is the number of degree days. For example, suppose that the average temperature for a 24-hour period is 40°F (5°C). Subtracting 40 from 65 leaves 25. There are 25 heating degree days for that period.
3. Total up all the degree days for a heating or cooling season to get the total degree days for the region's climate.

INSTALLING INSULATION

Insulation is manufactured from different materials and in different forms. Their R values are listed in Fig. 15-9.

Loose fill is used in ceilings or in space you cannot reach with batts or blanket insulation, Fig. 15-10. Rigid insulation is not normally used for floors or ceilings except in homes with cathedral type ceilings or in mobile homes. It is the preferred type, however, when

Fig. 15-10. Installing loose insulation requires special equipment as well as protection against breathing in insulating particles and dust. It is the preferred insulating method in areas difficult to reach. (Owens-Corning Fiberglas Corp.)

insulating basement walls. The sheets can be placed between thin furring strips.

Batt and blanket insulation are suited for use between standard spaced joists and rafters. On new construction it is also used between wall studs. Batts are sold in precut sections or in long rolls, Fig. 15-11. Vapor barrier is always faced toward the heated space. Fig. 15-12 shows insulation being applied between ceiling joists in new construction. When adding additional batt or blanket insulation to ceilings of older buildings, you

THERMAL RESISTANCE OF VARIOUS INSULATING MATERIALS

MATERIAL	R VALUE PER IN.	INCHES NEEDED FOR VALUE INDICATED				
		R-11	R-19	R-22	R-34	R-38
Batts/Blankets						
Fiberglass	3.14	3.5	6	7	11	12.5
Rock Wool	3.14	3.5	6	7	11	12.5
Loose/Machine Blown						
Glass Fiber	2.25	5	8.5	10	15.5	17
Rock Wool	3.125	3.5	6	7	11	12.5
Cellulose	3.75	3	5.5	6	9.5	10.5
Hand Poured Loose Fill						
Cellulose	3.7	3	5.5	6	9.5	10.5
Rock Wool	3.125	3.5	6	7	11	12.5
Glass Fiber	2.25	5	8.5	10	15.5	17
Vermiculite	2.1	5.5	9	10.5	16.5	18
Rigid Board						
Polystyrene Beadboard	3.6	3	5.5	6.5	9.5	10.5
Extruded Polystyrene	5.41	3	5	5.5	8.5	9.5
Urethane	6.2	2	3	3.5	5.5	6.5
Glass Fiber	4	3	5	5.5	8.5	9.5

Fig. 15-9. The insulating value of different insulators vary greatly. Note the R value of Urethane.

Fig. 15-11. Blanket insulation is available with or without vapor barrier. (Owens-Corning Fiberglas Corp.)

Fig. 15-12. In cold climates insulation of ceilings should be 12 in. thick or up to R 38 as recommended in Fig. 15-7.

Fig. 15-13. When additional insulation is applied to upgrade R values for a ceiling it is advisable to install batts across the ceiling joists. (Certainteed)

Fig. 15-14. Loose insulation can also be added over batts. (Owens-Corning Fiberglas Corp.)

Fig. 15-15. Keep insluation several inches away from light fixtures recessed in the ceiling lest they become overheated. (Certainteed)

should lay them across the joists as shown in Fig. 15-13. Loose insulation can also be applied on top of batts, Fig. 15-14. To prevent overheating of recessed light fixtures, keep insulation at least three inches away from them. See Fig. 15-15.

INSULATING FOUNDATIONS

R values of foundation walls are relatively low. This means that warm air travels through the concrete too easily. Fig. 15-16 shows how walls, sills, and headers should be insulated for the most effective control of heat loss through conduction.

CUTTING ENERGY LOSSES THROUGH GLASS

Since a single pane of glass has an R rating of 0.9, heat is easily transmitted. Thus, large areas of glass can cause significant energy loss during either heating or cooling of buildings.

Double or triple glazing are the most effective methods of increasing insulating value of windows, Fig. 15-17. Other effective ways of increasing the R value of window areas are:

• Installation of storm windows. Besides the insulating

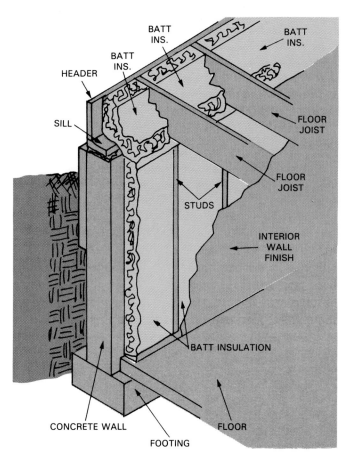

Fig. 15-16. In new or old construction, foundations can be insulated inside and out. Outside, rigid insulation should extend 2 ft. below grade.

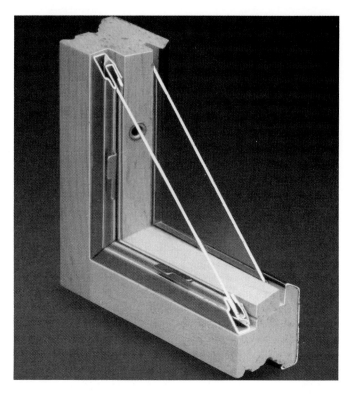

Fig. 15-17. Double and triple glazing add needed insulation to window areas. (Rolscreen Co.)

value of an additional pane of glass, the air space between the storm window and the regular window has an R value about equal to that of the glass.
- Installation of insulated shades.
- Installation of interior shutters that can be closed to reduce heat loss through the window during cold weather. Like the insulated shade, they could also be closed to block solar radiation in the hot summer months.

- Closing shades and/or drapes against either unwanted solar radiation or winter cold.

STOPPING AIR LEAKAGE

While buildings should be reasonably airtight, no structure can or should be made entirely airtight. Good ventilation within the building requires one complete change of air every hour. Air exchangers are often employed to draw fresh outdoor air into the building, heating it with warm stale air that is being exhausted outdoors. Fig. 15-18 is a diagram of a typical heat exchanger for a dwelling.

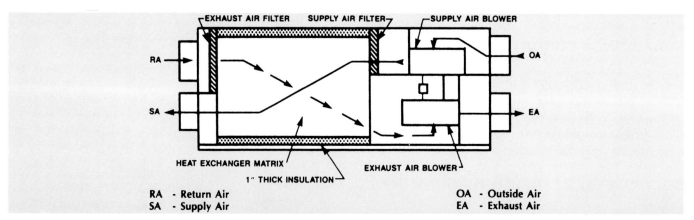

Fig. 15-18. Diagram showing how a heat exchanger works. Airstreams flowing close to each other allow transfer of heat from exhaust air to incoming outdoor air without actual mixing of the two airstreams. (Des Champs Laboratories Inc.)

FINDING THE LEAKS

Air leaks are usually found at these locations:
- Where sills rest on foundation walls, Fig. 15-19.
- Around doors and windows.
- Joints in a building's siding at corners, windows, or doors, Fig. 15-20.
- Cavities around switches and electrical outlets and around doors and windows that have not been filled with insulation.

It is easy to detect these leaks on a cold and windy day. Simply hold a lighted candle near a suspected leak. The movement of the air will cause the flame to flicker or even go out. More sophisticated methods are smoke testing and infrared photography.

SMOKE TEST

In smoke testing the air pressure inside the house is raised slightly. Then smoke is released around areas where leakage is suspected. Air pressure in the house will force the smoke out, pinpointing the source of the air leak, Fig. 15-21.

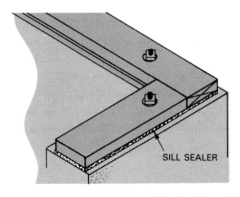

Fig. 15-19. The crack between a building's sills and its foundation can be sealed with caulk or with the sill sealer shown.

Fig. 15-20. Air leakage around windows can be sealed by caulking joints where siding meets the window and door trim.

Fig. 15-21. Smoke test is made with the house slightly pressurized. Smoke will escape through cracks.

INFRARED PHOTOGRAPHY

Infrared photographs taken outside of the building will also indicate location and extent of heat loss through walls and roof. A chemical in the film emulsion is sensitive to infrared waves (heat). Poorly insulated areas of the walls and roof heat the air and the heat will appear bright red on the photograph.

SEALING CRACKS

Buildings that are not properly sealed against air leakage can waste up to 40 percent of the heating and cooling energy.

Several commercial products are useful in sealing cracks. Included are caulking compounds, oakum, and weather stripping.

Caulk is a soft, puttylike material usually put up in tubes. The tube fits into a gun for application. It may also be sold in rope-like coils that can be pressed into cracks without benefit of tools.

Oakum is a tarred hemp product made into a light rope-like material. It is used to seal large cracks and is sometimes called caulk rope. The rope is peeled off the roll and pressed into the cracks.

Weather stripping can be made of thin strips of metal, rubber, vinyl, spring bronze, foam rubber, or felt. The strips can be tacked or glued around doors and windows.

To seal air leaks:
- In exterior walls. Inspect all joints where siding meets

corners, window frames, and door frames. Also check points where siding meets the foundation. Seal joints with a thin bead of caulk. If old caulk has dried out and cracked, remove it and recaulk. Caulk openings where pipes or ducts come through exterior walls, Fig. 15-22. Foam rubber gaskets are made to slip under switch plates, Fig. 15-23.

• Doors and windows. Weather strip edges where air leakage occurs. Reglaze sashes where glazing has cracked and fallen out. On old windows with sash weights, stuff small pieces of sponge rubber into openings around pulleys. Weather strip doors. Special gaskets are available to seal gaps between the bottom edge of the door and the threshold.

NEW CONSTRUCTION

New homes can be designed to reduce the amount of energy needed to heat and cool them. The savings come about in two ways:

• If the home is designed to use less energy, a smaller, less costly, heating/cooling system can be used.
• Tighter, super insulated homes and a smaller heating/cooling system will consume less energy to keep the home comfortable.

One house, known as the Arkansas Energy Saving Home, was part of a study which compared nine energy-saving homes with nine similar homes that were built to the specifications of the 1974 Minimum Property Standards (MPS) of the FHA. The energy-saving homes used 50 percent less energy over a three-year period.

Superinsulated, panelized factory constructed systems are now available. Outside wall sections and roofs are made up in panels consisting of a thick 5 1/2 to 7 1/2 in. sheet of polystyrene sandwiched between two sheets of wafer wood. (Waferwood is a construction sheet like plywood but made up from wood chips.) The system "wraps" the building in insulation, providing higher R values than conventional construction. Heating costs are said to be reduced by half from conventional, stick-built structures. Refer to Fig. 15-24. Fig. 15-25 is a list of construction practices that should be avoided.

EFFICIENT APPLIANCES

Appliances account for about 25 to 28 percent of the energy used in the average household. Fig. 15-26 lists estimated electrical power consumed annually by different appliances.

Appliances can be designed to do their job using less energy. Since 1976, the appliance industry has been efficiency labeling their different models. This information is displayed on most new appliances and is known as an "energy efficiency rating" (EER).

EER RATINGS

Following the energy crisis of 1973, the National Bureau of Standards set up an appliance labeling section. Its job is to sponsor a program of efficiency rating of household appliances. Appliances covered by the rating program include: room air conditioners, refrigerators, freezers, water heaters, clothes washers and dryers, ranges, ovens, furnaces, and central air conditioners.

EER labels are attached to the appliances and contain the following information:

• The energy efficiency rating of the appliance. (The higher the rating, the more efficient the appliance in its use of electricity, gas, or fuel oil.)

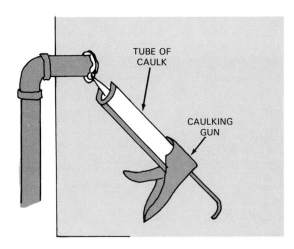

Fig. 15-22. Other leakage may occur around openings where ducts and pipes go through a wall. These, too, can be sealed with caulk.

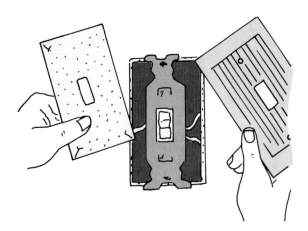

Fig. 15-23. A simple remedy like a foam pad under light switches and receptacles helps cut down air infiltration through walls.

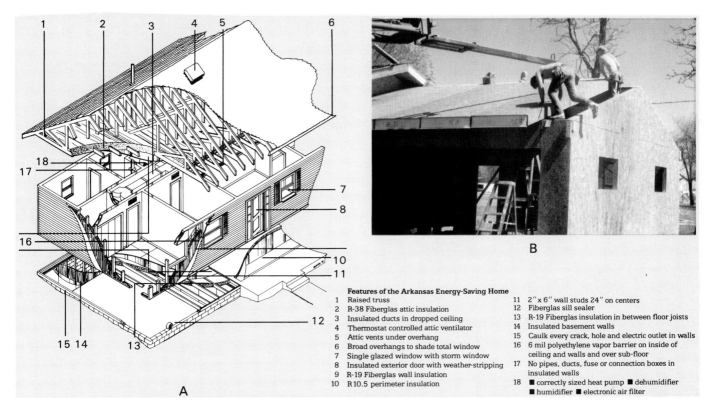

Features of the Arkansas Energy-Saving Home

1. Raised truss
2. R-38 Fiberglas attic insulation
3. Insulated ducts in dropped ceiling
4. Thermostat controlled attic ventilator
5. Attic vents under overhang
6. Broad overhangs to shade total window
7. Single glazed window with storm window
8. Insulated exterior door with weather-stripping
9. R-19 Fiberglas wall insulation
10. R10.5 perimeter insulation
11. 2″ x 6″ wall studs 24″ on centers
12. Fiberglas sill sealer
13. R-19 Fiberglas insulation in between floor joists
14. Insulated basement walls
15. Caulk every crack, hole and electric outlet in walls
16. 6 mil polyethylene vapor barrier on inside of ceiling and walls and over sub-floor
17. No pipes, ducts, fuse or connection boxes in insulated walls
18. ■ correctly sized heat pump ■ dehumidifier ■ humidifier ■ electronic air filter

Fig. 15-24. A—The Arkansas Energy-Saving home has 18 special features. (Owens-Corning Fiberglas Corp.) B—A super-insulated, panelized home under construction. Walls and roof are made up of polystyrene insulation with a wooden sheet bonded to each side. Channels are provided for wiring. (Ener-Cept Building Systems)

DON'TS

1. Avoid anything that will interfere with the insulated envelope, upset the humidity control, or create a need to readjust the thermostat controls once they have been balanced.
2. Avoid using sliding glass doors. They are hard to keep satisfactorily sealed, and the large glass areas present large hot and cold surfaces that can cause discomfort leading to higher- or lower-than-necessary thermostat settings. Substitute French doors that can be weather stripped to close tightly.
3. Don't use recessed lighting in the attic. Fire codes require a 3-inch uninsulated space around them, which leads to undesirable heat losses into the attic. Tracked lighting on the ceiling surface can provide similar aesthetic effects without the heat loss.
4. Avoid a fireplace unless the combustion chamber is supplied directly with outside air and has a tight-fitting glass screen and damper. (The exterior of the fireplace must be as well insulated as the other exterior walls.)
5. Don't allow anything in exterior walls, such as pipes, ducts, electrical distribution, or medicine cabinets, that will reduce insulating efficiency.

Fig. 15-25. For energy efficient construction, these practices should be avoided.

- The range of efficiency or energy consumption of other similar products on the market.
- An estimate of how much the appliance will cost to operate during the year at various levels of usage.

A typical label is shown in Fig. 15-27. Usually, they are referred to as "energy guide" labels.

FIGURING ENERGY EFFICIENCY RATINGS

The energy rating of an electrical appliance is calculated by dividing the energy output, in Btu per hour, by the electrical energy used (in watt/hr). For example, if a window air conditioning unit has a capacity of 7000 Btu/hr with an input of 840 watts of electricity per hour the EER is equal to 7000 divided by 840 watts or 8.4. This information can be found on the energy guide label.

Wise use of appliances is just as important as wise shopping for the most efficient ones. Refer to Fig. 15-28.

SPACE CONDITIONING EFFICIENCY

Furnaces are a major user of energy in the home. Let us assume that the efficiency of an older unit is around 50 to 60 percent. This means that for every 100 Btu of fuel energy consumed the furnace delivers 50 to 60 Btu of heat energy.

Much of the heat energy is exhausted up the chimney. Temperatures must usually be above 300°F (149°C) in

ANNUAL ENERGY REQUIREMENTS OF ELECTRIC HOUSEHOLD APPLIANCES*

*Source: Edison Electric Institute

Major Appliances	EST. kWh	Kitchen Appliances	EST. kWh		EST. kWh
				Heating Pad	10
Air-conditioner (room) (Based on 1000 hours of operation per year. This figure will vary widely dependong on geogrpahic area and specific size of unit)	860	Blender	15	Humidifier	163
		Broiler	100	**Laundry**	
		Carving Knife	8	Iron (hand)	144
		Coffee Maker	140		
Clothes Dryer	993	Deep Fryer	83	**Health & Beauty**	
Dishwasher including energy used to heat water	2100	Egg Cooker	14	Germicidal Lamp	141
		Frying Pan	186	Hair Dryer	14
Dishwasher only	363	Hot Plate	90	Heat Lamp (infrared)	13
Freezer (16 cu. ft.)	1190	Mixer	13	Shaver	1.8
Freezer—frostless (16.5 cu. ft.)	1820	Oven, Microwave (only)	190	Sun Lamp	16
		Roaster	205	Toothbrush	.5
Range with oven with self-cleaning oven	700 730	Sandwich Grill	33	Vibrator	2
Refrigerator (12 cu. ft.)	728	Toaster	39	**Home Entertainment**	
Refrigerator—frostless (12 cu. ft.)	1217	Trash Compactor	50	Radio	86
		Waffle Iron	22	Radio/Record Player	109
Refrigerator/Freezer (12.5 cu. ft.)	1500	Waste Disposer	30	Television Black & White Tube type Solid state	350 120
Refrigerator/Freezer—frostless (17.5 cu. ft.)	2250	**Heating and Cooling**		Color Tube type Solid state	660 440
		Air Cleaner	216		
Washing Machine—automatic (including energy, used to heat water) washing machine only	2500 103	Electric Blanket	147	**Housewares**	
		Dehumidifier	377	Clock	17
Washing Machine— non-automatic (including energy to heat water) washing machine only	2497 76	Fan (attic)	291	Floor Polisher	15
		Fan (circulating)	43	Sewing Machine	11
		Fan (rollaway)	138	Vacuum Cleaner	46
		Fan (window)	170		
Water Heater	4811	Heater(portable)	176		

Note:
When using these figures for projections, such factors as the size of the specific appliance, the geographic area of use, and individual use should be taken into consideration.

Fig. 15-26. This chart might be the basis for determining the annual electrical energy costs in your home. Electric rates will vary from one geographic area to another.

the chimney so it will draw. When the furnace is not running, additional warm room air is lost as it rises up the open chimney.

Automatic vent dampers are available which close off chimney flues when furnaces and other gas burning appliances are not in operation. Fig. 15-29 shows a vent damper system.

Fuel efficient furnaces

New furnace designs make better use of fuel, Fig. 15-30. Some of these systems are rated 91 to 96 percent efficient by the Department of Energy.

The high efficiency systems are designed with heat exchangers that extract more of the heat of combustion. Exhaust gases are in the 100°F (38°C) range. Therefore, the furnace does not require a chimney. Spent gases can be exhausted through a small plastic pipe.

A unique design is known as the Pulse furnace. The fuel charge is ignited by a spark plug. Fuel is burned in a series of small ignitions called "pulses."

Heat pumps

An appliance which may become an important energy saver in the future is the heat pump. The **heat**

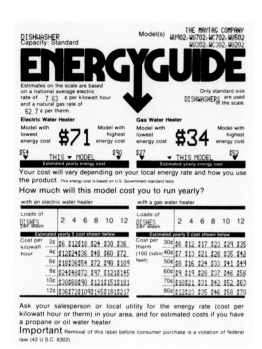

Fig. 15-27. Left. Typical Energy Guide found today on appliances. By comparing, shoppers can select models that are more energy efficient. Right. Annual cost of operation may rise as costs of energy rise. (GE and Irv Wolfson Co.)

About 8 percent of all the energy used in the United States goes into running electrical home appliances, so appliance use and selection can make a considerable difference in home utility costs. Buying an energy-efficient appliance may cost a bit more initially but that expense is more than made up by reduced operating costs over the lifetime of the appliance.

Energy efficiency may vary considerably though models seem similar. In the next few years it will be easier to judge the energy efficiency of appliances with the Government's appliance labeling program. (See page 31 for details.) In the meantime, wise selection requires a degree of time and effort.

You will find a number of tips on how to save energy when buying or using appliances in other sections of this booklet, but here are a few general ideas to consider.

- **Don't leave your appliances running when they're not in use.** It's a total waste of energy. Remember to turn off your radio, TV, or record player when you leave the room.

- **Keep appliances in good working order** so they will last longer, work more efficiently, and use less energy.

The list shows the estimated annual energy use of some household appliances. With this information, you should be able to figure your approximate energy use and cost for each item listed. You also should get a good idea of which appliances in your home use the most energy and where energy conservation practices will be the most effective in cutting utility costs.

- **Use appliances wisely;** use the one that takes the least amount of energy for the job. For example: toasting bread in the oven uses three times more energy than toasting it in a toaster.

- **Don't use energy-consuming special features on your appliances if you have an alternative.** For example, don't use the "instant-on" feature of your TV set. "Instant-on" sets, especially the tube types, use energy even when the screen is dark. Use the "vacation switch," if you have one, to eliminate this waste; plug the set into an outlet that is controlled by a wall switch; or have your TV service man install an additional on-off switch on the set itself or in the cord to the wall outlet.

Fig. 15-28. All appliances are more efficient with wise use. (U.S. Dept. of Energy)

pump is basically an air conditioner with a reversing valve so that it can function as either a heating or a cooling unit. In winter it pumps heat indoors; in summer it pumps it outdoors.

Using 60 percent less electricity than an electric heater, it makes use of three well-known principles of heat energy:

- Heat energy will always move from a warmer to a cooler area.
- A gas will heat up if it is compressed (squeezed together).
- When allowed to expand, the heated gas will become colder.

These statements are also known as the **principles of thermodynamics** (**thermo** for heat and **dynamic** for motion). They deal with movement of heat.

A heat pump system

Heat pumps operation can be modified to heat or cool as seasons change. The basic system includes:

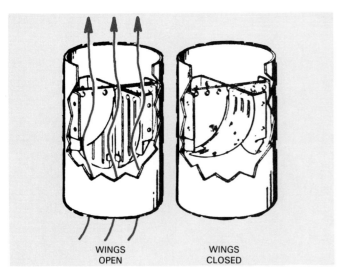

WINGS OPEN   WINGS CLOSED

Fig. 15-29. Automatic dampers open when heat from fuel-fired appliance strikes it, close when the appliance shuts down. (Ameri-Therm)

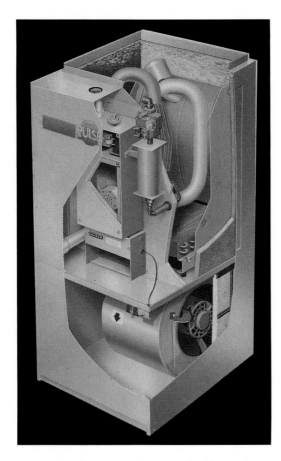

Fig. 15-30. New furnace designs "wring" more heat out of their fuel. Manufacturers provide AFUE (annual fuel utilization efficiency) ratings for their units.   (Lennox Industries Inc.)

• A pump for compressing a gas. This pump is usually called a compressor.
• Two chambers or heat exchangers. One collects heat and is called an evaporator. It is where the gas ex-

pands. The other gives off heat and is called a condenser.
• Pipes or tubing which connect the evaporator and condenser with the compressor.
• A reversing valve which changes direction of the gas flow for either heating or cooling.

How the heat pump works

The heat pump's system is filled with a gas called a **refrigerant**. Its purpose is to absorb heat and transport it to other parts of the system. When the reversing valve is in the "heat" position, Fig. 15-31, the chamber that is located outdoors is like an evaporator. The refrigerant expands into it. Being cooler, the refrigerant absorbs heat from the warmer outdoor air. Moved along by the compressor, the refrigerant flows to the chamber inside the house which is the condenser. Here the gas gives up its heat to the indoors. When the valve is reversed, the heat pump reverses and acts like an air conditioner. Now it draws heat from the building and moves it outdoors, Fig. 15-32.

HEAT PUMP EFFICIENCY

Using the same amount of energy, a heat pump will deliver up to three times as much heat as an electrical resistance heater. Since it heats as well as cools, it is cheaper to purchase than both a furnace and an air conditioner.

In the heating mode, the heat pump's efficiency drops as outside temperatures drop. Below 20°F (-7°C) it usually is not able to provide enough heat for a building. An auxiliary heating system must be used to assist the pump.

This problem can also be solved by placing the outdoor coil in a better heat source. One method is to bury the coil in the ground where the temperature is always higher in the cold season. Fig. 15-32 shows a system with a ground coil. Another solution that works well is to place the coil down a well, Fig. 15-33, or on a lake bottom where it can collect heat from the water. A third solution is to install the outdoor coil in the heat storage tank or bin of a solar collector. (Solar collectors are described in the chapter on Solar Energy.)

REDIRECTING HOME HEAT ENERGY

Because of the way homes are designed and constructed, heat generated by appliances in warm weather adds to the heat load and increases air conditioning costs.

Similarly, much of the heat carried through uninsulated duct work and pipes is lost. In 1976 the federal

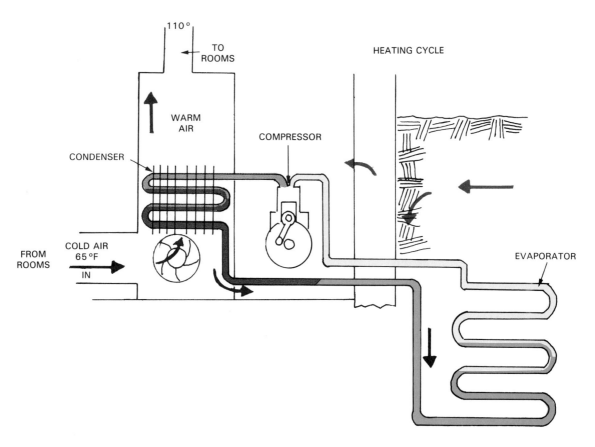

Fig. 15-31. Diagram of a heat pump set to heat. An outside ground coil collects heat that the refrigerant carries indoors and releases to air flowing through the condenser.

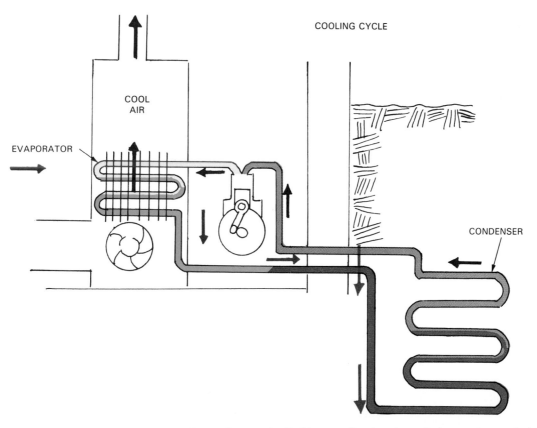

Fig. 15-32. A heat pump diagram showing the cooling mode. Refrigerant flowing through the evaporator indoors absorbs heat from the indoor air and carries it outdoors to be released through the ground coil.

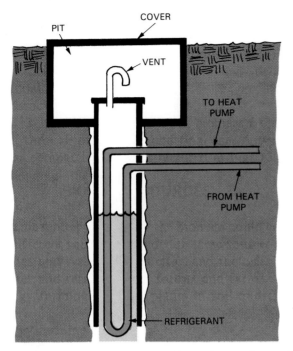 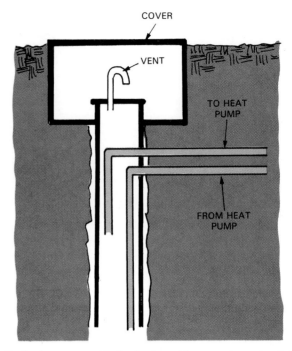

Fig. 15-33. A well is a suitable place to locate the outdoor coil of a heat pump. Left. A closed loop system circulates refrigerant. Right. An open system in which well water is the medium for carrying heat to or from the building.

government published a report on an experimental house in Hampton, Virginia. The home's designers sought to conserve energy in three ways: through use of an active solar collector system, through insulating its systems, and through relocation of its ducting system. All ducts were located entirely within the heated or cooled space. At the same time, heat from appliances could be ducted out of doors during the cooling season. During the heating season the appliance-generated heat could be distributed throughout the home.

Fig. 15-34 is a schematic for the heating system for a home that serves as a model for the Saskatchewan Conservahome Program.

The home was so weathertight and was so well insulated that it needed no furnace. It was estimated that the home's heating needs had been reduced by 78 percent. Even the smallest conventional furnace would have been too large and, therefore, inefficient. Instead, the home's water heater became the primary heating source. Hot water from the heater is pumped through a coil in the warm air distribution system. Warm exhaust from the clothes dryer is also fed into the heat distribution system. During hot months, the dryer is vented outside.

The hot water heater draws its combustion air from outside. Thus, it does not rob warm air from the house.

Hot water piping systems are another source of heat loss. Pipes carry heated water from the water heater to faucets in the kitchen, bathroom, and laundry. Some of the heat is lost to the cooler air surrounding the pipe. Insulating the pipe will prevent this heat loss. Insulating

the water heater itself will reduce heat loss by about 400 kWh a year of electricity or 3600 cu. ft. of natural gas. See Fig. 15-35.

Other methods of saving energy on water heating include:
- Lowering the temperature setting on water heaters to the lowest temperature setting acceptable for household needs.
- Fixing leaking faucets.
- Running the dishwasher only when fully loaded.
- Installing flow restrictors in shower heads.
- Laundering soiled clothing in cold or warm water when practical.
- Loading washing machines fully or adjusting the water level to the size of the load.

LIGHTING

According to the U.S. Department of Energy, 20 percent of the electricity generated in the United States is used for lighting. Energy savings involving use of lighting revolve around these three activities:
- Selection and purchase of more efficient bulbs.
- Careful selection of fixtures or placement of lights.
- Changing of personal habits regarding use of lights.

Selecting light bulbs

One of the first considerations in selecting bulbs is to determine its efficiency. The measure of lighting efficiency is the amount of light delivered for the energy used up by the bulb. Light from all sources is measured

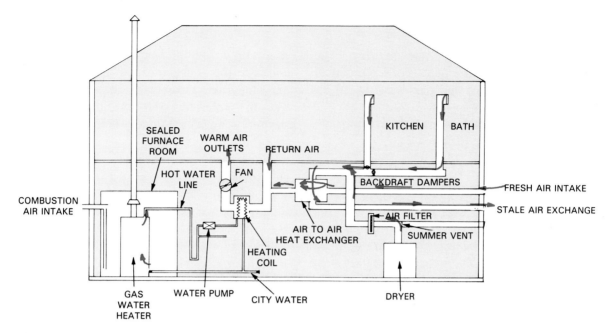

Fig. 15-34. An energy-saving heating system. This dwelling needs no furnace but heat is supplied by a gas-fired water heater. Hot exhaust from the clothes dryer is also fed into the heating system. (Saskatchewan Energy and Mines)

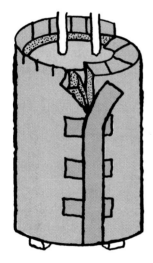

Fig. 15-35. Insulating blankets around hot water heaters conserve energy by insulating the tank.

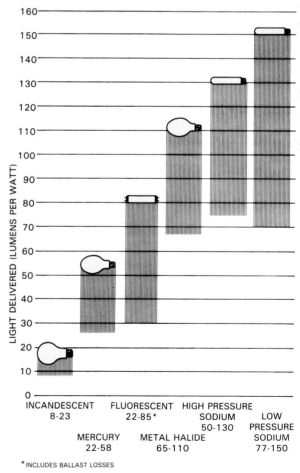

EFFICIENCY OF VARIOUS LIGHT SOURCES

LIGHT DELIVERED (LUMENS PER WATT)

| INCANDESCENT 8-23 | FLUORESCENT 22-85* | HIGH PRESSURE SODIUM 50-130 | LOW PRESSURE SODIUM 77-150 |
| MERCURY 22-58 | METAL HALIDE 65-110 | | |

*INCLUDES BALLAST LOSSES

Fig. 15-36. Types of artificial lighting vary greatly in their efficiency. Least efficient is the incandescent since it must first convert the electrical energy into heat.

in lumens. A **lumen** is the amount of light that falls on a surface. Electrical power usage is measured in watts. A **watt** is equal to a current of 1 ampere at a pressure of 1 volt.

The efficiency of a light source is determined by how much light (lumens) is produced by a watt of electrical energy.

The wattage of a bulb indicates the amount of energy the bulb will use in a given amount of time. Usually, the package will also indicate the lumens the bulb will deliver. The more lumens delivered per watt the more efficient the bulb. Fig. 15-36 is a Department of Energy chart that compares the efficiency of various types of bulbs.

Incandescent bulbs. Incandescent bulbs are least efficient because they produce light by first heating the tungsten filament until it glows. Ninety percent of the energy is used to produce the heat. For all practical purposes, the heat is wasted and passes off to the atmosphere.

Fluorescent lighting. Fluorescent lights do not depend upon heat buildup to produce light. They use a high-voltage electric charge to excite atoms of an inert gas trapped inside the tube. A small step-up transformer called a **ballast,** Fig. 15-37, causes an electric charge or spark and keeps a current moving through the tube. The atoms of the inert gas give off ultraviolet radiation that the phosphor coating on the inside of the tube absorb. The phosphor coating produces the visible light. Fluorescent lights are five times more efficient than incandescent bulbs. Fluorescent tubes last 20 times longer than incandescent bulbs. Fig. 15-38 is a guide to selection of fluorescent lighting.

Low-pressure sodium lights. Low-pressure sodium lights are highly efficient, delivering from 77 to 150 lumens of light per watt of electricity. They convert into light nearly 35 percent of the energy used. These lamps are used mostly in Europe and occasionally in North America for lighting streets and parking lots.

High intensity discharge (HID) lighting. Like low-pressure sodium lights, HID lighting is designed for outdoor use. High-pressure sodium lights are most often used for interior industrial lighting such as warehousing and manufacturing. Metal halide lights are sometimes used for commercial interior lighting. They are preferred also for stadium lighting where there are TV broadcasts. Mercury lamps are the least expensive outdoor lights.

Proper placement of lights

Placement of light fixtures in relation to surroundings affects their efficiency. The following practices will help increase the amount of light you can get from lighting:

- Light-colored walls and bright surfaces reflect more light than dark surfaces.
- A low-wattage lamp placed near a work area will give as much usable light as a larger general-area light such as a ceiling fixture.
- Lamps used for work or study areas should be placed to the side. This keeps the light from reflecting off the work into the viewer's eyes.

Maintaining lights

Frequent cleaning of lights, particularly the reflectors, helps to keep them delivering the light levels for which they are designed. A lamp producing 20 lumens per watt when installed may produce only half that when covered with dust. Fluorescent reflectors, especially, are known to collect dust. They need frequent cleaning.

Efficient use of lighting

Wise usage of lighting will produce more light for less wattage. The checklist in Fig. 15-39 suggests how to increase lighting efficiency through wise use.

Automatic controls

Lights can be controlled with automatic devices that will turn them off should they be forgotten. All of these

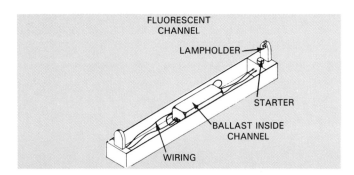

Fig. 15-37. The fixture required for fluorescent lighting is more expensive than incandescent lighting. One requirement is a step-up transformer called a ballast. On the other hand, fluorescent bulbs are more durable and efficient.

	EFFECT ON "ATMOSPHERE"	COLORS STRENGTHENED	COLORS WEAKENED OR GRAYED	REMARKS
Cool white	Neutral to fairly cool	Orange, yellow, blue	Red	Blends with natural daylight
Deluxe cool white	Neutral to fairly cool	All nearly equal	None appreciably	Simulates natural daylight
Warm white	Warm	Orange, yellow	Red, blue, green	Blends with incandescent light
Deluxe warm white	Warm	Red, orange, yellow, green	Blue	Simulates incandescent light

Fig. 15-38. Fluorescent tubes are manufactured for different qualities of light. (U.S. Dept. of Energy)

- Do not overlight an area. Try lower wattages and choose one that is adequate. Use daylight when you can.
- Larger bulbs use energy more efficiently. Try to replace two bulbs with a single bulb which delivers as much light as the two.
- Four-watt night lights have a clear finish and deliver almost as much light as 7 watt night lights. They use only about half as much energy.
- Put fixtures on individual switches. Use only as many lights as you need.
- Turn off decorative outdoor gaslights or replace them with low-wattage incandescent lamps.

Fig. 15-39. This checklist has suggestions for more efficient use of lighting.

devices, if used consistently, will reduce the amount of energy used. The savings more than offset their cost. Such devices include:

- Timers. These can be set to turn lights on and off at set intervals. For example, they can be made to turn the light on at dusk and off at daylight.
- Photocells. Natural light acts upon a photosensitive mechanism causing a small current of electricity. The current turns off a switch to put out the light. When light decreases, the photocell will turn the light on.
- Dimmers. These devices allow adjustments to the light level in a room according to requirements at different times of the day. They save energy because they reduce the amount of energy being used when light demands are low. There are two types: solid state and variable automatic transformers. A third device, a rheostat, is not an energy saver. Though it does cut down the light level, the excess energy is turned into heat and so it is wasted.

CONSERVATION IN TRANSPORTATION

There are many opportunities for energy conservation in transportation. Transportation accounts for 26 percent of the total energy used in the United States. Most of this energy comes from petroleum. Transportation fuels include gasoline, diesel fuel, and alcohol. Over 38 percent of petroleum products consumed is gasoline.

The energy conversion machine is primarily the internal combustion engine. These include the gasoline engine, diesel engine, gas turbine, and jet engine. Electric motors are used in a few types of transportation units such as commuter trains. In the future more electric motor vehicles will be seen on the streets and highways. Environmental concerns are mandating them in some areas.

Transportation has two purposes: to move people and things. Seventy percent of transportation energy is spent in moving people. By far the most-used mode of transport is the automobile. Next most popular for passenger service is the airplane and then the train and the bus. Trucking is the preferred mode for moving freight. See Fig. 15-40.

When we compare the costs of using each type of vehicle for moving people we can readily see that we pay a high price for convenience and speed when we travel. The automobile and the airplane, though most used, cost more in energy per passenger mile than other types of transportation, Fig. 15-41.

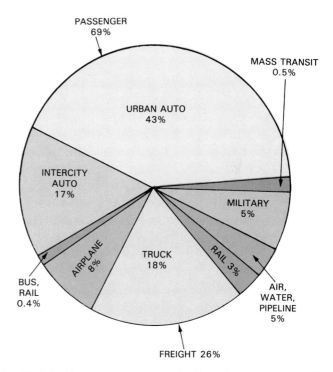

Fig. 15-40. How energy use is distributed among types of transportation. Percentages are approximate.

Mode	Inter-City		
	Average Passenger Capacity	Average Load Factor (percentage)	Energy Cost (BTU per passenger mile)
Railroad	69/car	35%	2900
Bus	41	45%	1600
Automobile	5	48%	3500
Airplane	106	50%	8400
	Urban		
Bicycle	1	100%	200
Walking	—	—	300
Rail (Trolleys, Subways)	?	?	2300
Bus	55	20%	3800
Automobile	5	28%	8500

Fig. 15-41. Comparison of modes of transportation and energy cost per mile passenger is transported. What does it tell you about efficiency of the automobile?

METHODS OF CONSERVATION

The automobile is not an efficient way to transport people in the city. However, we can use it more efficiently by joining a carpool to get to and from jobs. The Federal Highway Administration estimates that on a 30 mile round trip a driver can save from $280 to $650 per year by carpooling.

Other methods of conserving fuel might include:

• Planning shopping trips so that a number of chores could be accomplished at one time. For example, if your Saturday schedule includes trips to the supermarket, lumberyard, and post office, try to do it all in the same trip with the most efficient routing, Fig. 15-42.

• Using other modes of transportation. Consider walking four blocks to pick up the newspaper. Use public transportation for getting to and from work.

• Changing driving habits to get better gas mileage from the automobile. Consider the following: Accelerate at a smooth, steady pace. It saves on engine wear as well as tires and fuel. Keep up a steady pace if possible. Anticipate slowdowns and stops; hold braking to a minimum. Avoid long warmups in cold weather. A couple minutes should be enough. Keep air filters clean. Dirty filters cause an overly rich fuel mixture lowering gas mileage. Keep tires properly inflated. When buying new tires consider radials; they reduce energy consumption. Remove unnecessary loads. Extra weight reduces fuel efficiency. Keep windows rolled up, if possible, it reduces wind drag and improves gas mileage.

Improving vehicle efficiency

Internal combustion engines convert from 22 to 38 percent of the energy they use to useful motion. The rest escapes as heat or is absorbed as friction. It is not possible to improve this level of efficiency greatly because of the law of therodynamics. Still, some improvements are possible.

One suggestion is to reduce the size and weight of the vehicle. Less power is needed thus engines can be smaller. Another is to use manual transmissions with more efficient gear ratios. Automatic transmissions are also becoming more efficient. A new design, called the continuously variable transmission (CVT), uses belts and pulleys. The pulleys automatically change their ratios. See Fig. 15-43.

The variable pulleys in the CVT can be made larger and smaller through the use of weights in the hubs of one set of driven pulleys. Thus, as less power is needed the power pulleys become larger and the driven pulleys smaller. See Fig. 15-44.

Reducing wind drag

Auto and truck manufacturers are working on ways to reduce the wind drag on vehicles. Drag is due to the laws of aerodynamics. Aerodynamics is the study of the motion of air and gases and the forces on bodies moving through such gases or air. Vehicle body designs are placed in wind tunnels where the effects of wind drag on different shapes can be studied. Usually, a model of the body is placed in the tunnel while air is blown past it.

Three types of drag imposed by motion of the car body, Fig. 15-45.

• **Pressure drag.** Bicycle riders experience this. Air is forced against the front of the body, creating greater

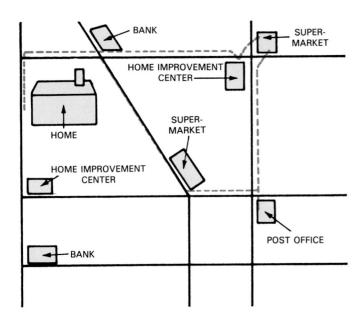

Fig. 15-42. Planning to take care of your shopping trips in the most direct and efficient way. Look at the proposed route. Is it the shortest?

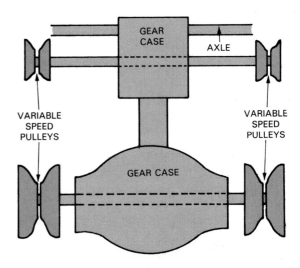

Fig. 15-43. New CVT (continuously variable transmission) design is an attempt to improve efficiency of the automobile.

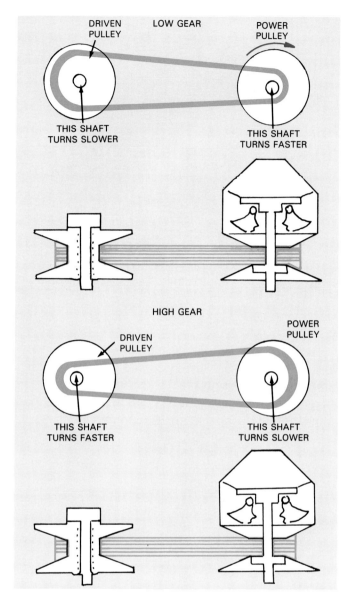

Fig. 15-44. How the CVT works. In low gear, power pulley is small for more power at low speed. In high gear, power pulley is large to drive driven pulley faster.

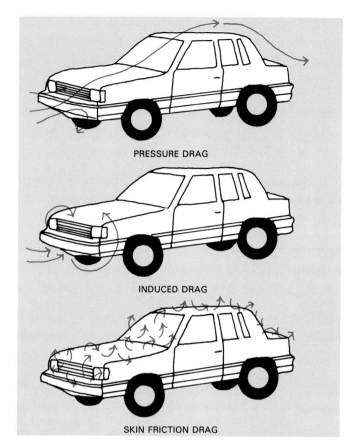

Fig. 15-45. These three types of drag reduce the efficiency of automobile as it speeds increases.

air pressure. Eddies and whirls form in the back of the body as the air flows past and absorbs energy. Streamline forms are shapes which reduce this drag. The classic streamline form is the teardrop. A cross section of an airplane wing takes this basic shape.

- **Induced drag.** This type does not concern automobile body design too much. It occurs mostly at the tips of airplane wings. Air compressed under the wing slips over the tip to the lower pressure on top of the wing. This creates eddies which drag on the wing.
- **Skin friction drag.** This is the tendency of the air to "stick" to the surface of a moving object. This is somewhat like the effect of pulling an object through thick molasses. Of course, the air is not as viscous, but, at high speeds, the effect is similar.

REFINEMENTS TO INTERNAL COMBUSTION ENGINES

Under EPA guidelines, many improvements are making automotive engines more fuel efficient. The most important of these advances currently in production are fuel injection, electronic ignition, and computers. On board modules control fuel mixture, ignition, timing, air pollution controls, and other engine functions.

Port fuel injection

Port fuel injection is a system of delivering air and fuel separately to the combustion chambers of the engine. It uses fuel more efficiently than a carburetor fuel system. In a carbureted system, the carburetor mixes air and fuel before they enter the intake manifold, a group of passages leading to the combustion chambers. Because of the design it is hard to give each cylinder the same amount of fuel. Port injection delivers precise amounts of air and fuel inside each cylinder.

Controlled combustion

Researchers for Texaco have been developing an improved internal combustion design which has been giving more than 35 percent improvement in gas mileage during tests. Known as the Controlled Combustion System, Fig. 15-46, it is adapted to a conventional gasoline engine. The system carefully controls fuel injection, ignition (spark), and air swirl in the combustion chamber. See Fig. 15-47.

STIRLING ENGINE

For a number of years, automotive manufacturers have been experimenting with different types of power units for automobiles. One of these is the Sterling cycle engine. The Stirling gets its energy from heat generated outside the cylinder. The cylinder of the Stirling engine is filled with trapped air or gas. The cool gas is heated by an outside source. The heat can be caused by any source, including solar. The gas is compressed and then heated. The hot gas expands, driving the piston which turns a crankshaft. After the gas has expanded, an outside source cools it and it contracts again. The piston returns and the cycle repeats. Greater efficiency is possible with the Stirling engine because it has a power stroke every revolution. Efficiency of the Stirling cycle engine is above 48 percent.

Fig. 15-46. Texaco Controlled Combustion System combines fuel injection (arrow) with carefully controlled ignition and air turbulence in the combustion chamber of a six cylinder engine. (Texaco, Inc.)

Hydrogen and helium have proven to be the best working gases for the Stirling engine. The gas must be contained at 200 atmospheres pressure.

Engineers are working on two problems. One is providing cooling for the great amounts of heat it

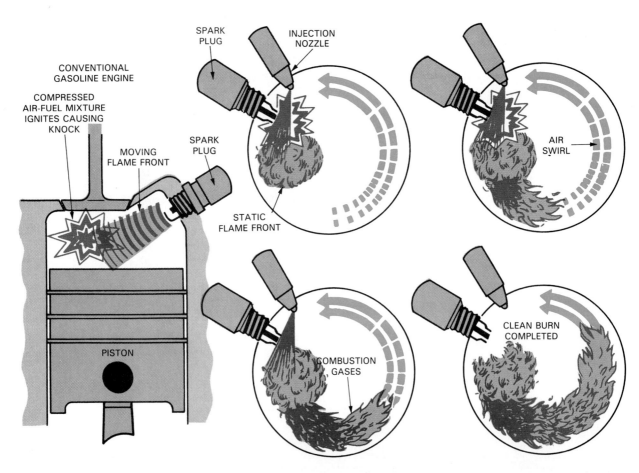

Fig. 15-47. This is how a controlled combustion system works.

generates. A second problem involves the use of hydrogen as fuel. At high pressure, the hydrogen is hard to contain. A seal leak could be a safety hazard.

Gas turbine

Gas turbines operate on the same principle as the water turbine. They have been used successfully to power airplanes, generators, helicopters, large trucks, and boats. See Fig. 15-48. Cool air is taken into the turbine and heated by compression and an exchanger to 1100°F (593°C). As hot air enters a combustion chamber, fuel sprayed into the chamber ignites. Hot gases at 1700°F (927°C) exhaust into a turbine causing it to rotate at high speed.

The turbine is more efficient than the internal combustion engine. Where the latter are at 22 to 38 percent, turbines are 42 to 48 percent efficient. Moveover, they have greater power for their size. Being expensive to manufacture, they have not been used in automobiles. As fuel costs rise, their cost will be less a factor. Another feature in their favor is that they will operate on almost any fuel, including powdered coal, grain alcohol, and biogas.

Electric automobile

The electric motor has many advantages as an energy converter supplying power for an automobile. It is efficient, makes hardly any noise, does not use hydrocarbon fuels, and does not directly pollute the air. Electric vehicles, Fig. 15-49, have been in limited use since 1920 and new designs are under development.

That the electric automobile has not been more popular is due to the lack of suitable batteries. Today's lead acid batteries cannot store enough energy and are too heavy. There are other disadvantages:

• They cost over $600 for a set.
• They do not last long. A set must be replaced every three years.

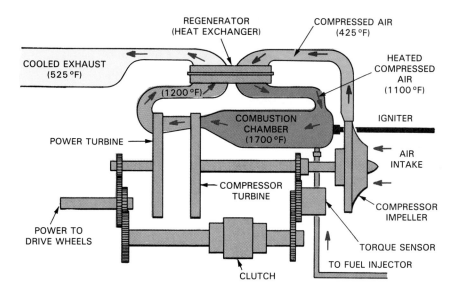

Fig. 15-48. Simple schematic of a gas turbine engine.

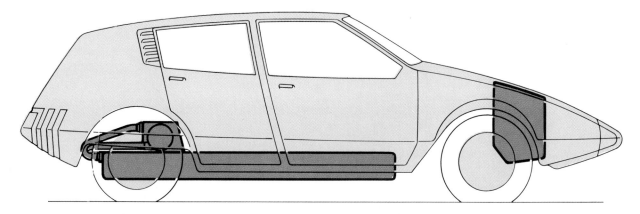

Fig. 15-49. Basic concept of an electric car. This design is for a four-passenger vehicle that would use a zinc-chloride energy storage system. Long container under passenger compartment holds storage batteries. Forward storage is for battery electrolyte. (Energy Development Associates, A Gulf & Western Co.)

- Constant charging and discharging shortens battery life.
- Too bulky. The batteries take up space that is needed for passengers or cargo.

Engineers are working on battery refinements and various automakers are developing electric vehicles. New types of batteries are promising to provide more storage capacity, which means longer driving ranges between recharges. Fig. 15-50 lists the new battery types, their ranges, and expected life. Other electric car technology under development includes:

- **Regenerative braking.** This is simply using the motor for braking by running it as a generator to recharge the batteries.
- "Super flywheel" technology. To store electric power the flywheel would be brought up to top speed by an electric motor plugged into house current at night. The flywheel would drive a generator. The current would be fed to individual electric motors at each wheel. Again, regenerative braking would feed electricity back to the motor driving the flywheel.

LONG RANGE PLANNING

Looking into the future, it is considered possible that the need to conserve fuel will affect the planning of communities and change the modes of travel from what we know today.

It is possible that cities would be much smaller and laid out in a circle. Private vehicular traffic would be prohibited. Bicycles would become the most-used conveyance. Shopping and other necessary community services would be within walking distance. Small neighborhood stores would spring up. Truck traffic would be permitted to service stores and other businesses. Automobiles would be used for intercity traffic within 500 miles (805 km). Aircraft would be used for more distant travel.

CONSERVATION IN INDUSTRY

Industry—manufacturing, mining, agriculture, forestry, and fishing—consumes 40 percent of U.S. energy. Just six industries use half of this energy. They include: metals, chemicals, petroleum and coal products, stone, clay, glass, pulp and paper, and food processing.

ENERGY MANAGEMENT FOR INDUSTRIAL CONSERVATION

In some respects, conservation in industry is like residential conservation. One area of similarity is the confinement of heat so it is not wasted. Thermostats can be lowered. Offices and factories can be better insulated. Air infiltration can be controlled through sealing up cracks and installing doors and windows with better thermal insulation qualities.

A second area where industrial conservation is an entirely different problem is in improving machines and processes to get more product for the same input of energy. Control of heat and improvement of energy usage is called **energy management.**

HEAT RECOVERY

Much of the energy used by industry is spent to generate heat. About 39 percent is used to produce process steam. Another 26 percent is used directly for space heating.

We have already suggested the need for better insulation and reduction of air infiltration. Another important method of conservation is recovery of heat that is now being carried away by smokestacks, exhaust systems, and water cooling systems. There are several systems for recovering this heat.

COGENERATION

Cogeneration is the production of electric power and use of the same heat for industrial operations or other purposes. See Fig. 15-51. Spent steam from electric power generation has lost all its mechanical energy. Since it can do no more work, it is normally released to the atmosphere. In cogeneration it is captured and used in manufacturing processes or for the condition-

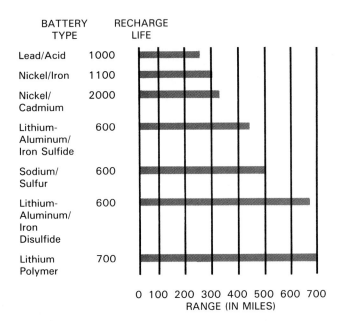

BATTERY TYPE	RECHARGE LIFE
Lead/Acid	1000
Nickel/Iron	1100
Nickel/Cadmium	2000
Lithium-Aluminum/Iron Sulfide	600
Sodium/Sulfur	600
Lithium-Aluminum/Iron Disulfide	600
Lithium Polymer	700

0 100 200 300 400 500 600 700
RANGE (IN MILES)

Fig. 15-50. New technology promises to build electric car batteries with greater range and greater durability than any now on the market.

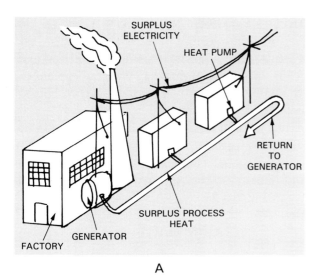

Fig. 15-51. Cogeneration system. A—Mechanical energy of heat generates power. At same time exhaust heat is used to heat buildings and to provide heat for manufacturing processes. B—This 220 MW cogeneration plant is part of the processing plant at a refinery in New Jersey. It will sell electricity to a utility. (The Coastal Corp.)

ing (heating and cooling) of buildings. It is believed that energy savings of more than 2 quad annually are possible if all capabilities were to be developed.

The captured waste heat would mean savings in both energy sources and the cost of installing conventional heating units. One heat input is made to produce two outputs with the same amount of energy.

SYSTEMS FOR COGENERATION

Several different systems can be used for cogeneration. One system uses a gas turbine; the other uses the conventional steam turbine.

The gas turbine uses hot, high-pressure gases resulting from combustion of some type of fuel. Pressure from the expanding gases spins the turbine which is connected to an electric generator. Spent gases from the turbine are piped into a boiler system or a heat exchanger. From 62 to 90 percent of the energy can be recovered.

In a steam system, the turbine is driven by high-pressure steam produced by a boiler that burns either fossil fuel or municipal wastes. Like the gas system, the mechanical energy of the steam produces electric power while the remaining heat energy provides climate conditioning for buildings or process heat for manufacturing.

RECYCLING

Recovery or separating of metals and minerals from the waste stream and making them available for remanufacture is called **recycling.** Since 1992, federal guidelines have required municipalities to recycle at least 25 percent of their waste.

Recycling saves energy since reprocessing recycled metals takes less energy than the processing of ores. These savings are significant:
- 200 million Btu per ton for aluminum.
- 12 million Btu per ton of steel.
- 42 million Btu per ton of copper.

According to the EPA 60 to 75 percent of the aluminum and 90 percent of the steel used in the U.S. could be recycled.

A high percentage of municipal waste is paper. Now essentially banished from landfills, it could be a valuable resource. When remanufactured into new paper and other paper-based products it uses only about one-quarter of the energy needed to process virgin wood pulp.

A new process for recycling used rubber automobile tires into a purified rubber powder has been developed by the French company CIMP (Comptoir Industriel des Metaux et Plastiques). The tires are reduced to rubber powder by this four-stage process. The powdered rubber can be used as surfacing for school playgrounds, running tracks and sports facilities, patios and terraces, as well as noise free road surfaces. Toy car wheels and shoe soles are other uses.

In the first stage of the process, the tires are fed into two high-capacity choppers which reduce the tires to half-inch sized pieces. See Fig. 15-52. The cutting principle is similar to that of scissors. The reversible rectangular blades are spaced around a rotor driven by an inertia flywheel. The rectangular blades contact a row of fixed blades to shear the pieces of tire to the half-inch size. The pieces of rubber are then discharged to a perforated plate which retains the oversized pieces

Fig. 15-52. This recycler turns tires into powdered rubber to be used in making asphalt. (CIMP)

INFRARED HEATING

Most space heating is done through convection and conduction. Infrared heat moves by way of radiation. This is the type of heat thrown off by an old-fashioned potbellied stove.

Factories, warehouses, and public buildings can be more efficiently heated with infrared systems than by conventional systems. The radiant heat travels through space and releases its heat energy when it strikes solid matter. See Fig. 15-53.

SUMMARY

Energy conservation means the efficient use of energy through careful selectioin of energy-using devices. It also means not wasting it through carelessness. Wasteful use of energy has serious consequences because the world and its people are experiencing shortages of energy. This shortage will only become more serious in the years ahead.

Some energy losses are unavoidable. They are the natural result of conversion. However, it is possible to reduce energy consumption through: choosing energy efficient housing, appliances, and automobiles; maintaining and upgrading housing so energy is con-

but allows the granulated pieces (smaller than a half-inch) to go to the operation's second stage.

In the second stage the pieces move over a magnetic separator. A horizontal moving belt vibrates to keep the pieces in a fluid state. A magnetic over-belt, equipped with permanent magnets, separates the ferrous materials from the rubber pieces.

The remaining synthetic and rubber materials continue to the third stage which consists of high speed finishing granulators which produce granules, ranging in size from one to four millimeters (4/100 to 16/100ths inch), in order to separate the rubber from the fibers. These granulators are designed for quick replacement of the blades and grids. The reduced size materials are moved by an air transporter to the finishing stage. In addition to moving the material between the third and fourth stages, the air transporter provides cooling air for the granulators and also dries the material.

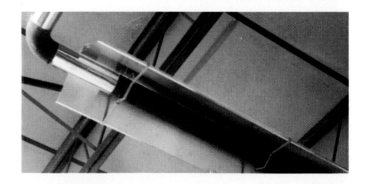

Fig. 15-53. Radiant heating system. A—Heat pipe and reflector. B—Exhaust is vented through the wall.

served; developing habits that save energy; recycling of products and materials.

DO YOU KNOW THESE TERMS?

ballast
cogeneration
conduction
conservation of energy
convection
degree day
energy management
heat pump
induced drag
insulation
lumen
pressure drag
principles of thermodynamics
R value
recycling
refrigerant
regenerative braking
skin friction drag
U value
watt

SUGGESTED ACTIVITIES

1. Survey your community to see what your neighbors may have done to conserve energy with respect to any of the following:
   a. Their automobile.
   b. Their commuting arrangements.
   c. Their house or apartment.
   d. Their business.
2. Have a class discussion on how to plan a city that would save energy resources. Assume that you are designing a new community and can place recreational, business, residential, and industrial areas for maximum convenience.
3. Plan and build a device which will be useful in recycling some type of household organic waste or mineral waste.
4. Plan a new energy-saving residence and produce a scale model to demonstrate your plan. Ask your instructor to recommend reference books on energy saving construction.
5. Contact a company which conducts energy audits of housing. Collect literature and interview them on how an energy audit is done. Report to the class.

TEST YOUR KNOWLEDGE

1. Conservation of energy means to _____ any energy source by using it _____.
2. List five purposes of conservation of energy.
3. Give four personal actions anyone can take in energy conservation.
4. Cite the thermodynamic principle behind a dwelling loss or gain of heat.
5. An R value is:
   a. Unit for conductance of heat through matter.
   b. Unit of resistance to transfer of heat through any matter.
   c. A symbol for insulation.
6. Severity of climate in a certain locality is measured in _____ _____.
7. If temperature averaged 32°F (0°C) over a 24 hour period what would be the degree days for that period?
8. Warm air rising in a closed space to leak out of cracks or into unheated upper levels is called the chimney effect. True or False?
9. What is an EER rating and how is it calculated?
10. A _____ _____ is basically a refrigerator which is capable of either heating or cooling.
11. Why are fluorescent lights more efficient than incandescent lights?
12. Automotive internal combustion engines convert to useful motion from _____ to _____ percent of the energy they consume.
13. How does fuel injection differ from a carburetor in feeding fuel to the automotive engine?
14. _____ is the more efficient use of heat energy to produce both electrical power and space or process heat at the same time.
15. Explain why use of radiant energy systems is more efficient than using convection and conduction systems.
16. Explain how recycling saves energy.
17. Briefly describe cogeneration.

Modern technology, whether directed at more efficient products, new energy sources, or problems of pollution, is working hand-in-hand with science. (Union Pacific)

16

CHAPTER

ENERGY IN THE FUTURE

The information given in this chapter will enable you to:
- *Discuss the future direction of energy technology in the light of environmental concerns and the prospect of limited supplies of fossil fuels.*
- *List some pollution problems directly related to the extraction and processing of energy supplies.*
- *Cite new technologies that will be developed in the 1990s and beyond.*

A constant, reliable supply of energy for the world is a growing challenge and concern. It will continue to occupy people's energies for the foreseeable future. At the same time, the world community is greatly troubled by global warming. This phenomenon, also known as the greenhouse effect, is to some extent, normal. It benefits all life on earth up to a point. Greenhouse gases have always been in the atmosphere. They allow most of the sun's visible radiation to pass through to warm the earth. However, the gases trap a certain amount of the infrared heat reflected from the earth's surface. The gases, in this way, are acting like the glass panels of a greenhouse. The warmth the gases trap provides the warmer temperatures needed to support life on earth. Without them, the earth would be one massive iceberg, lifeless, with an average temperature of $-3°F$ ($-19°C$).

What is worrisome is that since the industrial revolution there has been a dramatic increase in these heat-trapping gases. The result appears to be a warming up of the earth that could cause major climate changes. Fig. 16-1 lists the greenhouse gases and shows how they act in the atmosphere.

The Greenhouse Effect

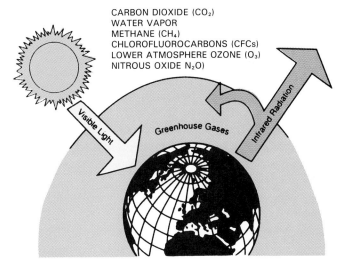

CARBON DIOXIDE (CO₂)
WATER VAPOR
METHANE (CH₄)
CHLOROFLUOROCARBONS (CFCs)
LOWER ATMOSPHERE OZONE (O₃)
NITROUS OXIDE N₂O

Visible Light

Greenhouse Gases

Infrared Radiation

Fig. 16-1. The greenhouse gases are beneficial in certain quantities. They trap infrared heat needed for a comfortable climate. However, too-heavy concentrations of the gases could warm the earth too much. The result: serious and disastrous climate changes.

Burning of fossil fuels is thought to be one of the major causes of the buildup of greenhouse gases. Industrialized nations make the heaviest demand on fossil-fuel energy supplies and contribute heavily to the levels of carbon dioxide in the atmosphere. As a group these nations use three times as much commercial energy as developing countries. At the same time, each person in an industrialized country uses 10 times more energy than persons in developing countries.

Not all of the blame for global warming can be laid on industrial activity, however. Natural decay in the world's swamps also releases carbon and methane into the atmosphere. Activities connected with clearing of land for other uses are another source. For example, satellites monitoring greenhouse gases show large amounts of pollution coming from equatorial, underdeveloped areas where there is large-scale cutting and burning of rain forests. Deforestation also reduces the number of trees available to absorb the carbon dioxide.

The amount of methane in the atmosphere has doubled in 300 years. It will likely double again in 100 years given current rates. There are several sources of methane. One is a direct result of global warming increasing the rate of decay of organic matter. Another source is from food production. Other sources are emissions from landfills, coal mines, and leaking of fossil fuel gases. Adding to the problem are carbon monoxide emissions from engines. It is known that the emissions partially block nature's process for removing the methane from the atmosphere.

The immediate challenge to people of the world is to use our energy resources more wisely and efficiently. This will, in turn, reduce the pollutants that are creating global warming. At the same time, it will conserve dwindling fuel supplies. A second challenge is to develop alternate (renewable and inexhaustible) energy resources.

Stiffer laws curbing pollution are one response of governments to the problems. This has encouraged research and development in new energy and energy-conservation technology. One of these is the development of nonpolluting electric cars. American auto manufacturers as well as some foreign companies have been developing cheaper, more durable, and more efficient batteries to power the new cars. See Fig. 16-2. (Batteries for electric vehicles are discussed in Chapter 15.)

California, with one of the worst urban pollution problems, has passed legislation curbing use of polluting gas engine autos. By 1998, at least 2 percent of the new

Fig. 16-2. Electric vehicles are being legislated in some parts of the United States. (GM)

cars sold in the state must be pollution free. (This leaves no choice but electrics; all other vehicles directly pollute the air.) By 2003, the sales ratio must be 10 percent.

Los Angeles' program goes even farther. The city's Department of Water and Power, along with Southern California Edison, has set up a $17 million fund to sponsor development of electric cars. It is the city's intention to bring 10,000 electric cars into the metropolitan area by 1995. Two companies (out of 200 submitting proposals) have been selected to produce the cars. One will produce four passenger car models with driving ranges of 150 miles (240 km). The second firm will produce a mid-size van with a driving range of 120 miles (about 190 km).

Other states are studying regulations to encourage sale of low-emission or zero-emission vehicles. Several auto manufacturers are close to production of electrics. The greatest challenge facing engineers is a cheaper more efficient battery with faster recharge time. At the same time the batteries must be lighter and more durable than those currently in use.

PHOTOVOLTAIC CELL DEVELOPMENT

The photovoltaic cell, usually known as the "PV," should become increasingly important as an alternate source of electric power. Cheaper methods of production are reducing their cost. At the same time, efficiency, long a problem, is being improved.

As demand for PV arrays grows, production costs will continue to drop. In turn, the cost of the electricity they generate will drop. Energy firms, encouraged by lower costs and greater efficiency, will develop photovoltaic "farms" on nonproductive land. The electricity produced could be used to supply residential electric power or as a cheap source of energy to produce other energy supplies. In 1989, following encouraging developments in cheaper PVs, a New Jersey firm announced plans to develop the world's largest PV station. It is to be located on desert land in the Southwest United States. It would be capable of producing 50 peak MW of electricity. This is enough to supply the electrical power needs of 25,000 homes.

FROM WATER TO HYDROGEN

The development of the photovoltaic cell is directly related to development of another alternate energy—hydrogen. New technology for cheaper mass production of PVs means that sunnier regions of the world could use cheap electricity produced by the sun to extract hydrogen from water.

Such a development is possible by the year 2000. At a cost of about two cents a kilowatt for the electricity, hydrogen fuel would solve pressing fossil-fuel-related problems, including urban air pollution, acid rain, and the previously mentioned global warming.

Hydrogen is cleaner burning than any fossil fuel or synfuels (such as methanol). It does not emit carbon monoxide, particulates, or sulfur dioxide. Its only pollutant is nitrogen oxide which can be controlled and kept to very low levels. Though more expensive than methanol derived from fossil fuels, its environmental costs are much less. If these "societal" costs are included in the cost of methanol, experts say, then PV-generated hydrogen should be clearly preferred.

HOUSING AND ENERGY CONSERVATION

Energy efficient construction of the future is likely to stress super insulation. Earth-protected designs also save energy but are not cost effective because of higher construction costs.

This is the conclusion of a study by the Oak Ridge National Laboratory. Underground construction costs, ranging from 30 to 49 percent more than above-ground construction are almost impossible to recover. Their study indicates that the best choice is an energy-efficient, superinsulated house. See Fig. 16-3.

POWER BEAMING

Power beaming is the sending of energy through space on electromagnetic waves. It is similar to sending low-energy radio waves through the atmosphere to distant receivers. Energy could be sent from earth to satellites or from space to earth receivers. Satellites receiving power beams from earth could redirect them anywhere in space or on earth. Air and space vehicles could use electric motors rather than atmosphere-polluting internal combustion engines.

Does this seem like a Buck Rogers story for the far-away future? Perhaps so, but there are already plans to employ power beams by the year 2000. Alaska has submitted a proposal to the Department of Energy of the United States. Their plan eliminates the danger of oil spills and does away with the pollution of burning fuels in urban areas. Hawaii sees power beaming as a way to avoid costly underwater power lines linking their volcanic islands.

Eventually, power generators on the moon could beam power back to earth. Too fantastic? Scientists don't think so.

Power satellites are basically satellites placed in geostationary (moving at same speed as earth rotation) orbit. Their purpose is to collect solar energy and convert it into electricity. Power beaming would allow the satellites to transmit the electrical energy to ground-

Fig. 16-3. Superinsulated structures are being designed and built to conserve energy. Some firms are specializing in factory-built construction panels that are superinsulated. Top. The roof panels are being laid for a commercial building. Bottom. A panelized home under construction. Panels which are a sandwich of rigid insulation and 1/2 in. wood sheets, are insulated to R 28. (Enercept, Inc.)

based solar collectors. Photovoltaic cells located in space can intercept at least four times more solar energy than those placed in the sunniest spot on earth. There are no clouds, stormy weather, or nights to block out the sun.

The power satellites would work much like present-day communication satellites. From their stationary positions above the earth they would always be in contact with receivers on earth. Power would be transmitted as microwaves. The receiver arrays on earth would convert the microwaves back to electricity with high efficiency. It is believed that we could have power satellites at work around the year 2010.

A power satellite, Fig. 16-4, would generate electric power in one of two ways:
- Using the traditional turbine. Concave mirrors of reflective plastic film would concentrate solar heat on helium gas. The expanding helium would drive a turbine harnessed to a generator. Four reflector units 11 miles long could deliver as much electric power as ten land-based power stations.
- Using photovoltaic (solar) cells. Radiation would be

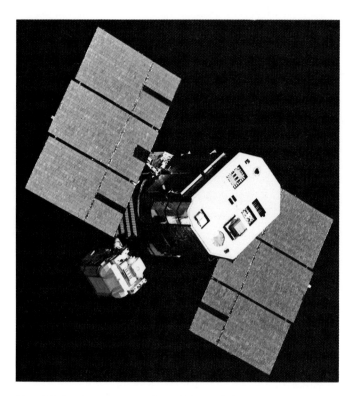

Fig. 16-4. A power satellite might look like this artist's concept. Photovoltaic arrays would capture solar energy that could be beamed to earth or to space vehicles.   (NASA)

converted directly to electricity. An array of 10 billion cells covering 20 sq. miles, could deliver as much electric power as five land-based stations.

MOTOR FUELS OF THE FUTURE

Nearly half of all North Americans live in areas that do not meet clean air standards. Contaminated air causes many ailments. The air they breathe can cause a chronic cough, shortness of breath, chest pains, or even cancer. The major air contaminant is vehicular exhaust emissions. Of the culprits, the private automobile is the worst.

Fuels of the future will need to be less polluting. Something is being done to make gasoline less polluting and research is underway to find cleaner burning alternatives.

Today's gasolines combine as many as 100 different hydrocarbons. Then there are additives to maintain octane levels. These changes have added to airborne pollution. Refineries can add oxygenated compounds to allow more complete combustion. The most common are methanol (wood alcohol), ethanol (grain alcohol from corn or sugar cane), and ethers made from either methanol or ethanol. The oxygenates lower carbon monoxide emissions but can increase smog formation. Major oil companies, the American Petroleum Institute, and three domestic automakers are involved in a research project to produce cleaner gasoline. The

results of the research are still undetermined but will immediately reduce pollution when the fuel is available.

HYDROGEN AS A FUTURE FUEL SOURCE

Hydrogen promises to be an unlimited energy source. Furthermore, it has none of the pollution problems of fossil fuels. Its major by-product is water vapor.

Using present-day electrolysis technology, hydrogen can be extracted from water in large central plants. The main obstacle to its use is cost. Most conversion processes make hydrogen more expensive than gasoline. Still, if we were to consider the hidden costs of gasoline — damage to environment, military actions to assure a constant supply — hydrogen might be the cheaper of the two.

Solar cells, some believe, may soon offer a cheaper way to extract hydrogen from water. A leading manufacturer of cells predicts that PVs (solar cells) will one day be the primary energy source for electrolysis of water. West Germany is working on such a research project. German scientists would like to put hydrogen-producing photovoltaic "farms" in African desert areas. The hydrogen would be piped to Germany. Saudi Arabia has built a solar-hydrogen plant near Riyadh as a testing and training facility.

Quebec and countries in Europe have begun a study to explore use of hydroelectric power to produce hydrogen. It would then be shipped in liquid form to Europe. There is would be used as a fuel source for fuel cells and vehicles.

Other fuel and energy sources that are likely to be developed include: methanol and ethanol which are already being used as additives to gasoline, compressed natural gas, electricity, and solar extracted hydrogen fuel. The chart in Fig. 16-5 summarizes the "pluses and negatives" presented by these alternative sources.

FUEL CELLS

A promising power source that has been around for a long time is the fuel cell, Fig. 16-6. Discussed briefly in an earlier chapter, it is a device somewhat like a car battery. It controls a chemical reaction that produces electricity from a fuel and oxygen. The fuel cell, however, differs from the battery in two respects:
- It needs a continuous supply of fuel to maintain production of electricity.
- The cell does not need to be recharged. Its chemical composition is always the same.

HYDROX FUEL CELL

The principle of the fuel cell has been known for a long time. The idea dates back to 1839. A British scien-

FUEL	ADVANTAGES	DRAWBACKS
ETHANOL	105 octane rating compared to 87 for regular unleaded and 93 for premium. Renewable resource. Exhaust emits less carbon dioxide than gasoline but total CO_2 impact must consider distillation and crop cultivation/harvest. Added to gasoline, lowers CO emissions.	Delivers less energy per gallon. Requires more frequent fill-ups. Expensive. Added to gasoline, may promote formation of smog.
METHANOL	High octane rating (105). Hydrocarbon emissions lower by 35% for M85; up to 90% lower for M100. Emits less CO_2 from exhaust (10% less if produced from natural gas; 100% more, if produced from coal) Airborne toxics 30-40% lower than gasoline's.	Less energy per gallon of fuel. Toxic: Can blind or kill if swallowed. Corrosive. Difficulty in starting engines on cold (−20 °F) days. Emits formaldehyde, a suspended carcinogen at a rate 4 to 8 times higher than gasoline.
ELECTRICITY	Quiet. Almost no emission from vehicle.	Technology at least 2 years away. Batteries extremely expensive. Limited range (no more than 250 miles/400 km). Recharge time 6 to 8 hours. Indirect pollution source, depending on source of electric power.
SOLAR-HYDROGEN FUEL	Renewable source. Practically emission-free. Adds nothing to global warming.	Technology feasible but not readily available; probably 8-10 years off.
COMPRESSED NATURAL GAS	Supplies abundant. Inexpensive. Low hydrocarbon emissions. Low carbon monoxide emissions. Lower carbon dioxide emissions. Good distribution system in place.	Frequent refueling (every 100 miles/160 km). Bulky fuel tank. Lengthy refueling time.

Fig. 16-5. These are likely to be the fuels for tomorrow's transportation vehicles.

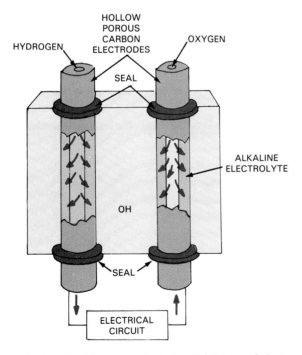

Fig. 16-6. A simplified drawing of a fuel cell. Like a car's battery, it produces electrical energy as a result of chemical reaction. However, unlike a battery, it never needs to be recharged. It will continue to produce electricity as long as it receives fuel and oxygen.

tist, Sir William Grove, while decomposing water through electrolysis, found that reuniting the hydrogen and oxygen atoms produced an electrical charge. Little was done with it from its discovery early in the 19th century until the exploration of space. Then engineers needed a different method of supplying electricity to spacecraft. The hydrox fuel cell was the answer. While fine for outer space, the fuel cell was just too expensive to attract land-based users. Electric power plants could produce electricity 2 1/2 times cheaper. Using a fuel cell did not make economic sense. That picture is changing, however, as tighter regulations begin to drive up electricity costs. Today, the cost of fuel-cell electricity is only 25 percent higher.

FUEL SOURCES

Fuel cells developed for spacecraft use hydrogen as a fuel and oxygen as an oxidizer. (An oxidizer is a substance which supports combustion or a chemical reaction.) Cells can operate on many different types of fuel. In fact, any hydrocarbon fuel can be used, such as natural gas or biogas from garbage or sewage. A special processor breaks down the fuel to produce hydrogen.

HOW A FUEL CELL WORKS

A single hydrox fuel cell is made up of a container which holds the electrolyte, and a solution of phosphoric acid. On either side of the electrolyte are two carbon plates called electrodes. (An electrode is a terminal or conductor of electricity. The electrodes are part of the path through which electricity moves.)

Hydrogen and oxygen are fed into the cell. The fuel flows down the positive electrode called the anode. The fuel loses electrons to the anode. The anode becomes negatively charged.

Meanwhile, the cathode (negative terminal) is supplied with oxygen, the oxygen collects electrons from the cathode leaving it positively charged. This sets up the fuel cell for the next operation.

When a load, such as a light, is connected across the anode and the cathode, free electrons on the anode travel through the conducting wire, through the light bulb to the cathode, Fig. 16-7. While this is happening, positive ions (particles) of hydrogen leave the cathode, travel through the electrolyte, and combine with positive ions of oxygen. This reaction forms water.

There are other waste products besides water. Included are nitrogen (from the air) and carbon dioxide from the fuel. Another by-product from the process is heat. Usually there is enough heat to produce steam during the process. This heat, though often passed off to the atmosphere, could be used for other purposes.

Modern fuel cell designs use two porous electrodes. The electrolyte can be either phosphoric acid or various solid compounds or membranes. The cell draws its oxygen from the air. Catalysts coated on the electrodes help the cell form water. This causes the cell to release electricity and heat.

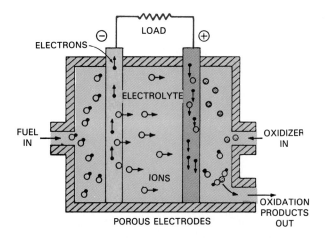

Fig. 16-7. Schematic of a fuel cell. The cell will provide a steady current across its terminals to drive electrical motors or provide lighting.

The shape of a cell can vary. Those using phosphoric acid are 2 1/2 ft. square and 1/4 in. thick. Hundreds of cells packed in a cabinet 24 ft. long, 10 ft. wide, and 10 ft. high could light up 150 homes. The fuel cell continues to work as long as it is supplied with fuel. The fuel cell requires little maintenance. Every five years the cell stack (where the reaction takes place) must be replaced.

Before the year 2000 we may see fuel cells supplying electric power to hospitals and community centers. An extra bonus: heat generated by the cells can be used to preheat water or provide space heat.

THE MHD GENERATOR

The MHD (for magnetoheterodyne) generator produced electricity from hot gases. The hot gases (called plasma) at a temperature of 4000 to 5000°F (2200 to 2800°C) are passed across a magnetic field at right angles to the magnetic lines of force. Electrodes attract the free electrons in the gas. The collected electrons will flow through the conductors connected to the electrodes. Fig. 16-8 shows a simple diagram.

THERMOELECTRIC CONVERTERS

Thermoelectric converters are other devices that turn heat directly into an electric current. The oldest device of this type is the thermocouple. In 1821 a German physicist discovered that if two dissimilar (unlike) metals were joined and heated, a voltage or electromotive force could be measured across the free ends.

For over 100 years, no practical use was made of this knowledge. Then the semiconductor was invented. It is a better thermoelectric material. Semiconductors have impurities built into them without which they would be insulators.

One type of semiconductor has impurity atoms with some free electrons. The extra electrons are free to move about. This is known as an n-type (negative type) semiconductor.

A second type of semiconductor has impurity atoms with too few electrons. This leaves positive holes in the semiconductor. These holes can move around just like positive charges.

Suppose that you connected a p-type and an n-type conductor as in Fig. 16-9. Then, if you applied heat to the plate marked "hot junction" this is what would happen:

• Heat at the hot junction drives the loose electrons and holes toward the cold junction. You can think of them as gases being driven through the semiconductor material by the temperature difference.

• The holes will naturally travel up one semiconduc-

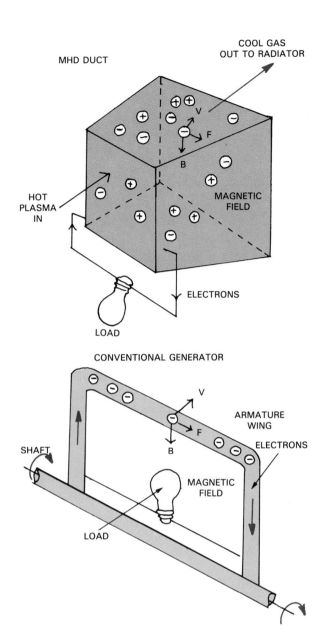

Fig. 16-8. Diagram of a magnetohydrodynamic (MHD) generator. It uses superheated gases to produce electric power.

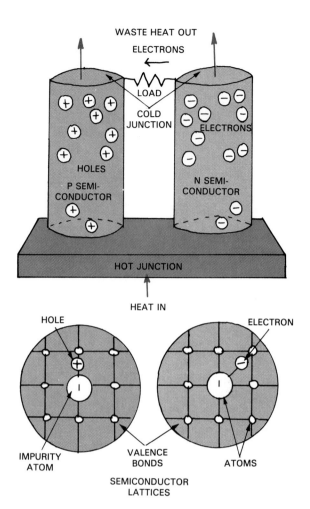

Fig. 16-9. Diagram of a thermoelectric device. Top. Thermoelectric couple is made from pair of semiconductors linked by a hot junction and a load. Bottom. Impurity atoms (in the centers) are different in each leg. One has too many, the other too few electrons. Heat drives both holes and electrons toward the terminals.

tor and the electrons up the other.
- The holes create a positive terminal and the electrons a negative terminal.
- The flow across the load connected to the terminals is driven. This sets up the flow of electrons which causes the voltage.
- If a load such as a light bulb is connected across the terminals, the voltage will turn on the light.

THERMOELECTRIC POWER GENERATORS

Thermoelectric generators can be located in remote areas where it is not possible to attend them. They will operate quite satisfactorily without maintenance. Small units operating with nuclear power have been launched into space aboard satellites. They are also used in polar regions and beneath the ocean.

Small propane units are being manufactured for use on campsites, ocean buoys, and in remote places where only a small amount of electricity is needed. See Fig. 16-10.

The efficiency of the thermoelectric generator is about 5 percent. This is low compared to the 35 to 40 percent efficiency of the steam-powered power plants. Efficiencies would have to become much higher before thermoelectric power replaces steam powered units.

OTHER ENERGY CONVERSION DEVICES

Other conversion devices are under study and development. They include:
- Thermionic converter. This device consists of two metallic plates. When one of the plates is heated it causes electrons to "boil off" and create an electric current.

Fig. 16-10. Small thermoelectric generator. It is powered by a small propane heater.
(Global Thermoelectric Power Systems, Ltd.)

- Ferroelectric converter. A nonconducting material is made to absorb heat by producing change in molecules of the substance.
- Thermomagnetic converter. This device uses heat to make a change in the magnetic field of a magnet core.

These devices are still in the development stage. However, they may provide important energy conversion capabilities for the future.

HEAT ENGINE TECHNOLOGY

Engines with many parts made of silicon nitride, a ceramic material, are able to run without a coolant. Tests on a compression-ignition engine have shown that the ceramic parts can withstand extreme heat as well as the stresses placed on internal combustion engines. Further testing has proven that a diesel engine can run at temperatures up to 2200°F (1205°C). The engine has no radiator, fan, or water pump, thus, reducing its weight. Also, there are fewer parts to fail. Ceramic parts include: cylinder sleeves, pistons, piston pins, heads, rocker arms, push rod ends, tappets, and turbochargers. Its advantages: cheaper to manufacture, less polluting due to higher operating temperatures, and a better power to weight ratio.

COAL GASIFICATION

Gasification of coal is the process of turning it into a gas fuel. While not a new or emerging process it may have implications for the future as other fossil fuels become scarce.

Coal is almost pure carbon. It has only small amounts of hydrogen, oxygen, sulfur, and nitrogen. Gasification (or any of the processes which convert coal to other fuels) adds hydrogen atoms to the carbon atoms.

In coal there are 16 carbon atoms for every hydrogen atom. Many more hydrogen atoms are found in gas or liquid fossil fuels, Fig. 16-11. Producing any of these fuels from coal requires large amounts of hydrogen.

ELEMENTS IN THE PROCESS

There are four primary ingredients in the gasification process:
- Carbon. This is supplied by the coal.
- Hydrogen. Some is supplied by the coal. The rest usually comes from breaking down natural gas (methane is mostly hydrogen) or by introducing steam which then reacts with the coal. (Water, from which the steam is made has two hydrogen atoms for every oxygen atom.)
- Oxygen. This is supplied either in its pure form or from air.
- Heat is supplied by burning coal or a gas which is a by-product of the process. Fig. 16-12 illustrates the basic chemical reaction. The gas produced by this reaction, carbon monoxide and hydrogen, will burn but has only a third of the heat value of methane. It can be burned to carry on the process but it is not worth moving long distances by pipeline.

Instead, it is put through another process or step called **methanation**. The hydrogen and carbon monoxide are combined with a catalyst (something that triggers a chemical reaction) to make methane. Several steps are necessary to produce methane. First the carbon monoxide reacts with steam to produce carbon dioxide (CO_2) and hydrogen. The carbon dioxide is then separated along with another gas, hydrogen sulfide (H_2S). The latter is formed when the hydrogen combines with the sulfur impurities in the gas. At present,

RATIOS OF CARBON TO HYDROGEN IN SEVERAL FUELS		
Fuel	Number of Carbon Atoms	Number of Hydrogen Atoms
Coal	16	1
Heavy Fuel Oil	6	1
Gasoline	1	2 or 3

Fig. 16-11. Comparison of coal with liquid fuels. Coal has a great many carbon atoms in its molecular structure.

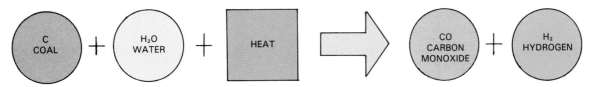

Fig. 16-12. Basic chemical reaction for coal gasification.

the process is being refined and developed for possible commercial use, Fig. 16-13.

LIQUID FUELS FROM COAL

As with gasification, liquefication of coal is aimed at reducing the high carbon-to-hydrogen ratio (16 to 1) to at least 6 to 1. This is the carbon-hydrogen ratio of fuel oil. There are three different processes which will accomplish this:

• Hydrogenation. Pure hydrogen is added. It produces a heavy oil that can be used in power plants.
• Pyrolysis. Coal is heated in the absence of oxygen.
• Catalytic conversions.

Pyrolysis produces three fuels: a pipeline gas, a synthetic crude oil, and a carbon residue called **char**. The char can also be burned as a fuel if it does not contain too much sulfur.

Catalytic conversion produces the same reaction as gasification described earlier. The carbon monoxide and the hydrogen are then combined with a catalyst to make a liquid fuel.

Fig. 16-13. A coal gasification plant that was established to provide research data on the process. It converts coal to synthetic gas for use in steam boilers and gas turbines. (Texaco Inc.)

PRODUCING SOLID FUELS

Making solid fuel from coal is called **solvent refining**. The solid fuel produced by this method is called solvent refined fuel or SRC.

First, the coal is crushed and mixed with a solvent such as light oil. This mixture is heated at high temperature and pressure. As the coal dissolves, the ash and the impurities can be separated from it. The fuel resulting is like a solid tar and can be crushed or heated and melted. Its heat value is higher than coal and it is a much cleaner fuel.

COAL CONVERSION AND THE ENVIRONMENT

Gasification and liquefication have some serious drawbacks. Among the most serious is the heavy use of water which not only must provide cooling but becomes a raw material in the production of hydrogen. About 1 1/2 to 3 lb. (roughly 2 qt.) are needed for every pound of coal processed. This is double the normal electric power plant usage.

Then too, most of the coal which will be gasified is located in areas where water is scarce. (Heavy coal deposits are found in the southwest, and the north central regions of the Dakotas, Montana, and Wyoming.)

Some pollution problems are also present. Air pollution controls will be needed to collect particulate pollution and SO_x. There will be solid wastes also. Some of these, of course, could be returned to the mined out areas.

An added health concern comes with production of synthetic fuels: presence of carcinogens. These are chemical elements that produce high risk of cancer for those who work with or come in contact with them. The high temperatures used in synthesizing coal produces molecules known as polycyclic aromatic hydrocarbons (PAH). There is plenty of evidence that these materials can cause cancer.

Experts say that it is difficult to predict how extensively synthetic fuel conversion will be practiced. Developing the huge commercial plants needed will be costly and the process itself is expensive. Gas companies are reluctant to undertake the development of such plants since they would place a serious strain on their assets.

THE FUTURE OF FUSION ENERGY

Many engineering problems need to be solved before nuclear fusion can be a success. So far, a self-sustaining fusion reaction has not been possible. For fusion, three conditions must occur at the same time. There must be exact:

• Temperature.
• Confinement time.
• Plasma density.

Only then will fusion reaction continue on its own. Despite almost 30 years of work on fusion, scientists and engineers have not been able to produce all three of these conditions in one machine. Should these researchers succeed tomorrow in demonstrating a method of sustaining a fusion reaction, other problems would have to be solved. One of them would be the construction of a chamber that could withstand the fusion reaction for a long period of time without disintegrating. Another problem would be the tremendous cost of designing and building such a reactor. Should engineers and scientists succeed in overcoming the obstacles, humankind will have an energy source for millions or even billions of years.

FUSION "TORCH"

It is possible that once developed, fusion power could be put to other uses than for propulsion and generation of electric power. One proposal is the development of the "fusion torch." The plasma coming from the exhaust of the fusion reactor might be used to break down solids or liquids. The extremely high temperature would vaporize the material and then ionize it. (This means to break up the particles into atoms.) Fig. 16-14 shows a simplified schematic for a fusion torch.

The torch could be used for disposing of all kinds of waste materials, processing of ores, and production of liquid fuels. It is believed that the fusion torch could also be used to generate ultraviolet or X rays. This would be done by putting small amounts of heavy atoms into the plasma exhaust.

Energy created in this way could then be used for various purposes:

• Desalinization (taking salt out) of seawater.
• Production of hydrogen.
• Processing of chemicals.

Because plasmas, not chemicals, would be used to separate elements there would be less pollution of the environment.

The idea of a fusion torch, being untested, may not be practical in every case. It does appear that exciting industrial uses may be found for fusion energy if the scientific and engineering problems can be solved.

SUMMARY

One of the challenges facing all of us in the years ahead is to use our energy resources wisely and frugally. At the same time, our cooperation is needed in efforts

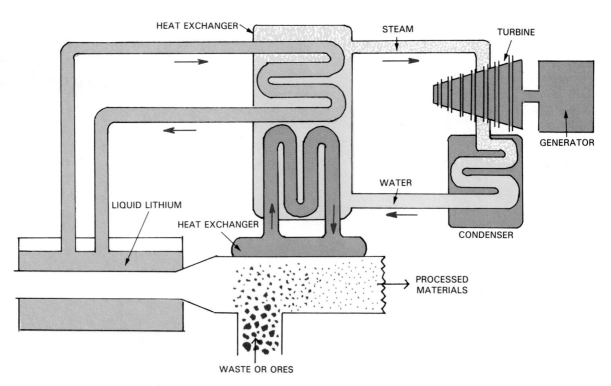

Fig. 16-14. This simplified diagram shows how a fusion torch might work. Some of the energy from the superheated plasma would be used to vaporize waste material or ores and reduce them to atoms. Then the atoms could be combined to produce useful raw materials.

to reduce pollution caused by production and use of energy.

One of the major impacts of pollution, particularly from burning of fossil fuels, is the "greenhouse" effect. We are uncertain of the long-term effects of global warming.

Industrialized nations are responsible for much of the carbon dioxide and carbon monoxide released into the atmosphere. However, cutting of rain forest to produce wood fuel, or simply burning the trees to make room for grazing and farmlands also contributes heavily to greenhouse gases.

These concerns are bringing about change in energy usage. With or without government assistance, research is underway to find alternate energy sources that are economical and less polluting. There is also research to find ways to make fossil fuels less polluting.

New laws will provide an incentive to manufacturers, designers, and we, the users of energy, to work toward pollution reduction. Especially critical will be the effort to conserve a resource already in short supply. Running parallel will be a greater effort to preserve our environment.

DO YOU KNOW THESE TERMS?

char
methanation
power beaming
power satellites
solvent refining
thermoelectric converters

SUGGESTED ACTIVITIES

1. Build and operate a working model of a fuel cell.
2. Research power beaming and prepare a report with sketches on how the system operates.
3. Select one of the "fuels of the future" listed in Fig. 16-4 and research your local libraries for additional information. Prepare a report on your findings.

TEST YOUR KNOWLEDGE

1. Why is the burning of fossil fuels a major concern today?
2. What factor is driving research and development for less polluting energy technology?
3. What two factors are giving photovoltaic cells greater potential as a technology for generating electric power?
4. _____ is the sending of energy through space with the use of a satellite.
5. What two methods could be used to generate electric power in space?
6. Discuss the advantages of producing hydrogen from water using photovoltaics.
7. Tell in your own words what might be significant advantages of the fuel cell as an energy converter.

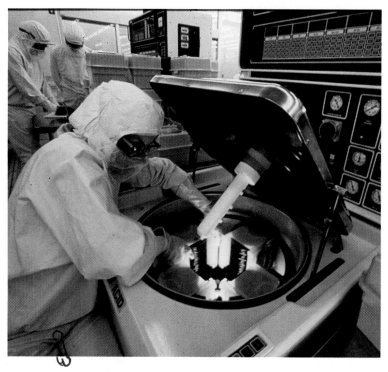

Above. Many manufacturing industries offer energy-related careers. The manufacture of integrated circuits, which are the heart of every computer, is one example. The woman pictured operates equipment used in the manufacture of semiconductors. The semiconductors will end up on computer systems that control operations as widely diverse as electric power stations, automobile electronic controls, and energy systems in a building.   (Harris Corp.) Below. A fuel technician services automobiles adapted to run on LPG fuel.   (The Coastal Corp.)

17

CAREERS IN THE ENERGY FIELD

The information given in this chapter will enable you to:
- *Discuss energy job opportunities in different industries.*
- *List some types of jobs offered by energy industries today.*
- *Describe the kinds of work done by energy industry employees.*
- *Discuss the effects of conservation and pollution control efforts on workers in energy jobs.*

Energy companies in North America are one of the largest groups of employers. Every year the United States alone uses a third of the energy consumed in the world today.

As you learned in Chapter 4, energy is consumed by various sectors of society. Some industries, such as steelmaking, drug manufacturing, and plastics fabrication, use petroleum, natural gas, or coal as feedstocks. (That is, these products are the raw materials from which other products are manufactured.)

Other fuel users include farms and factories which consume energy to operate machines and produce process heat. Each of us represents two other groups which are the heaviest users of gas and liquid fuel. Our residences, offices, schools, and automobiles consume the lion's share of these energy sources to help us live, ride, study, and work in relative comfort.

THE JOB MARKET

It is not easy to predict what the job market will be for energy workers in the years ahead. A study by Congress once predicted that 2.1 million jobs could be created in energy-related fields. The study added that

we would have to work hard at conservation and development of solar energy. In the years following the study, cheap oil slowed development of the alternate energy industry.

We do not expect that energy jobs will continue to be the same as they have been. The energy industry is entering a period that will see many changes.

NEED FOR CONSERVATION

As mentioned before, there is a need to conserve scarce fossil fuels. We need also to develop renewable resources as well as the know-how to harness them for our energy needs.

There will continue to be jobs in traditional energy fields, and coal mining will, without question, continue to increase. See Fig. 17-1. The need to conserve precious fossil fuel reserves will have an effect on every person working in the industry. Anyone entering the field today will need certain additional understandings and skills:
- Why conservation is necessary.
- How we can cut down our use of fossil fuels.
- Technical know-how for conservation.
- Why continued pollution of our environment is not acceptable.
- Skills for developing and finding ways to use alternate energy sources.

FOSSIL FUEL MARKET

As we said, U.S. consumption of energy is much greater than that of any other country, Fig. 17-2. Our per capita (each person) usage is four times that of all western Europe and 20 times that of Asia. Our autos alone use one out of every nine gallons of oil consumed in the world.

Fig. 17-1. Coal mining will offer long-term job opportunities. This coal miner is operating a longwall coal mining machine. (American Coal Foundation)

AREA	Percentages of Total									
	10	20	30	40	50	60	70	80	90	100
UNITED STATES	Population									
			Energy usage							
			Gross national product							
WESTERN EUROPE		Population								
				Energy usage						
				Gross national product						
JAPAN	Population									
	Energy usage									
	Gross national product									
REST OF WORLD									Population	
					Energy usage					
				Gross national product						

Fig. 17-2. Comparison of U.S. population and energy consumption with the rest of the world. With 16 percent of the world population, Americans use more than 30 percent of the world's energy.

Most of the job opportunities in the energy field will continue to be in the fossil fuel industries. This will be true for at least the next decade. However, new jobs are going to be opening up in alternate energy fields.

FOSSIL FUEL CAREERS

Opportunities will continue strong for careers in fossil fuel industries. It is expected that scientists will be in demand in the oil, coal, gas, and conservation industry. They will be concerned not only with technology for improving fossil fuel products but with new technology for producing synthetic fuels and the alternatives to fossil fuels. They will also be in demand to solve the problems of waste disposal and handling of environmental hazards.

JOBS IN ENERGY EXPLORATION

A number of skills are required for exploration of energy resources under the ground. One group of specialists are called **geophysicists**. Their jobs are to explore solid earth for the presence of oil, coal, and other mineral deposits. See Fig. 17-3.

Exploration geophysicists use seismic prospecting techniques to locate the oil or minerals. They use a seismograph to send sound waves into the earth and record the echoes that return from rock layers below. See Fig. 17-4. This information can be studied to see if rock conditions are favorable to the presence of oil. **Seismologists** work with geophysicists. They study and interpret the data created by the recorded vibrations.

Geology is another field related to the oil industry. **Geologists** study the makeup of the earth, identify rocks and minerals, make surveys, take measurements, and record data. Petroleum geologists work with teams exploring for oil.

Mining engineers find, extract, and prepare minerals for use. They design and construct pit and underground

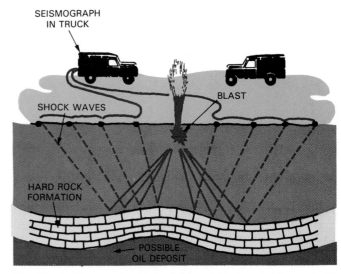

Fig. 17-4. Teams of geophysicists and seismologists set off small explosions and study the shock waves for presence of oil, coal, or gas deposits. (Shell Oil Co.)

mines. Some develop special mining equipment. Many specialize in coal mining and some may work only on solving pollution problems resulting from coal mining activities.

Petroleum engineers work mostly at exploring and drilling for gas and oil. They design and develop pumps and other devices to force the oil to the surface. Some supervise drilling operations.

Seismic observers work with seismologists. An electrical prospecting observer uses electrical apparatus for measuring earth resistance to electrical charges. Gravity prospecting observers record readings of meters and other gravity-measuring equipment for purpose of finding oil. They are assisted by observer helpers.

Mine surveyors do surface and underground surveys at coal mines to control the direction and extent of mining. They also estimate the volume of coal in parts of the mine using survey information. Sometimes they draw up maps of mine workings.

Oil well directional surveyors measure sonar, electrical, or radioactive characteristics of earth formations in oil and gas wells. They bore test holes to determine how productive the well will be.

JOBS IN ENERGY EXTRACTION, REFINING, AND DISTRIBUTION

Once oil, gas, or coal deposits are discovered, extraction and refining processes are required. Also, distribution of the energy products must be done.

The **derrick operator** works on oil drilling crews rigging the derrick equipment and operating mud pumps. See Fig. 17-5.

Fig. 17-3. Exploration for fossil fuels often takes geophysicists like this woman into rugged terrain. (Marathon Oil Co.)

Fig. 17-5. An offshore oil rig offers steady employment for many oil workers. (Dupont)

Fig. 17-6. Forklift operators work in warehouses handling energy products. (Caterpillar)

A

B

Fig. 17-7. A—Drivers are needed to deliver fuel to gas stations in larger tanker trucks. B—Others operate cranes that hoist trailers aboard trains in "piggyback" operations. (Union Pacific)

Acidizers operate equipment for treating oil wells with acid to make them produce more oil. They supervise the acidizer helpers in loading and mixing chemicals at the well site.

Industrial truck operators drive gasoline or electrically powered trucks or tractors equipped with a forklift, elevating platform, or hitch to move petroleum products in a warehouse, Fig. 17-6.

A **blaster** plants and sets off charges of explosives in mining operations.

Gasoline finishers control equipment for blending natural gasoline with chemicals to improve its fuel qualities.

Heavy truck drivers drive coal trucks in strip mining operations, haul gasoline, oil, and other petroleum products, Fig. 17-7. They usually lubricate their own trucks and perform minor repairs.

Pipeline workers are a classification of careers that are used in several industries. Many are employed in the petroleum industry. **Pipeline construction inspectors** check materials and work during pipeline construction to see that they are up to specifications. They supervise or observe preparation of right-of-way, check trenches for depth and pitch, examine welds, coatings, and wrappings, and advise supervisor of corrections needed. See Fig. 17-8. A **pipeliner** maintains and repairs pipelines, pumping stations, and tank farms. This person will drive equipment such as backhoes, bulldozers, and side booms to dig ditches. He or she will also lay pipe and backfill ditches. Pipeline workers bolt together pipe sections, working as a team. They pour sealing compounds over joints and wrap asbestos coverings to prevent corrosion and leaks.

Mine-car repairers are skilled at metalworking. They repair and replace damaged parts of underground mine cars. The work often includes welding, straightening parts, and refitting them.

Mine foremen supervise and direct the work of other mine workers. They also supervise the opening of new cuts, pits, or underground rooms and passageways. Another responsibility is the construction and installation of mining equipment.

Mine machinery mechanics repair and maintain mine machinery such as stripping and loading shovels, drilling and cutting machines, and continuous mining machines.

Fire bosses inspect underground mines for fire hazards, presence of toxic gases, and inadequate ventilation. They tour hallways, shafts, and work areas looking for hazards. They also keep logs of accidents.

Miner is a term given to all who work in mines. Jobs are classified according to the work performed. For example, cutter operators control self-propelled machines that saw channels along the bottom or side of the working face of the coal mine.

Stripping shovel operators control the shovels that remove overburden at strip mines prior to actual coal mining operations. Some also repair and replace shovel parts.

Timber framers cut, fit, and install supporting timbers and other framework in underground mines, using carpentry tools. Some may build forms and pour concrete supports.

Filling station attendants are a type of salesperson, Fig. 17-9. They service vehicles of all types with fuel and lubricants, pump the fuel into the fuel tanks, check and change oil, and lubricate the vehicles. Many also perform light maintenance and repair.

Salespersons are responsible for selling energy products such as coal, gasoline, and various petroleum products. Some sell the products directly to the consumer. Some travel to call on customers. They describe their products to customers, take orders, and keep records of their sales calls.

Natural gas utility companies provide many employment opportunities. These companies are largely concerned with supplying fuel for residential heating and industrial use. In 1990 U.S. customers consumed more

Fig. 17-8. Pipeline inspectors' jobs may take them anywhere in the world. This 24 in. pipeline connects Beruk oil field in central Sumatra with three new oil fields. (Texaco Inc.)

Fig. 17-9. Filling station attendants perform a variety of tasks such as checking oil, pumping gasoline, cleaning windshields, and operating a cash register. They work in small retail operations like the station shown. (Chevron Corp. and Dupont) than 18 trillion cubic ft. of natural gas.

Developing new markets for natural gas is a concern of all gas utilities, Fig. 17-10. It is important for marketing people to understand their customer's needs, especially those engaged in manufacturing, so the utility can serve them better.

Gas utilities also employ engineers to design and keep their equipment and systems in good operating condition. Engineering management people, Fig. 17-11, have as one of their tasks, the maintaining of thousands of miles of piping in a gas utility's distribution systems.

There are many other satisfying careers working for gas utilities. Two are shown in Fig. 17-12.

Some energy companies offer scholarships and summer internships to high achieving college students. In

Fig. 17-10. Public utilities, such as gas companies must work at anticipating the needs of customers. This man is an area manager for a natural gas utility. A degree in chemistry and numerous in-house workshops help him to understand the needs of the customer.   (Northern Illinois Gas)

Fig. 17-12. Utility companies employ people for various types of jobs. Top. Appliance repairs are made by gas company employees who come to the customer's home. Bottom. A meter reader travels from house to house to record gas usage by customers. This information is used by the company to bill for fuel used.

Fig. 17-11. As an engineer, her job is to help maintain 25,000 miles of gas pipeline so that it meets or exceeds all operating and safety standards.   (Northern Illinois Gas)

addition to financial assistance and a competitive salary they receive valuable experience in their chosen field as part of their education. See Fig. 17-13.

These are but a small sample of the occupations within the fossil fuel industry. You can find others through consulting the Dictionary of Occupational Titles, or the Occupational Outlook Handbook. Both are published by the U.S. Labor Department.

POWER GENERATION INDUSTRIES

Generation of electric power is a large industry involving many utility companies. Every state and prov-

Fig. 17-13. Some utilities have intern programs for management-level students. These interns often are hired on full-time after getting their degrees. (Northern Illinois Gas)

ince has a number of these companies, so career opportunities are available everywhere in the U.S. and Canada.

A number of career fields are open in the production of electrical power for fossil-fueled, hydropowered and nuclear-powered facilities. Some will require baccalaureate engineering degrees while others will require a high school diploma and specialized training.

WORKING AS AN ENGINEER

Those likely to succeed in engineering careers have an abiding interest in things scientific and technical and may enjoy mathematics and mathematical puzzles. Along with a serious interest in the field of engineering, the prospective engineer will need certain aptitudes and capabilities. One of these is a mastery of the various types of mathematics and the ability to apply mathematical principles to the solution of practical problems. She or he should also have an aptitude for stating a practical problem in abstract terms or mathematical equations without becoming confused.

In addition one interested in engineering as a career should do well in and be interested in related fields such as physics and chemistry. It is also helpful to a practicing engineer to be able to visualize what is being described in words and to sketch them out in detail.

Relating well to other people is another useful trait of the engineer. Often, he or she must work as a member of a large team where communication of ideas

goes on continuously as a project moves through the planning and execution stages.

Electrical/electronic engineer. As a group, electrical/electronic engineers are occupied with the application of the laws of electrical energy and the principles of engineering for the generation and transmission of electricity. Specializations related to the energy field are: power generation, power transmission and distribution, and nuclear power generation.

Electrical engineer. This person researches, develops, and designs systems or subsystems. She or he may specialize in power generation and distribution and may be involved in the design of power stations. Other specialists may direct the operation of a power station, or the maintenance and repair of an operating power station and its components.

Distribution field engineer. A distribution field engineer plans and remodels distribution facilities such as transmission lines and substations to overcome unsatisfactory conditions such as overloads. Other tasks include making field surveys or studying maps to locate relays, line intersections, overhead and underground connectors, feeder lines and so on. She or he also recommends installation of new or additional facilities or changes in existing facilities that cannot handle the increased loads.

Power engineer. She or he designs power systems and oversees construction and maintenance of power stations, transmission lines, and substations. Other tasks might include compiling of rates and supervision of maintenance. Some power engineers may specialize in areas such as: distribution lines, protection equipment, or substation design.

Civil engineer. This is an area of engineering concerned with planning, designing and constructing structures.

CAREERS IN NUCLEAR POWER

Between the United States and Canada there are about 129 nuclear power generating stations in operation, Fig. 17-14. Because of public concern over the dangers of radiation and the problems of disposing of nuclear waste, construction on nuclear reactors has slowed. The growth of nuclear power and the potential of nuclear power for careers is uncertain in the United States. However, there should be continuing employment in existing nuclear power stations. Also, engineering firms in the U.S. have been performing engineering design work in nuclear power for foreign countries.

Following are brief job descriptions for certain types of engineering in the nuclear power field.

Nuclear equipment design engineer. Designs, develops and tests nuclear equipment and monitors testing operations, and maintenance of nuclear reactors; prepares technical reports; may direct operations and maintenance of nuclear power stations.

Nuclear fuels reclamation engineer. Plans, designs, and oversees construction of nuclear fuels reprocessing systems; writes project proposals; studies safety procedures, guidelines, and controls; oversees nuclear fuels reprocessing system construction and operation.

Nuclear fuels research engineer. Studies behavior of various fuels to determine safest and most efficient usage of nuclear fuels; designs tests and tests fuels.

Fig. 17-14. Nuclear power generating station. There are more than 110 such stations in various parts of the U.S. and Canada. (Commonwealth Edison)

Nuclear criticality safety engineer. Researches, analyzes, and evaluates proposed or existing methods of transporting, handling and storing of nuclear fuel to ensure against accidents.

Waste management engineer, radioactive materials. Designs, implements, and tests systems for reducing volume and disposing of radioactive wastes; examines and analyzes sludge and effluents from nuclear operations to determine level of radioactivity; compares disposal costs; designs and sketches systems for disposal and oversees their construction; writes manuals on work procedures and safety procedures for disposal of hazardous, radioactive wastes; advises management on site selection for disposal of wastes and closure methods.

Radiation protection engineer. Supervises and coordinates activities of those working on monitoring radiation levels and conditions of equipment used to generate nuclear electric power to ensure safe operation of the plant facilities.

Sales engineer, nuclear equipment. Sells nuclear machines and equipment and provides technical services to clients.

Nuclear plant technical adviser. Monitors plant safety status and advises operators; prepares reports on operation; checks instrumentation and also checks with workers to ensure safe procedures.

Trained personnel are needed for operation of the nuclear power stations now in use. It is difficult to determine what growth might occur in the field since new construction has been halted for several years. New designs, concern over the greenhouse effect of fossil-fueled power stations, and need for more electric power, may produce a new spurt of construction. The following occupations do not require college degrees. However, it will be necessary to have a high school diploma and additional training or studies up to and including an associate degree in some engineering technology.

As with engineer, person aspiring to these positions should have a strong interest in principles of physics, and better-than-average ability to understand and apply math principles.

Trained technical workers are needed for the following jobs:

Nuclear materials processing technician. These technicians help to fabricate (make), handle, reprocess, transport, and store nuclear materials and fuels, Fig. 17-15. Candidates are selected with great care by the firms who employ them. Two years of study in technical schools is a good background for a trainee. They are employed by nuclear facilities that prepare nuclear fuels and by research organizations and chemical industries.

Fig. 17-15. Inspecting a fuel rod assembly. Those who do this kind of work are carefully selected. (Siemens/Advanced Nuclear Fuels.)

Nuclear instrumentation and control technician. These persons are specialists in maintaining and repairing instruments and control systems. They must have a broad background in electronics, mechanics, and mathematics. They find employment with electric utilities, manufacturers of nuclear instruments and controls, handlers and processors of nuclear fuels, reactor research, nuclear power plant builders, and many electronic and instrument industries.

Radiation protection technician. These men and women use special instruments to check radiation levels and keep records on intensity and duration (how long it lasts) of radioactivity. They also identify types of radiation. They may also train others in the proper use of instruments. Employers are the same as for nuclear instrumentation and control technicians.

Nuclear reactor operator trainee. This is an entry level job and requires special training in the operation of a nuclear power plant or a reactor used in research. The trainees may learn the fundamentals of reactor operation at a two-year vocational-technical school. Employment and additional training comes from nuclear power stations, nuclear reactor research facilities, or from federal agencies that regulate nuclear activities.

Alternate energy research and development is being carried on in numerous places. Private enterprises, universities, large oil companies and the federal government are all involved in research and development projects. This work requires various types of scientists and engineers. Working with them are technicians who can translate ideas and theories into practical energy applications. Generally, applicants for these technician's jobs need two years of education beyond high school or specialized on-the-job training. To be considered for a technician's job, applicants must have taken math and science in high school. A technician's duties in the alternate energy field can range from basic wind research involving tasks as simple as measuring wind velocities or as complex as doing fusion research.

JOBS IN SOLAR ENERGY

New jobs in the solar energy industries will be closely tied to conservation and the construction industry. Some of the job areas are:
- Insulating older buildings to make them more energy efficient.
- Building and remodeling houses to use solar energy and cogeneration for space and water heating. (Cogeneration is using waste energy from one operation to do another job. For example, spent steam from a power generator may be used to heat buildings.)
- Legal and engineering jobs related to solar construction and conservation.
- Collecting and recycling metals and organic wastes which still have some energy in them. New solar and conservation industries will provide jobs for sheet metal workers, carpenters, plumbers, pipefitters, construction workers, insulation workers, painters, welders, air conditioning, heating, and refrigeration technicians, masons, and electricians. See Fig. 17-16.

Fig. 17-16. Workers fabricate tubes for boiler of solar thermal electric pilot plant. Tubes 45 ft. long are welded at the rate of 5 ft. per minute. (Rocketdyne Div., Rockwell International)

Careers will be available in law, sales, solar engineering, architecture, waste management, zoning, and consumer protection. Development of equipment for solar and conservation technology will create jobs for process, production, design, and chemical engineers.

It could be that in 20 years solar energy systems in homes and businesses could save 3 million barrels of oil a year. The solar industry has been growing. Sales and installation of solar heating products have grown from $25 million to $260 million in just two years.

Many of the persons needed to install solar heating systems will be trained by two unions. The Sheet Metal and Air Conditioning Contractors National Association and the Sheet Metal Workers International Association have set up 150 training programs all over the United States. They could train up to 11,000 solar installers a year.

A two-year training program for those working or planning to work in solar system installation might include courses in: refrigeration fundamentals, electrical fundamentals, related math, soldering and brazing, electrical controls and circuits, refrigeration systems components, airflow characteristics, domestic heating and cooling systems, blueprint reading, sheet metal fabrication, physics of heat transfer, computer programming, statics and fluids, industrial electronics, solar energy system analysis, energy economics, system design, related solar principles, and business management.

JOB CATEGORIES

The following job descriptions are based upon solar careers listed in the Occupational Outlook Handbook, a publication of the U.S. Labor Department:

- **Solar energy systems installer** plans the systems installation and tests the completed system.
- Solar energy systems installer helper helps install and repair solar energy systems.
- Solar technician performs all functions of a solar mechanic and can also design active solar systems for dwellings and light commercial use.
- **Solar engineer** designs all types of active solar systems; must have a degree from an accredited four-year engineering program.

CAREERS IN KINETIC ENERGY

Between 1850 and 1930 the windmill industry provided many jobs. It died out with the coming of rural electrification in the 1940s.

It is about to be reformed again. Large companies have begun serious work in research and development on generators capable of large-scale production of electricity. Small wind generators that supply 5 to 10 kW are already working. They are big enough for ranches, farm, and home use. The government is looking at the possibility of tying many such small units together. Then large amounts of electricity could be generated for whole towns and cities.

Another plan being researched by Westinghouse would create whole "farms" of large wind generators in the ocean to use the breezes that are always present offshore. Many problems must be solved with this plan. Among them are the design of floats to support the towers and power lines to carry the electricity to shore. Other questions to be worked on are what effect the towers will have on recreation, shipping, and fishing. Besides engineers and technicians, this development will require welders, steel workers, operating engineers, machinists, and electricians.

WATER POWER JOBS

The water power industry in the U.S. has been developed to the point where hydroelectric plants supply 14 percent of our electricity. Many workers will be needed to maintain and operate present plants.

Additional jobs will be available as we begin to expand hydroelectric facilities. Three-fourths of the nation's water power resources have not been tapped. A survey by Army Corps of Engineers has marked 50,000 locations where hydroelectric plants could be built. This will require engineers, technicians, environmentalists, and surveyors. Construction will require a variety of skilled workers such as welders, concrete workers, carpenters, and pipefitters. See Fig. 17-17. Once these power plants are built, many technical and highly skilled jobs will open up.

Fig. 17-17. Highly skilled workers install a generator rotor into a stator at a new hydroelectric power station.
(Power Authority of State of New York)

BIOMASS CAREERS

Careers in biomass industries are expected to grow slowly at first. Rapid growth of this area is unlikely because of the need for better technology. At the same time, the problems of collection need to be solved. However, small industries are already at work.

WASTE PRODUCT INDUSTRIES

At least 20 companies are involved in bioenergy. Products include:
* Entire systems for collecting fuel derived from municipal waste.
* Silvicultural biomass (tree) farm.
* Plants to produce methane by anaerobic digestion.
* Plants to convert wastes into fuel by pyrolysis process.

FORESTRY

Forestry technicians or forestry aids assist foresters in care and management of forest lands and forest resources. Their work involves forest protection, improvement of forests, and harvesting of trees for forest products.

Duties can be varied. For example, in forest production they may estimate yields in certain areas. Other tasks could include supervising, surveying, and building of roads.

Most foresters work for either lumber companies or the federal government in the Forest Service of the U.S. Department of Agriculture. Some work for state governments. A few are employed by mining and oil companies on reforestation projects.

OTHER CAREER AREAS

There are many other professional and skilled people whose services are required in energy fields. For example, certain engineering specialties common to other industries are also employed in energy-related jobs. In general, engineers apply science and mathematics to solve technical design problems. They design industrial machines and equipment. They work at testing, production operations, or maintenance. They may supervise manufacturing processes in refining of fuels. Some will work only in sales.

Energy related engineering includes:
* Agricultural engineering. Part of their work is to design systems to improve conservation of energy and soils.
* Chemical engineering. Since petroleum is the raw material for many chemical products, this type of engineering is important in the oil industry. Chemical engineers may help design synthetic fuel plants or systems to prevent pollution.
* Electrical engineering. These engineers may design electrical equipment needed for power generating and transmission.
* Mechanical engineering. Many engineers of this type are engaged in the design and development of energy producing machines. Some may specialize in producing machines for the petroleum industry.

Other skilled occupations important to the energy industry include carpenters, electricians, machinists, auto mechanics, plumbers and pipefitters, concrete workers, and masons. Many are employed by the energy companies for maintenance work, Fig. 17-18.

A

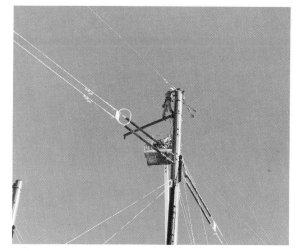

B

Fig. 17-18. Maintenance of all types of energy and power systems is very important. A—Transportation vehicles need frequent adjustment and repair. (BC Transit, Mark Van Manen) B—Line crews keep power lines in good repair. (Nevada Power Company)

BUSINESS OWNERSHIP

There are opportunities for many people to start their own business in the energy field. There are many small service and retail businesses connected with the transportation industry. Those who own or start their own business are known as **entrepreneurs.** They are people who are willing to risk their financial resources on the success of a business venture.

TYPES OF OWNERSHIP

There are three major forms of business ownership. Each type has advantages and drawbacks. The types are:
- Single proprietorship.
- Partnership.
- Corporation.

In a **single proprietorship** one person owns the entire enterprise. A gas station or a tire store are typical single proprietorship enterprises. In this type of ownership, the owner has complete control of the company's operation. It is easier to make business decisions because all authority rests with one person. On the downside, the owner is personally responsible for all debts of the company.

A **partnership** is a business owned by two or more persons. These are usually larger operations than a small retail store. Examples might be an automobile dealership or a freight company. Partnerships have several advantages. Costs of starting up or operating the business are shared by the partners. Another plus is that the partners often have a variety of skills that help in making good management decisions. Of course, the need to make joint decisions makes management tasks consume more time. As in the single proprietorship, the partners are liable for all company debts.

A **corporation** is a business which by law has the rights and obligations of a person. Many of the energy companies are corporations, for example, Amoco (transportation fuels) or Commonwealth Edison (electric power). The owners of corporations are persons who purchase stock in the corporation. **Stock** is a share in the ownership of the company. Many shares of a large company may be offered to the public.

Corporations need a license to operate. Known as a charter, it lists the company name and its officers. General rules for the corporation are also stated in the charter. A more detailed list of rules is known as the corporate bylaws.

Corporations do not have some of the weaknesses of other forms of ownership. Owners have limited liability. They are only responsible for debts up to the limits of their investment. Owners need not fear losing personal property, such as a home, because of business debt. Profits of the corporation are distributed to stockholders as dividends.

CHARACTERISTICS OF ENTREPRENEURS

A successful entrepreneur must possess certain characteristics that helps her or him to be successful. The following summarize the major qualities:
- Good health—long hours make heavy demands on owners of businesses.
- Knowledgeable—to have a well-run company, he or she must know all aspects of the business. In addition the entrepreneur must understand the industry and the products being marketed.
- Good planner—running a successful business means that nothing is left to chance. She or he must be able to foresee difficulties as well as plan how to take advantage of opportunities.
- Willing to take calculated risks—once a plan has been conceived that takes into account events likely to occur, the person must have the courage to risk money and future on making the plan work.
- Innovative—successful in finding ways to improve and find ways to produce better service and thus gain the confidence of customers.
- Responsible—willing to accept the consequences of decisions whether good or bad. This includes paying debts, keeping promises, and accepting the responsibility for mistakes of employees.
- Goal oriented—likes to set goals and works hard to achieve them.

SUMMARY

The recovering of energy and converting it into power to operate our technological systems, provide process heat for manufacturing, comfort condition our buildings, and transport goods and people requires the work of many people with a variety of skills and knowledge. As we near the 20th Century and become more conscious of the need to conserve energy resources and protect the environment, other people are finding work in conservation and environmental sectors.

While we may look more and more to alternate energy sources and how to extract that energy, there will continue to be excellent opportunities for jobs in the fossil fuel industry.

While controversy and concern over nuclear power continues, there will continue to be careers in nuclear electrical power generation. New designs for nuclear reactors now on the drawing boards or being tested promise to make nuclear power more efficient, safer, and less expensive. If this is the case, then nuclear energy may become an acceptable alternative to fossil-fueled power stations and there will be additional career opportunities in nuclear power.

It is harder to predict what job opportunities there may be in alternate energy industries. These have not shown the growth anticipated in the 1980s. Solar energy continues to show great promise for development as does biomass (with processes that produce ethanol and methanol as a gasoline "extender.")

There should be career opportunities in environmental preservation as people and governments become more concerned with preserving the planet. Environmental engineers and technicians should find rewarding careers in the 1990s and on into the 20th century.

DO YOU KNOW THESE TERMS?

acidizers
blaster
civil engineer
corporation
derrick operator
distribution field engineer
electrical/electronic engineer
electrical engineer
entrepreneurs
gasoline finishers
geologists
geophysicist
heavy truck driver
industrial truck operators
miner
mine surveyors
mining engineers
nuclear criticality safety engineer
nuclear equipment design engineer
nuclear fuels reclamation engineer
nuclear fuels research engineer
nuclear instrumentation and control technician
nuclear materials processing technician
nuclear plant technical adviser
nuclear reactor operator trainee
oil well directional surveyors
partnership
petroleum engineers
pipeline construction inspectors

pipeliner
pipeline workers
power engineer
radiation protection engineer
radiation protection technician
sales engineer, nuclear equipment
seismic observers
seismologists
solar energy systems installer
solar engineer
single proprietorship
stock
timber framers
waste management engineer, radioactive material

SUGGESTED ACTIVITIES

1. Interview someone in your community who works for an energy company. Report to your class on the skill requirements, educational requirements, duties, and responsibilities of the job.
2. Study the publications in the guidance counselor's office or library on career opportunities. Select an energy-related career and research it. Prepare a written report.
3. With your instructor's permission and assistance, invite a speaker from the energy industry to talk to your class.
4. Visit a utility company or an energy company of any type and observe people at various jobs. Arrange for a representative to speak to your group while you are there.

TEST YOUR KNOWLEDGE

1. Congress once predicted that as many as _____ jobs could be generated in energy-related fields.
2. There (will, will not) continue to be jobs in traditional energy fields.
3. What do exploration geophysicists do?
4. Name one source of additional information on energy careers.
5. What are the likely growth areas for energy-related jobs?
6. What courses might a trainee in solar heating systems need to take to become knowledgeable in that field.

Appendix A
SAFETY IN TECHNOLOGY EDUCATION LABORATORIES

SAFETY WITH MATERIALS

1. Gasoline, lubricants, and other liquids used in the power laboratory are extremely flammable. Use care in their use. Gasoline has more explosive energy than many explosives.
2. Close the fuel container immediately after filling a fuel tank.
3. Study the use of fire extinguishers. Know what fire extinguisher should be used with different types of fires.
4. Be careful with compressed air. It can cause flying debris that can injure you or those around you. Fuels and lubricants vaporized by the compressed air can be easily ignited causing painful and serious burns.
5. Read manufacturer's instructions carefully before attempting to use special fuels, solvents, or lubricants.
6. When uncertain of the nature of fluids in containers do not inhale fumes directly from the opening in the can. Check with your instructor.

PERSONAL SAFETY

1. Dress appropriately while engaging in energy or power laboratory activities. Avoid wearing loose clothing. Roll up long sleeves, remove rings and other jewelry. If you are wearing a necktie, remove it before starting your activity. Never wear oily clothing. It could easily catch fire.
2. Wear safety glasses whenever there is danger from airborne chips, dirt, or from splashing liquids.
3. Use ear protectors when operating noisy engines or other equipment or devices.
4. If unsure of a procedure, check with your instructor before attempting it.
5. Ask your instructor to repeat any safety

Fig. A-1. Safety glasses should be worn when working around batteries since there is always danger of an explosion splashing acid into a technician's eyes. Care should also be taken when working around belts and fans of running engines.

demonstration that you do not understand or that you have forgotten.

LABORATORY/SHOP SAFETY

1. Never use the power laboratory or shop as a "playground." Running or horseplay is dangerous and could cause you or those around you serious injury.

2. Keep work areas clean. Always place soiled or oily cleaning rags in a closed container so they are not a fire hazard.
3. Wipe up spilled fuel and lubricant immediately. Use rags rather than sorbents which become highly combustible when saturated with fuels.
4. Provide adequate ventilation when operating engines or where activities require materials that give off hazardous fumes. If available, attach an exhaust hose to the exhaust pipe of an engine. Be sure to turn on the exhaust system.
5. Do not smoke in an automotive or power/energy lab. Fuels are present and a serious fire or explosion could result.
6. Store gasoline and solvents in safety cans designed for that purpose. Keep the container in a special safe storage area away from flames or sparks.
7. Get help when attempting to move heavy items or equipment.
8. Never use a piece of equipment before you have been instructed in its proper, safe use.
9. Compressed air pressures in the laboratory or shop are usually around 100 to 150 psi. Be careful! This is enough to cause severe injury or even death.
10. Know where the fire extinguishers are located in the shop. Learn how they are used. Also note the location of the fire alarm located in the lab and how it operates.

SAFETY WITH TOOLS

1. Never use hammers or other striking tools with loose heads.
2. Hammer faces are extremely hard. Striking two hammerheads together may cause chips of metal to fly off and cause injury.
3. Keep fingers clear of the cutting edges of pliers.
4. Never push on a wrench when loosening or tightening a fastener. The wrench may slip and cause knuckles to strike the work causing painful injury.
5. Avoid use of a pipe section to get more leverage on a wrench. Use a larger wrench.
6. Only use files with handles on them. The tang may injure your hand.
7. Never test sharpness of tools by running a finger across the edge.
8. When using metal cutting tools remove chips with a brush, never your hand.
9. Avoid use of damaged or broken tools.
10. Before use of power equipment, make certain that all safety guards are in place.
11. Never use a steel hammer on hardened parts. Airborne metal chips could cause injury. Use a brass or softheaded hammer.

12. To prevent flying chips, keep chisel heads in good condition. Grind away and chamfer a chisel head that has become mushroomed.
13. Never use a dangerous tool unless you have had instruction in its safe use.

ELECTRICAL SAFETY

1. Use pliers with insulated handles when working around electricity.
2. Disconnect a tool from electrical power when you are working on it.
3. Disconnect a circuit from its power source before working on it.
4. Always disconnect a battery when indicated by a service manual. Hot (having electrical current in them) wires can cause fires upon making accidental contact with a ground (such as an automobile frame).

VEHICLE SAFETY

1. Drive slowly and carefully within the shop area. There are other students and vehicles in the area and serious damage and injury could result from careless or fast driving.
2. Use care when working on running engines. Fan blades can inflict serious cuts. A tool dropped in a fan can become a dangerous missile.
3. Diesel fuel systems and fuel injection systems carry high pressure. Fuel under high pressure can puncture the skin or eyes. Blood poisoning or blindness may be the result of such injuries. Wear eye protection when working on such fuel systems.
4. Set parking brakes on vehicles and block the wheels if the engine must run during tests.
5. Keep leads of test equipment well away from fans or belts and hot engine components.

FEDERAL/STATE SAFETY REGULATIONS

By the late 1960s legislators were beginning to worry about the high rate of accidents in the workplace. Every year some 2.2 million workers were killed or disabled by accidents at work. In addition to the pain and suffering of these workers there was the costs to the country and the economy of deaths, disablement, and illnesses. The cost had been placed at a half million dollars a year. This represented a huge dent in the gross national product.

These statistics, though sobering, were only part of the story. A Senate Committee Report pointed out a startling fact which told the missing story: countless individuals fell ill each year because of occupationally induced diseases. Industrial hygienists had known for more than 40 years the effects some industries had on its workers. Among these are asbestos, coal mining, cotton textile, chemical, pesticide and fungicide industries — to mention only a few.

After assessing the testimony and documentary evidence, Congress deliberated through 1969 and into 1970 and finally passed the Occupational Safety and Health Act of 1970. President Nixon signed it into law on December 29.

The purpose of OSHA was and is to assure safe and healthful working conditions to every employee in America and to preserve this country's human resources. The ways in which the Act tries to do this include:

1. Encouraging employers and workers in their efforts to make the workplace safer and healthier.
2. Giving the Secretary of Labor the power to set safety and health standards.
3. Creating a commission to adjudicate (judge) cases where companies were in violation of the Act.
4. Providing for research in occupational safety and health.
5. Developing new methods, techniques, and approaches to dealing with safety and health problems on the job.

6. Exploring ways to discover latent (hidden) diseases and uncovering connections between disease and work in environmental conditions.
7. Conducting research relating to health problems induced by work environment.
8. Providing safety and health standards for industry.

Since OSHA became law, standards have been set for work safety and health. These affect every company with one or more employees. With the encouragement of the federal government, states assume varying degrees of responsibility for enforcing their own safety and health laws and for bringing them into accord with OSHA standards.

STATE SAFETY LAWS

When Congress passed the Occupational Safety and Health Act, state safety laws were preempted. The Act took precedence over existing state safety laws. The Act does, however, invite the states to step in and take on the responsibility of enforcing the occupational and safety health laws within the state. Many of the states have reshaped their occupational health and safety laws to be the same as the federal Act. States with federally approved plans are entitled to federal funds. These funds help them enforce their laws.

Appendix B
STATISTICS ON RENEWABLE ENERGY

Table 108. **Solar Thermal Collector Shipments by Type and End Use, 1989**
(Thousand Square Feet)

End Use	Low-Temperature Collectors	Medium-Temperature, Special, and Other Collectors	Total
Application Total	**4,283**	**7,199**	**11,482**
Pool Heating	4,283	405	4,688
Water Heating	0	1,374	1,374
Space Heating	0	205	205
Other [1]	0	5,215	5,215
Market Sector Total	**4,283**	**7,199**	**11,482**
Residential	3,915	1,889	5,804
Commercial	368	56	424
Industrial	0	42	42
Other	0	5,212	5,212

[1] Includes collectors for electricity generation by independent power producers, process heating, and space cooling.
Note: Sum of components may not equal total due to independent rounding.
Source: Energy Information Administration, *Solar Collector Manufacturing Activity 1989* (March 1991), Tables 6 and 7.

Table 109. **Photovoltaic Module and Cell Shipments, 1989**

End Use	Amount Shipped (peak kilowatts)	Percent of Total
Application Total	**12,825**	**100.0**
Health	5	(1)
Water Pumping	711	5.5
Transportation	1,196	9.3
Communication	2,590	20.2
Consumer Goods	2,788	21.7
Electric Generation		
Grid Interactive	1,251	9.8
Other Remote	2,620	20.4
Cells to Original Equipment Manufacturers	1,595	12.4
Other	69	0.5
Market Sector Total	**12,825**	**100.0**
Residential	1,439	11.2
Commercial	3,850	30.0
Government	1,077	8.4
Industrial	3,993	31.1
Electric Utility	785	6.1
Transportation	1,130	8.8
Other	551	4.3

[1] Less than 0.05 percent.
Note: Sum of components may not equal total due to independent rounding.
Source: Energy Information Administration, *Solar Collector Manufacturing Activity 1989* (March 1991), Tables 17 and 18.

Figure 107. Solar Thermal Collector Shipments by Type, 1974–1984 and 1986–1989

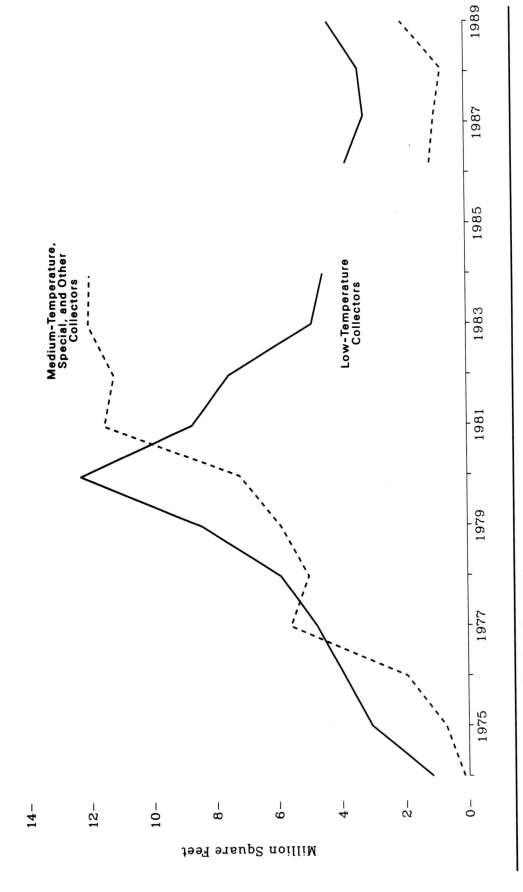

Note: Data were not collected for 1985.
Source: See Table 107.

Sources: Number of Manufacturers: Energy Information Administration (EIA), "Annual Solar Thermal Collector Manufacturers Survey." **Other Data:** •1974 through 1976—Federal Energy Administration, *Solar Collector Manufacturing Activity,* semi-annual. •1977—EIA, *Solar Collector Manufacturing Activity, July through December, 1981,* March 1982 (semi-annual). •1978 and forward—EIA, *Solar Collector Manufacturing Activity,* annual.

Table 104. Renewable Energy Consumption Estimates, 1988-1990
(Quadrillion Btu, Except as Noted)

	1988	1989	1990
Reported Renewable Energy Consumption	**2.88**	**3.10**	**3.15**
Hydroelectric Power	2.64	2.88	2.94
Electric Utilities	2.28	2.74	2.89
Industrial	0.03	0.03	0.03
Imported Electricity	0.40	0.27	0.21
Exported Electricity	0.07	0.17	0.19
Geothermal Energy at Electric Utilities	0.22	0.20	0.18
Wood and Waste Energy at Electric Utilities	0.02	0.02	0.02
Wind Energy at Electric Utilities	(1)	(1)	(1)
Additional Renewable Energy Consumption Estimates	**3.43**	**3.19**	**3.55**
Biofuels	3.15	2.90	3.23
Residential, Commercial, and Industrial Use	3.09	2.83	3.15
Transportation Use	0.06	0.07	0.08
Geothermal Energy (Non-Electric Utilities)	0.13	0.14	0.16
Solar Energy (Non-Electric Utilities)	0.09	0.09	0.08
Wind Energy (Non-Electric Utilities)	0.03	0.03	0.04
Hydroelectric Power (Additional Industrial Use)	0.03	0.03	0.04
Total Renewable Energy Consumption Estimates	**6.31**	**6.29**	**6.70**
Adjusted Total Energy Consumption Estimates [2]	**83.63**	**84.54**	**84.99**
Total Renewable Energy Consumption Estimates as Share of Adjusted Total Energy Consumption Estimates (Percent)	**7.5**	**7.4**	**7.9**

[1] Less than 0.005 quadrillion Btu.
[2] Adjusted Total Energy Consumption Estimates is Total Energy Consumption from Table 3 plus Additional Renewable Energy Consumption Estimates from this table.
Sources: **Reported Renewable Energy Consumption:** Table 3. **Additional Renewable Energy Consumption Estimates:** Energy Information Administration, Office of Coal, Nuclear, Electric and Alternate Fuels.

Table 106. Households That Burn Wood, 1980-1982, 1984, and 1987 [1]

	1980	1981	1982	1984	1987
Households That Burn Wood					
Number of Households (millions)	21.6	22.8	21.4	22.9	22.5
Percent of All U.S. Households	26.4	27.4	25.6	26.6	24.8
Number of Cords Burned (millions)	42.7	44.0	48.6	49.0	42.6
Average Number of Cords Burned per Household					
Mean	2.0	1.9	2.3	2.1	1.9
Median	0.7	1.0	1.0	1.0	0.7
Wood Energy Consumed (trillion Btu)	854	881	971	981	853
Households That Burn Wood as Main Heating Fuel					
Number of Households (millions)	4.7	5.3	5.6	6.4	5.0
Percent of All U.S. Households	5.8	6.4	6.7	7.5	5.6
Number of Cords Burned (millions)	22.4	24.7	28.7	29.4	23.5
Average Number of Cords Burned per Household					
Mean	4.7	4.6	5.1	4.6	4.7
Median	3.3	3.0	4.0	4.0	4.0
Wood Energy Consumed (trillion Btu)	448	493	574	589	470
Households That Burn Wood as Secondary Heating Fuel and for Other Purposes					
Number of Households (millions)	16.9	17.4	15.9	16.5	17.4
Percent of All U.S. Households	20.6	21.0	18.9	19.1	19.3
Number of Cords Burned (millions)	20.3	19.4	19.9	19.6	19.2
Average Number of Cords Burned per Household					
Mean	1.2	1.1	1.3	1.2	1.1
Median	0.3	0.5	0.5	0.5	0.5
Wood Energy Consumed (trillion Btu)	406	388	397	392	383

[1] Data are for the heating season beginning with the latter part of the previous year shown.
Note: Consumption estimates are based on respondent reports and may be subject to reporting biases.
Note: No data are available for 1983, 1985, or 1986.
Source: Energy Information Administration, Form EIA-457, "Residential Energy Consumption Survey."

Table 110. Electric Utility Net Summer Capability [1] and Net Generation of Electricity by Renewable Energy Resource, 1949-1990

Year	Geothermal Net Summer Capability [3] (thousand kilowatts)	Geothermal Net Generation (million kilowatthours)	Wood and Waste Net Summer Capability [3] (thousand kilowatts)	Wood and Waste Net Generation (million kilowatthours)	Wind and Other [2] Net Summer Capability [3] (thousand kilowatts)	Wind and Other [2] Net Generation (million kilowatthours)
1949	(4)	(4)	13	386	0	0
1950	(4)	(4)	13	390	0	0
1951	(4)	(4)	13	391	0	0
1952	(4)	(4)	37	482	0	0
1953	(4)	(4)	37	389	0	0
1954	(4)	(4)	37	263	0	0
1955	(4)	(4)	37	276	0	0
1956	(4)	(4)	37	152	0	0
1957	(4)	(4)	64	177	0	0
1958	(4)	(4)	64	175	0	0
1959	(4)	(4)	64	153	0	0
1960	11	33	64	140	NA	NA
1961	11	94	64	126	NA	NA
1962	11	100	64	128	NA	NA
1963	24	168	64	128	NA	NA
1964	24	204	64	148	NA	NA
1965	24	189	72	269	NA	NA
1966	24	188	72	334	NA	NA
1967	51	316	72	316	NA	NA
1968	78	436	72	375	NA	NA
1969	78	615	72	320	NA	NA
1970	78	525	72	356	NA	NA
1971	184	548	77	311	NA	NA
1972	290	1,453	77	331	NA	NA
1973	396	1,966	77	328	NA	NA
1974	396	2,453	77	251	NA	NA
1975	502	3,246	77	191	NA	NA
1976	502	3,616	77	266	NA	NA
1977	502	3,582	77	481	NA	NA
1978	502	2,978	78	338	NA	NA
1979	667	3,889	78	498	NA	NA
1980	909	5,073	78	433	NA	NA
1981	909	5,686	79	368	(5)	NA
1982	1,022	4,843	212	321	6	NA
1983	1,207	6,075	321	379	6	3
1984	1,231	7,741	350	886	17	12
1985	1,580	9,325	343	1,383	18	16
1986	1,558	10,308	401	1,177	19	18
1987	1,549	10,775	421	1,477	25	14
1988	1,667	10,300	465	1,674	7	10
1989	1,606	9,342	465	1,965	4	3
1990[6]	1,606	8,581	465	2,061	4	3

[1] See Glossary.
[2] Includes photovoltaic and solar thermal energy.
[3] At end of year.
[4] No geothermal capability prior to 1960.
[5] Less than 500 kilowatts.
[6] Previous-year data may have been revised. Current-year data are preliminary and may be revised in future publications.
NA = Not available.
Sources: Net Summer Capability at End of Year: •1960 through 1984—Energy Information Administration (EIA) estimates. •1985 and forward—EIA, Form EIA-860, "Annual Electric Generator Report." Net Generation: •1949 through September 1977—Federal Power Commission, Form FPC-4, "Monthly Power Plant Report." •October 1977 through 1981—Federal Energy Regulatory Commission, Form FPC-4, "Monthly Power Plant Report." •1982 and forward—EIA, Form EIA-759, "Monthly Power Plant Report."

Appendix C

LOCATIONS OF U.S. NUCLEAR POWER STATIONS

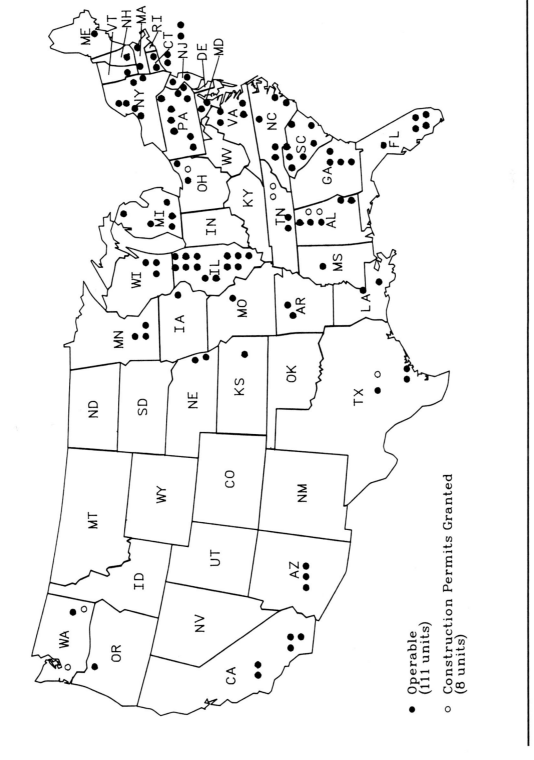

Figure 101. Nuclear Generating Units, December 31, 1990

• Operable (111 units)

○ Construction Permits Granted (8 units)

Note: Due to space limitations, symbols do not represent actual locations.
Source: See Table 101.

Appendix D

HOW THE U.S. USED THE PETROLEUM CONSUMED IN 1990

(Million Barrels per Day)

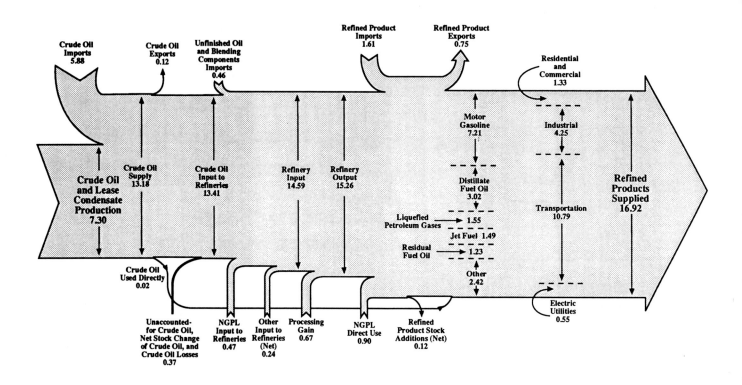

Note: Data are preliminary.
Note: Sum of components may not equal total due to independent rounding.
Source: *Petroleum Supply Montnly.*

Annual Energy Review 1990
Energy Information Administration

Technical Terms

A

Absorber: flat surface of a solar collector that absorbs the heat of solar radiation.

Acceleration: a change (speeding up) in velocity over a period of time.

Accumulator: a storage place for liquid that is under pressure in a hydraulic system. It acts as a damper or "shock absorber" as pressures change or movement stops in the system.

Acidizers: workers who operate equipment for treating oil wells with acid to make them produce more oil. They supervise acidizer helpers in loading and mixing chemicals at the well site.

Air gasification: process that uses small amounts of air and steam to convert char to a low energy gas used for fuel.

Alternator: generator in which the electric current alternates, changing direction many times a second.

Amperes: the measure of the rate of electron flow (current).

Anaerobes: cells that are able to oxidize organic matter and to maintain their life processes without air or free oxygen.

Anaerobic digestion: the controlled decay of organic matter without oxygen. The process produces methane gas.

Anemometer: an instrument that measures wind speed. A recording anemometer will keep a record of the wind speed over a period of time.

Anode: the positive terminal of a battery.

Anthracite coal: coal that is older and much harder than bituminous. It burns with a hot, blue, smokeless flame. However, it is not very plentiful.

Aquifers: underground rock formations holding large quantities of water that has filtered down from the surface. Gas is pumped in for storage, displacing the water.

Asphalt: semisolid form of petroleum used for waterproofing as early as 3000 B.C.

Associated gas: gas found in the same reservoir as crude oil.

B

Barrel: measure used for crude oil, equal to 42 gallons.

Base: one of three layers of doped (impure) silicon in a transistor.

Biofuels: fuels made from plant materials.

Biogas: gas produced by anaerobic digestion of biomass.

Biology: study of living organisms.

Biomass: live organic material such as trees, farm crops, seaweed, or algae. It is also wastes such as manure, trash, sewage, and garbage.

Bituminous (soft) coal: the most plentiful and most used coal.

Black body radiation: reverse radiation that occurs at night as substances which have absorbed heat during the day give up heat to the atmosphere.

Blaster: technician whose job is to plant and set off charges of explosives in mining operations.

Blowout preventers: valves that prevent uncontrolled flow from an oil well.

Bottom hole pressure: the pressure of fluids and gases at the bottom of an oil well being drilled.

Breaker: building where anthracite coal is prepared for use.

Breeding: the process of making new fuel in a nuclear reactor.

Btu (British thermal unit): unit of measure for heat. One Btu is the amount of heat needed to raise the temperature of one pound of water by 1°F.

Bubble caps: placed above holes in a fractionating tower, they collect and condense vapors.

Bulk freighters: vessels that transport dry and liquid bulk (loose) material on water.

Buoyant: capable of floating in water.

C

Cable tool string: all of the tools that go into an oil well, including a drill bit, a stem, connecting links called jars, sinkers for weight, and a cable for getting the tools down the hole to the drilling face.

Calorie: a unit of heat measure.

Cam: device that employs the principle of the inclined plane to achieve a lifting motion. As the cam rotates, its sloping surface exerts pressure on a rod to move it up or down.

Carbonate: sedimentary rock that came primarily from the shells of sea life.

Cathode: the negative terminal of a battery.

Cells: three distinct areas of wind patterns both north and south of the equator.

Chain reaction: the continuous splitting of atomic nuclei. Heat is released and can be put to use.

Changeable energy source: an energy source of uncertain availability.

Christmas tree: set of valves that controls distribution of the oil and gas and prevent unwanted flow from a well.

Circuit: path used to route and control electrical power to lights, machinery, appliances, and other loads.

Circuit breakers: electrical devices placed in circuits to protect them from overcurrents by "tripping" and opening the circuit. They can be reset when the overcurrent problem is resolved.

Clutch: a simple type of connect/disconnect device used in power transmission.

Coal gasification: the process of turning coal into a gas fuel.

Cogeneration: the production of electric power and use of the same heat for industrial operations or other purposes.

Coke: fuel converted from coal by pyrolysis.

Collector: one of three layers of doped (impure) silicon in a transistor.

Compound turbine: a series of turbines that steam passes through one after the other.

Computer: an electronic device that accepts data, processes it according to a stored program of instructions, and produces various outputs.

Concentrators: focusing lenses or mirrors that intensify the sunlight on the solar cells.

Conduction: process in which heat moves through materials by increased molecular activity.

Cone bits: special drilling bit for cutting through certain kinds of rock.

Conservation of energy: saving an energy source by using it efficiently and not wasting it.

Constant energy source: one that is always available when needed.

Container ships: water transporters designed to handle products preloaded into large containers.

Continuous mining: system in which special mobile (moving) machines mine and load coal in a nonstop operation. No explosives are used.

Control rods: movable rods loaded with neutron-absorbing elements that can be inserted into the nuclear fuel core to reduce the number of reactions taking place. They are removed to increase the level of nuclear reaction.

Convection: air in motion.

Core: the earth's iron center, roughly 2175 miles (3500 km) in diameter.

Corporation: a business which by law has the rights and obligations of a person. Many energy companies are corporations. The owners of corporations are persons who purchase stock in the corporation.

Cracking: the process of converting heavy fuel oils into high octane gasoline.

Crude oil: a fossil fuel, the natural state of liquid hydrocarbons.

D

Degraded: term used to describe energy that has been weakened to a point where we are not able to use it.

Degree day: a measure of how much temperature varies above or below a standard temperature of 65°F (18°C).

Densified biomass fuel: biomass compressed into pellets or cubes for easier handling and use.

Derrick: open steel tower that supports the oil drilling tools.

Derrick operator: worker on oil drilling crews who rigs the derrick equipment and operates mud pumps.

Deuterium: also called heavy hydrogen, is an isotope of hydrogen which has twice the mass of an ordinary hydrogen atom.

Development well: one which is drilled in an area that is already producing oil.

Diffused radiation: solar rays scattered by atmospheric conditions.

Direct current generator: one in which the current always moves in the same direction.

Direct gain solar heating system: system in which the space to be heated receives direct sunlight that falls on floors and/or walls constructed of materials which can do a good job of storing the heat.

Displacement digester: a long horizontal cylinder used to digest organic wastes.

Downwind rotor wind machine: one that has the rotor located behind the tower and rotor housing. That is, the wind blows over the tower and housing before striking the rotor.

Drag: frictional force that tends to hold an aircraft back.

Drill pipe: a length of steel pipe to which the bit is connected in rotary drilling for oil.

Drive (lift): force needed to bring oil to the surface from several thousand feet down.

Drive gear: the gear connected to the power source.

Driven gear: the gear that receives the power from the drive gear.

E

EER: energy efficiency rating. System used to rate an appliance on how efficiently it uses electricity, gas, or fuel oil.

Efficiency: ratio between the amount of energy put into a power system and the energy output.

Electric motor: device that changes electrical energy back into mechanical energy.

Electrical/electronic engineer: engineer occupied with the application of the laws of electrical energy and the principles of engineering as they apply to the generation and transmission of electricity.

Electrical power system: one that transfers and controls electric current.

Electrodrilling: drilling method that takes its power from an electric motor located just above the bit.

Electromagnetic radiation: the scientific term for radiant energy.

Electrons: negatively charged particles which travel around the nucleus of an atom in elliptical or circular paths.

Emf (electromotive force): another term for voltage.

Emitter: one of three layers of doped (impure) silicon in a transistor.

Energy: the ability to do work.

Energy conversion: the changing of energy from one form to another. Also, the process of harnessing energy or beasts to do work.

Energy converters: devices or beasts harnessed to do work.

Entrepreneurs: persons who own or start their own business. They are people who are willing to risk their financial resources on the success of their venture.

Entropy: the principle that all energy, as it is used, becomes so diffused or scattered (random) that it loses the ability to do work.

Entries: in coal mining, hallway-like passages that give mining machinery access to the coal.

Enzyme: type of protein that causes chemical action to take place within plant and animal cells.

Ethanol: ethyl alcohol.

Evacuated tube collectors: more expensive than flat plate solar collectors, but they produce higher temperature heat.

Exploratory well: one drilled to find a new pool, to extend an existing oil field, or to find an entirely new oil field.

External combustion converters: engines that burn their fuel outside of the cylinder or combustion chamber where the conversion from heat to motion takes place. Steam engines are the most common external combustion engines.

F

Feedback: any signal or information that might adjust the process or initiate it.

Fermentation: anaerobic process for producing alcohol through the action of enzymes on the starches and sugars in fruits, grains, and other biomass forms.

Fishtail bit: drilling bit best for soft clay and sand.

Fissile: able to go into fission.

Fission: process that changes the nuclei of uranium.

Fixed path vehicle: one that can only move along one path which has been constructed for its use, such as a train or a pipeline. The vehicle cannot leave the path.

Flame drilling: method in which jets of burning liquid rocket fuel burn a hole through rock and soil.

Flat plate collectors: solar collectors that consist of a closed, insulated box with a clear covering. These are best suited to residential use.

Fluid power: system that uses a fluid in transmitting power from one place to another and for altering power to do certain types of work.

Fluidized bed combustion: a process that has proven successful in cutting down sulfur and nitrogen oxide (NOx) emissions and fly ash from burning coal.

Foehn: a hot dry wind in mountainous regions.

Foot-pound: the unit of measure for work, equal to the amount of force needed to move a 1 lb. load a distance of 1 ft.

Fossil fuels: those that developed in the earth as a result of heat and pressure acting on formerly living material over a long period of time. Coal, oil, and natural gas are all fossil fuels.

Four-way switch: one that is used when a light must be controlled from three locations.

Fractionating (bubble) tower: tall tower with an arrangement of trays at different levels to collect the products of distillation.

Fractions: products refined from crude oil.

Free piston engine: somewhat of a hybrid, it combines elements of both reciprocating and gas turbine engines. The pistons are called "free" because they are not connected to a rod or crankshaft and they have no ignition system to ignite fuel. The free piston engine will run on almost any fuel, but is noisy, has low power, is hard to start, and violent forces of combustion soon cause breakdown.

Frequency: the time lapse between electromagnetic waves reaching a receiver.

Fuel assemblies: bundled rods of uranium pellets that make up the core of a nuclear reactor.

Fuel cells: devices, somewhat like batteries, that produce electricity from a fuel source and oxygen.

Fumaroles: wisps of steam emerging from the ground in volcanic areas.

Fuses: protective devices placed in circuits to protect them from overcurrents by opening and interrupting current flow.

Fusion: process that taps the energy of the atom by combining two light atoms into one heavy atom.

Future reserves: energy stocks that have been estimated to exist with a lesser degree of certainty than proven reserves. There is a possibility that they can be recovered sometime in the future with technology not now available.

G

Gas turbine: engine that has large, finned rotors which absorb the high energy of pressurized gas and spin at a high rate of speed. Unlike the steam turbine, which is designed primarily for stationary operation, the gas turbine is found in vehicles such as aircraft, boats, locomotives, and large trucks.

Gasoline finishers: workers who control equipment for blending natural gasoline with chemicals to improve its fuel qualities.

Gear: a wheel with notches called teeth around its rim. The teeth of one gear mesh with (fit between) the teeth of another gear to transmit force.

General cargo freighters: large vessels designed to operate on the open seas, carrying such cargo as cars, heavy equipment, palletized products, lumber, fruit, vegetables, or manufactured goods.

Geologists: scientists who study the makeup of the earth, identify rocks and minerals, make surveys, take measurements, and record data.

Geophysicists: scientists who explore solid earth for the presence of oil, coal, and other mineral deposits.

Geothermal energy: heat energy that comes from deep within the earth.

Giant excavator: open-pit mining machine with a cutting wheel at the end of a long boom. The wheel rotates, cutting and scooping up the overburden with bucketlike claws.

Gravimeter: instrument that measures gravity and indicates the density of rock deep in the ground.

Gravity: force pulling all objects downward (toward the center of the earth.)

Greenhouse effect: heat-trapping process caused by carbon dioxide and other gases in the atmosphere. It is feared that excessive amounts will cause global warming, upsetting weather patterns and causing coastal flooding.

Ground thermal energy: energy that is partly the result of solar radiation storage and partly the result of the heat locked in the earth when it was first formed.

Gushers: wells in which oil sprays out violently and uncontrollably.

H

Head: distance that water falls to the point where its energy is used to generate electric power.

Heat engines: devices that can contain the energy of heat and make it do work.

Heat exchanger: device used in a heating or cooling system to move or transfer heat energy from one medium to another.

Heat pump: basically an air conditioner with a reversing valve so that it can function as either a heating or a cooling unit. In winter it pumps heat indoors; in summer it pumps it outdoors.

Hole flow: in electronic devices, hole flow is always the opposite of electron flow.

Horse latitudes: high pressure belts where the air is sinking to the surface.

Horsepower: the unit of measure for power, equal to the energy needed to lift 33,000 lbs. a distance of 1 ft. in 1 minute.

Hovercraft: vehicles that depend on a cushion of air for propulsion and to support them above the surface. Such craft are capable of operating both on water and on land.

Hydrocarbons: substances that are a mixture of the two elements, carbon and hydrogen.

Hydrofoils: high-speed people movers that operate in inland waterways. With the aid of winglike structures on their hulls, they skim over the water at high speeds.

Hydrogenation: process used to produce a gaseous or liquid fuel, usually by extracting oil or gas from coal.

Hypergolic: term for rocket fuels and oxidizers that ignite simultaneously when brought in contact with each other in the combustion chamber.

I

Impulse turbine: one in which high pressure steam is piped to the turbine blades. It escapes through nozzles and is moving at very high speed as it strikes the blades.

Indirect gain solar heating: system in which sunlight strikes some kind of thermal mass first. It is absorbed by the mass and then transferred to the living space.

Industrial sector: energy users that include all types of mining, manufacturing, and farming businesses. They use energy in all its forms: gas, oil, coal, and electricity.

Inertia: tendency of a body to stay in motion or resist forces exerted to stop it.

Inertial storage: the storing of energy in flywheels. Inertia is the tendency of a moving or spinning body to continue moving or spinning in the same direction.

Input: system component (also called a resource) that is available to be processed.

Input well: well where gas is injected to help maintain the pressure and keeps the oil moving into the well.

Insolation: shortened form of the term, "incident solar radiation." It is a term for the amount of radiation falling on the earth's surface at any given time.

Insulation: a material that is highly effective at reducing heat passage.

Integrated circuit: a circuit containing resistors, transistors, and capacitors. It is a complete electronic circuit contained in one tiny package.

Internal combustion engine: a heat engine designed to burn fuel inside itself, rather than in a separate unit such as a boiler.

Inverter: electrical device which steps up voltage and converts direct current to alternating current.

Isobutane: a hydrocarbon liquid that vaporizes at low temperature.

Isomerization: process that converts chemical compounds in refinery gases into its isomers. (An isomer is an ion or molecule of any substance.)

Isotopes: atoms of the same element that have different numbers of neutrons in their nuclei.

J

Jet engines: aircraft power sources in which hot gases rushing out the rear of the engine at a high pressure move the airplane forward. Jet engines are powerful for their size and weight.

Joule: the unit of power measurement in the metric system.

K

Kinetic energy: the ability of objects that are moving to do work. The word "kinetic" comes from a Greek word that means motion.

L

Langley: metric unit used for measuring solar radiation. The langley equals 1 calorie per square centimeter (cm²).

Laws of Thermodynamics: basic behavior characteristics of energy.

Lift: an upward force that keeps an aircraft in the air.

Lignite: a soft, brown fibrous material formed from peat.

Lithosphere: thin crust of rock that makes up the outer layer of the earth.

Loader: an electrically powered machine with clawlike arms and an attached conveyor to carry coal to a shuttle car.

Local winds: winds caused by local temperature differences and topography (lay of the land).

Longwall mining: underground mining method, two parallel entries are driven through the coal seam and joined at the far end by another entry called a crosscut. The coal face formed by the crosscut is the longwall, where the coal is removed.

Lumen: measure of the amount of light that falls on a surface.

M

Magma: molten rock.

Magnetometer: device that measures the strength of a magnetic field.

Mantle: layer of rock beneath the lithosphere. The mantle is composed of rock that is molten at its upper edge.

Mechanical advantage: in mechanical power systems, a means of obtaining either greater force or greater speed than that applied.

Mechanical energy: term used for kinetic energy whenever the energy is harnessed to do work.

Methane: see "Biogas."

MHD (magnetoheterodyne) generator: device which turns hot gases directly into an electric current. It is an adaptation of a simpler device known as a thermocouple.

Microprocessor: a single-purpose computer with integrated memory units and controller units.

Microspheres: tiny particles of radioactive material formed into fuel pellets.

Mine surveyors: workers who do surface and underground surveys at coal mines to control the direction and extent of mining.

Miner: term given to all who work in mines.

Mining engineers: specialists who find, extract, and prepare minerals for use. They design and construct pit and underground mines.

Mistral: the strong cold wind that blows down the Rhone Valley in France.

Moderators: any material that can slow down the neutrons in a chain reaction.

Modes: term used to mean transportation environments. It is also applied to the systems or vehicles designed to suit each environment.

Module: group of interconnected solar cells.

Moments of force: the term used for torque in SI metric. The unit of measure is the newton meter (N.m).

Mud: special mixture of water, clay, and chemicals that cools the bit and carries away the rock pieces broken loose by the drilling.

N

Natural gas: the gaseous portion of the petroleum.

Neap tides: the lower tides in the tidal range.

Neutrons: uncharged (electrically neutral) particles in the nucleus of an atom.

Newtons of thrust: measurement of force developed by an aircraft engine.

Nonassociated gas: free natural gas which is found alone in its natural reservoir.

Nonenergy sector: group of industries that includes all those who use fossil fuels as a raw material in other products or goods.

Nuclear fusion: joining of the nuclei (centers) of atoms to create the heat and light given off by the sun.

Nuclear reactor: a strong, closed container (pressure vessel) in which nuclear reactions take place.

Nuclear reactor operator trainee: entry level job that requires special training in the operation of a nuclear power plant or a reactor used in research.

Nucleus: cluster of positively charged and neutral particles making up the center of the atom.

O

Ocean thermal energy conversion (OTEC): method of using stored heat from ocean surfaces in and near the equator as an energy source. A Rankine cycle heat engine floating on the ocean surface uses the heat to generate electricity.

Ohm: unit used to measure resistance to electron flow in a circuit.

Oil well directional surveyors: technicians whose job is to measure sonar, electrical, or radioactive characteristics of earth formations in oil and gas wells. They bore test holes to determine how productive the well will be.

On-site transport: means of transportation that do not leave the building or area where they are located, such as elevators and escalators.

Open-pit mining: recovery method used when coal is near the surface.

Output: in a system, the result of processing.

Overburden: the soil and rock covering a coal seam.

Overload: an overcurrent caused by an unusually high demand put on a circuit for electric power.

Overshot waterwheel: one in which the water flowed over the top of the wheel on its downward side. The weight of the water, added to its motion, made the wheel much more powerful.

Oxidizer: a substance which supports combustion or a chemical reaction.

Oxygen gasification: process carried on in an oxygen atmosphere which produces a medium energy gas made up of carbon monoxide (CO) and hydrogen (H^2). It makes a good fuel.

P

Parabolic concentrators: dish-shaped solar collectors capable of producing very high temperatures.

Parallel circuit: electrical circuit that provides more than one path for the current to take.

Particle accelerator: device used in fusion research to hurl deuterium particles against a stationary target made up of deuterium or tritium.

Partnership: a business owned by two or more persons.

Passenger liners: large ships designed to move passengers from one land base to another.

Passive solar heating system: one that uses no mechanical methods for bringing in heat and storing it for later use.

Peak output: the maximum output of a photovoltaic cell at noon.

Peat: the first stage of the coal-forming process, a product that looks like decayed wood.

Penstock: in a hydroelectric plant, the large water inlet tube.

People movers: means of transport generally employed in moving people short distances at relatively slow speeds.

Petroleum engineers: engineers who work mostly at exploring and drilling for gas and oil.

Phase-change materials: salts which can be made to change from a solid state to a liquid state as they store heat. When the heat is drawn off, they become a solid once more.

Photons: particles or bundles of energy that travel through space and give up their energy upon striking a surface through which they cannot pass.

Photosynthesis: process in which plants collect solar radiation, along with nutrients from the soil, and store it in matter which makes up leaves, stems, and stalks.

Photovoltaic cell: a small crystal of silicon that has been "doped" with other elements such as boron and phosphorous. Such cells can convert light energy into electrical energy.

Phytoplankton: name given to a huge number of plants and animals that drift near the surface in large bodies of water.

Pipeliner: worker who maintains and repairs pipelines, pumping stations, and tank farms.

Pipe still: a special type of furnace with many coils of tubing running through it, used to refine oil into various products.

Planetary wind: mass movement of air from the polar regions to the equatorial zone, and vice-versa.

Plasma: also known as ionized gas. A fourth state of matter.

Plenums: empty sections of a heat storage medium for air movement.

Plutonium: a fissile isotope of uranium.

PN junction: a neutral area in a diode that does not conduct electricity; it insulates the P and N layers from each other.

Pneumatic systems: power transfer systems that use air as the transfer medium.

Polycyclic aromatic hydrocarbons (PAH): materials produced as a byproduct of synthetic fuels production that can cause cancer.

Polymerization: a way of making gasoline from refinery gases.

Pond: thermal mass in the form of water on the roof of a building. Sunlight heats the water during the day. At night, the heat of the water will be transferred to the space beneath it.

Potential energy: energy that is waiting to do work. It is at rest, in storage, inactive.

Pounds of thrust: measurement of force developed by an aircraft engine.

Power: the amount of work done in a given period of time.

Power engineer: one who designs power systems and oversees construction and maintenance of power stations, transmission lines, and substations.

Power transmission system: a method of transferring, controlling, and adapting the output of an energy converter.

Pressure vessel: a steel pressure tank containing the fuel rods that make up the core of a nuclear reactor.

Primary loop: piping system that carries boiling water to a steam generator in a pressurized water reactor.

Principles of thermodynamics: natural laws that deal with the movement of heat.

Process: the change produced or the work required to give value to what is produced.

Process heat: heat used by a manufacturing operation to process raw material into a product.

Protons: positively charged particles.

Proven reserves: fossil fuel supplies which have been located and estimated with reasonable certainty and which can be recovered with available technology.

Pyrolysis: a breakdown of matter such as biomass by heat.

Q

Quad: unit used to measure large amounts of energy. A quad is 1 quadrillion Btu.

Quantum theory: energy theory that considers radiation as particles of energy.

R

R value: the value given a material's resistance to passage of heat.

Rad: unit measuring the radiation absorbed per gram of body tissue.

Radiant energy: energy that travels as a wave motion. Also called light energy.

Radiation: in the nuclear sense, is the result of unstable atoms (also called radionuclides) that break up as they give off particles of energy.

Radiation of energy: the method by which solar energy travels.

Radiation protection engineer: engineer who supervises and coordinates activities of those working on monitoring radiation levels and conditions of equipment used to generate nuclear electric power to ensure safe operation of the plant facilities.

Ramjet engine: an aircraft and missile engine noted for its simplicity. In fact, it is little more than an open-ended cylinder, with no moving parts, in which fuel is burned to develop thrust.

Random path vehicle: one that can be directed onto various routes. A bicycle's path or that of a truck or automobile is random.

Rankine engine: a reversible heat engine that uses hot water to heat up a volatile (turns to vapor easily) fluid.

Reaction turbine: turbines that work well with low-head hydroelectric plants.

Reaction-type turbine: one in which the steam passes through several sets of blades, all on the same shaft. The result is a stronger "push."

Receiver: a storage tank used to hold compressed air in a pneumatic system.

Recycling: recovery or separating of metals and minerals from the waste stream and making them available for remanufacture.

Refining: the process of breaking down crude oil into various usable liquids and semisolids.

Refrigerant: gas that absorbs heat and transports it to other parts of a refrigeration, air-conditioning, or heat pump system.

Relay: a switch that is operated by an electric current. It controls on and off switching with an electromagnet.

Rem: unit used to measure radiation dosage.

Reservoir beds: porous rock structures where oil spreads throughout the porous rock.

Residential/Commercial sector: group of energy users that includes all private and public buildings.

Rocket engines: reaction engines with few or no moving parts. The rocket engine, unlike a jet that depends on atmospheric air for combustion, can operate in outer space because it carries both its own fuel supply and the oxygen needed for combustion. Rockets are either liquid- or solid-fueled.

Roller bits: bits designed to cut through certain kinds of rock.

Room and pillar: underground mining system in which large pillars of coal are left at regular intervals to act as roof supports.

Rotary engine: an internal combustion engine in which the traditional piston is replaced by a triangular rotor.

Rotational force: force exerted in a turning or twisting manner around a center point.

Run-of-the-mine coal: unprocessed coal that contains rock and other impurities which must be removed.

S

Sandstone: sedimentary rock used to store natural gas.

Secondary loop: piping system that absorbs heat from the water in the primary loop of a reactor.

Seismic analysis: one that uses shock waves to determine the type and thickness of underground materials.

Seismic observers: technicians who work with seismologists.

Seismograph: instrument that records shock waves used to determine the type and thickness of underground materials.

Seismologists: scientists who study and interpret the data created by the recorded vibrations of controlled sound waves as a means of oil exploration.

Semiconductor: electronic component that sometimes acts as a conductor and sometimes like an insulator.

Series-parallel circuit: electrical circuit that combines series and parallel elements.

Shale: a rock made up of mud, clay, or silt.

Short circuit: an accidental, overly strong current in the circuit, due to a fault in the conductors.

Single proprietorship: business type in which one person owns the entire enterprise.

Slurry: a soupy mix of organic waste.

Solar cells: devices which turn sunlight directly into electricity.

Solar collectors: portions of active solar heating systems which absorb the sunlight.

Solid-state: semiconductor devices that replaced vacuum tubes.

Solvent refining: process of making solid fuel from coal.

Sonic drilling: system in which the bit vibrates at the same level as a high-frequency sound generated at the surface. This makes the bit cut faster.

Source beds: places where petroleum first formed.

Specific heat: capacity of a material to hold heat.

Spoil areas: places for dumping the excavated earth from a surface mine.

Spring tides: the higher tides in the tidal range.

Sprocket: a thin, flat gear with teeth.

Stirling cycle engine: engine in which the combustion process takes place in an external chamber, with heat transferred to a fluid such as gas or air through an exchanger. The exchanger heats up the compressed gas or air, which, in turn, expands, pushing on pistons to create mechanical energy.

Stock: a share in the ownership of a corporation.

Stripping: removal of the overburden above a coal seam.

Subbituminous coal: material with a heat content somewhat higher than that of lignite.

Sun: a luminous celestial body. The earth and other planets revolve around the sun and receive light and heat from it.

Supernatant: the spent liquid of the slurry from a digester.

Switches: disconnect devices that stop and start current in an electrical circuit.

T

Tankers: water transporters with tank-shaped holds designed to contain petroleum products and other liquid fuels, wine and molasses, and various kinds of liquid chemicals.

Thermosiphoning: flow of air between glazing and masonry of a Trombe wall, caused by differing densities of heated and cool air.

Three-way switch: switch that is used along with a second three-way switch when current to a light must be controlled from two different locations in a building.

Thrust: the force that moves an aircraft forward.

Tidal range: the difference between low tide and high tide.

Tides: periodic change in the water levels of seas, oceans, and large lakes.

Tipple: plant where bituminous coal is prepared for use.

Topography: the way the land lays.

Torque: a twisting force.

Towboats: small, powerful boats employed to push or pull barges loaded with cargo. Towboats are found mainly in inland waters.

Town gas: gas made from coal (also called "water gas") that was used in the cities for cooking and lighting in the 19th and early 20th century.

Transformer: an electrical device that is able, through use of a magnetic field, to change the voltage in an electrical current.

Transistors: key semiconductors in electronic circuits. They can switch current on and off, which makes them important to the operation of computers.

Transportation environments: the different mediums in which transportation vehicles travel.

Transportation sector: group of energy users that includes all methods of moving people and materials: automobiles, trucks, trains, ships, planes, conveyors, and pipelines.

Traps: nonporous rock strata that lie next to the reservoir rock and prevent the petroleum from moving farther.

Tugboats: small vessels with powerful engines that are used to tow or push larger vessels around in harbors or in close quarters.

Tungsten carbide: a hard metal used especially for drill bits.

Turbodrilling: in this drilling method, the force of the mud being pumped into the well powers a small turbine just above the bit.

Turbofan engine: turbojet engine to which a fan has been added in front of the compressor. This arrangement gives the engine extra power.

Turbojet: continuous-combustion aircraft engine in which superheated exhaust gases provide both thrust and the force (on turbine vanes) needed to compress air for the combustion process. Commonly used on both commercial and military aircraft.

Turboprop engine: a turbojet engine which is similar to the turbofan design, but with a propeller replacing the fan.

U

U value: the measure of a material's ability to conduct heat.

Underground mine: one that is deep beneath the surface of the earth.

Undershot waterwheel: one designed to let the water flow push against paddles from below.

Upwind rotor wind machine: a type that is designed to head into wind with the rotor and housing behind it.

Uranium: a silvery radioactive metallic element.

V

Vane: the fan-shaped metal piece keeping the rotor of a wind machine into the wind.

Velocity: the speed at which an object is moving.

Vertical axis wind machine: one in which the rotor spins on a vertical spindle. No vanes are needed since the rotor can pick up the wind from any direction.

Vertical-mixing digester:. one in which organic materials are digested in vertical chambers.

Vessels: water transporters made to move cargo and people.

Voltage: pressure that moves electricity through conductors. Another name for it is electromotive force.

Volts: see Voltage.

Volumetric heat capacity: measurement of the heat a given volume of a substance can hold. It is found by multiplying the specific heat of a substance by its density (number of pounds per cubic foot).

W

Walking dragline: a large bucket suspended from the end of a boom, used to strip overburden in surface mining.

Wankel engine: see Rotary engine.

Watt: measure of the power produced from an electrical device or system.

Wave theory: theory of energy based upon the activity of electrons in atoms of matter.

Wavelength: the distance from one crest to the next of an electromagnetic wave.

Wet gas: term for oil that goes to petroleum products such as propane.

Wildcat well: those drilled in search of more oil.

Work: that which causes change.

Y

Yellow cake: stage of uranium processing for nuclear fuel.

INDEX